CW01003729

Devon Great Consols

A Mine of Mines

R. J. Stewart

This book is dedicated to the memory of Richard Pepper (1946 - 2002), friend, colleague, and mentor.

Published by The Trevithick Society
for the study of Cornish industrial archaeology and history

ISBN
Paperback 978-0-904040-98-2
Hardback 978-0-904040-98-2

Printed and bound by Short Run Press
Bittern Road, Sowton Industrial Estate, Exeter EX2 7LW

Typeset by Peninsula Projects
c/o PO Box 62, Camborne, Cornwall TR14 7ZN

Contents

Foreword

What a pleasure it is to be asked to write a foreword to this splendid history of Devon Great Consols mine. It is a story that has long been waiting to be told and one which hitherto has been surprisingly difficult to piece together. Now at last, Rick Stewart has provided us with this much needed account and what a fascinating story it is.

From the earliest workings, through to the wonderful copper discovery in 1844 and the fabulous piles of glistening ore; thence to the realisation that even this, the richest sulphide lode in South West England was not inexhaustible. The mine's transformation into the world's greatest arsenic producer was followed by meticulous exploration for more ore reserves. The ultimately unsuccessful search for deep tin beneath the copper is described, together with the character of the directors (including the well-known William Morris), the mine school and the miners' labour disputes. Remarkably very few photographs of this great enterprise in its heyday have survived and this makes the written account all the more valuable.

After the mine's closure at the beginning of the twentieth century, the extent to which the dumps and shallow ore reserves were worked by the Bedford estate will come as a surprise to many. This account also fits into place the pieces of the jigsaw of twentieth century mining and arsenic production which were recorded by Barclay & Toll and Lt. Cdr. P. H. G Richardson in the inter-war years and for which some photographs survive. The dumps and halvans were still sporadically providing ore right up to the late 1970s and prospecting was still in progress in the early 1980s. It is surprising to think that this secondary period of sporadic activity lasted as long as the life of the original mine.

When I first visited the mine in 1959 few people were interested in it, but in 1964 a paper by John Goodridge *Devon Great Consols: A study in Victorian mining enterprise* was published and this was followed in 1967 by Frank Booker's best-selling *Industrial Archaeology of the Tamar Valley*. By 1970 the Dartington Amenity Research Trust had re-opened Morwellham as an historic port, and so began a revival of interest which has continued to the present time.

Now part of the Cornwall and West Devon World Mining Heritage Site and with the opening of the Tamar Trails which provide breath-taking views of Devon Great Consols, this interest is likely to increase rapidly.

It is marvellous to have this full history and Rick Stewart and his publisher The Trevithick Society are to be congratulated for making it available

Owen Baker

Author's foreword

My fascination, some might say obsession, with Devon Great Consols can be traced to a specific date: 1st November, 1997. Prior to that date I had had an interest in the mine however the 1st November was the first time I went underground at Devon Great Consols. As I dangled from a flimsy electron ladder the Main Lode stretched away beyond the light of my Oldham miner's lamp, I began to get an inkling of the enormity of the mine. Returning to grass nearly eight hours later I was hooked. Whilst I have explored the mine on numerous occasions since with colleagues from the Tamar Mining Group and latterly the Plymouth Caving Group the magic of that first trip still remains strong.

As my interest grew I read every thing I could about the mine, I devoured Frank Booker's standard work on the Tamar Valley, Hamilton Jenkin, John Goodridge's 1964 paper and P. H. G. Richardson's book on later operations in the valley.

As my understanding increased I began to feel a growing dissatisfaction with the standard works on Tamar Valley industrial archaeology and mining history. I also found it frustrating that no one had taken the subject much beyond the initial work of pioneers like Goodridge and Booker. This was a view shared by my long term collaborator Robert Waterhouse and we came to the somewhat arrogant conclusion that if we did not move the subject forward no one else would. In a delightfully arbitrary manner we decided that Robert would tackle the Tavistock Canal whilst I would take on Devon Great Consols. Much to Robert's credit he made an immediate start on the Tavistock Canal whilst I prevaricated. Over the following years Robert systematically worked towards his goal whilst I skirted around the edges of my chosen subject. My constant refrain being "I really must get around to writing the DGC book".

Circumstances change; in September 2009 my employers, the Morwellham and Tamar Valley Trust, went into administration. The future of Morwellham Quay Museum was seriously in the balance as was my job as mine manager. I decided that if the worst did come to the worst and my job disappeared I would take a "career break" and write the book. As it happened I managed to keep my job, however it did give me the impetus to make a start on the book after a long gestation.......

A work of this nature is by no means a sole venture and it would certainly not have been written without the generous help of family, friends and colleagues. I would like to extend my warmest thanks to the following:

Aditnow, Owen Baker, Rick Beament, the Duke of Bedford and the Trustees of the Bedford Estate, Big Gay Bob, Prof. Florence Boos, Tony Brooks, Colin Buck, Cornwall Record Office, Cornwall Studies Library (Redruth), Tony Clarke, Charlie Daniel, Bob Dersley, Devon Heritage Service, Devon Record Office, Ian Doidge, Barry Gamble, John Garner, Dr. John Goodridge, Dr. Tom Greeves, the late Roger Harrisson, Phil Hurley, Brian Jones, Pete Joseph, Chris Kelland, John & Cheryl Manley, The Mining History List, Morwellham Quay, Morwellham and Tamar Valley Trust, Alasdair Neill, Dr Patrick O'Sullivan, Plymouth Caving Group, Plymouth Local Studies Library, Plymouth Mining & Mineral Club, Jasen Quick, Steve Roberts, The residents of Devon Great Consols, Kim Sado, Audrey Stevenson, Tavistock Library, Graham Thorne, Tamar Mining Group, Tavistock Woodlands, Trevithick Society, Steve Wadlan, Dave Warne, Robert and Cathryn Waterhouse, Colin West, the late Richard Williams

Rick Stewart
Calstock, Cornwall
March 2013

Introduction

"It is not an ordinary mine, but a mine of mines". Peter Watson addressing shareholders at the May 1880 Annual General Meeting. Mining Journal 29 May 1880

Mining on the ground that included Devon Great Consols has a long history. Tinners worked the area from at least the sixteenth century; the last tin prospectors (for the time being) only leaving in 1985. However what most people think of when the name Devon Great Consols is mentioned is the massive nineteenth century copper and arsenic producer.

In spite of its stellar success the nineteenth century mine was comparatively short lived by Cornish standards: From the commencement of activity at Gard's Shaft in August 1844 until the pumps were stopped in 1902 was a mere fifty eight years. However during that fifty eight years Devon Great Consols' prosperity set the mining world alight, eclipsing even such Cornish copper giants as Carn Brea & Tincroft and South Caradon Mine. It is probable that the mine was, in its hey day during the 1850s, the largest copper producer in the world; it was certainly the largest arsenic producer. Between 1844 and 1902 737,401 tons of copper ore were produced realising £3,360,163 and 71,607 tons of refined arsenic realising £662,285. That the mine generated great wealth is beyond question, £1,232,105 being divided between the shareholders whilst successive Dukes of Bedford received £286,372 in dues as Lords of the soil.

The phenomenal wealth of the mine was built on two distinct but interrelated factors. First of which was the richness and extent of the Main Lode; exploited over a length of two miles the Main lode has aptly been described as the "largest sulphide lode in the west of England". Whilst the Main Lode did yield high grade copper ore it would be wrong to assume that high values were found throughout the ore body. From the 1850s much of the mine's wealth came from lower grade ores. This was made possible by the vision of Devon Great Consols' management who, on the back of rising copper prices, invested heavily in state of the art dressing floors allowing them to successfully exploit lower grade ores or halvans as they were known.

Arguably it was the quality of the mine's management, as much as the richness of the ground that fuelled the mine's prosperity. It was the vision and ability of men such as William Alexander Thomas, Thomas Morris, Josiah Hitchins, James and Isaac Richards which drove the mine in its early years. They were responsible for the introduction and development of sophisticated ore dressing facilities, the integrated water management system which was equalled nowhere in British mining, they built railways and erected a foundry and, not content to rest on the prosperity of the Main Lode, they continually invested in exploration discovering lodes such as South Lode and New South Lode.

It was these men who guided the mine through the precarious days of the mid to late 1860s characterised by falling copper prices and labour unrest. Nothing daunted they sought to reinvent Devon Great Consols as both an arsenic producer, erecting the largest works in the west of England, and as a tin producer expending much time and expense in, ultimately unsuccessful, deep exploration in emulation of the Cornish mines such as Dolcoath which had made the transition from copper to tin.

If the mine failed to make the much desired transition from copper to tin production it was spectacularly successful in making the transition from copper to arsenic production during the 1870s. By the 1880s the income from sales of refined arsenic had overtaken those from copper ore. Whilst the heady days of the 1850s copper boom would never return, the revenue from arsenic significantly prolonged the life of the mine, albeit a mine in terminal decline.

The transition from copper to arsenic in the 1870s also saw a transition in the management. The pioneers of the 1840s and 50s had grown old in the service of the mine and the late 1870s saw significant changes, most notable of which was the increasing influence of Peter Watson who, ably aided by Moses Bawden and William Clemo, would steer the mine through the hard times of the 1880s and 1890s when calls on shareholders were more common than dividends. Even a man of Watson's undoubted ability could not fend off the inevitable forever; the pumps stopping, without ceremony, in late September 1902. Never a man to keep his opinion to himself it is appropriate that, in addition to giving the title to the current work, Peter Watson's words should serve as an epitaph for the great mine:

This mine has not been stopped on account of the poverty of the mineral resources.

Watson's words were no empty bombast; the mine was reworked throughout much of the twentieth century albeit on a smaller scale than in the great days of the nineteenth century. During these latter days the old staples of copper and arsenic were joined by both wolfram and tin.

It is somewhat ironic that the mine which spent so much time and energy attempting to reinvent itself as a tin producer during the 1870s and 1880s should have started life as a post medieval tin producer and ended its life recovering tin from the nineteenth century dumps.

Today the only economic activity on the site of the mine is forestry. However with increasing metal prices one wonders how long it will be before the wealth still undoubtedly contained in the mine's dumps attracts the attention of a new generation of miners? Perhaps, just perhaps, the giant is only sleeping.........

List of illustrations

Chapter 1

Activity before 1844

"I would not give much for any ground where the ancients had not found the "backs" of the lodes and tried them considerably": Moses Bawden, 1892.

In 1892 Devon Great Consols' Purser Moses Bawden commented that:

> I have heard a great deal about virgin ground, but what do they mean by it? There have been no lodes discovered that I am aware of in Devon or Cornwall, certainly not in my lifetime, the "backs" of which have not been worked by the ancients, who have gone down to a certain depth prospecting. Even the "backs" of the Devon Great Consols' lodes had been laid open in many places, and shafts sunk nearly to the course of ore, for many years before the great discovery of 1845. I would not give much for any ground where the ancients had not found the "backs" of the lodes and tried them considerably.[1]

When operations commenced at Devon Great Consols in August 1844 the ground covered by the new sett was certainly not virgin ground and there was some knowledge of what the sett contained. For example the 1844 lease makes reference to lodes at Grenofen and Frementor.[2] The "lode in Grenofen" soon became known as the Main Lode.

There had certainly been some working on the Main Lode prior to the granting of the 1844 lease. In the 1870s Josiah Hitchins described the scene as he first saw it which included "some old workings, and a half and half sort of shode pit and trial shaft about 20ft deep".[3]

The initial 1844 development work at Wheal Maria took place in the "trial shaft" noted by Hitchins, which had been sunk to a depth of fourteen fathoms. Who originally sunk the shaft is a moot point. John Benson, the Bedford Estate's agent at Tavistock wrote to Christopher Haedy, the Estate's London agent, on the 19th March 1845:

> the place where the shaft is sunk there was an old one which had been given up – when that work was done we do not know.[4]

Writing in 1863 J. Y. Watson, a well known mining commentator, suggested that the original shaft was sunk by Josiah Hitchins' father:

> Many years ago an old miner, the father of Messrs. Josiah and Jehu Hitchins, commenced sinking a pit in a game preserve of the Duke of Bedford. The pheasants, however, did not like the "pick and gad;" they got disturbed on their roosts, and the Duke did not like that; and so the old miner was ordered to knock his incipient "bal," and for several years it was abandoned and neglected.[5]

Interestingly Josiah Hitchins' father Josiah (Jehu), as Secretary to the Tavistock Canal Company in the early years of the nineteenth century, was involved in the mining operations associated with the canal. However there is no contemporary, that is to say "canal period", evidence to suggest that any work was carried out on the Main Lode, although adjacent setts such as Capel Tor had received some attention.

The image of the worthy miner coming within feet of a phenomenally rich copper lode only to be thwarted at the eleventh hour by an overbearing aristocrat more concerned with the welfare of his game is picturesque. If one adds in the fact that his son managed to successfully vindicate his father's judgement, one has the stuff of legend.

Unfortunately the truth is probably more mundane. In the Tamar Valley a shallow tin zone overlies a deeper copper zone. Throughout the valley tinners had exploited this upper tin zone from at least the mid sixteenth to eighteenth centuries, some latterly developing into copper mines at depth during the eighteenth and nineteenth centuries (see Note 1). It is arguable that the old shaft where work commenced in 1844 was part of an earlier tin work. When abandoned the shaft had reached a depth of fourteen fathoms, requiring the 1844 miners to sink only another three and a half fathoms before cutting rich copper.[6] Presumably the point at which the tinners abandoned the work marked a transition zone between the shallow tin zone and the deeper copper. Moses Bawden's 1892 comment that a number of shafts had been sunk "nearly to the course of ore" would suggest that sinking to the bottom of the tin zone and not beyond was common practice.[7] If this was the case it is highly probable that the tinners were well aware of the copper which lay under their feet (see Note 2). Although they may have been aware of its presence it would have been of little interest to them, certainly prior to the 1690s when the British copper and associated brass industries were more or less non existent (see Note 3).

Both archaeological and documentary evidence supports the fact that prior to becoming Devon Great Consols in the nineteenth century the area was extensively exploited for tin at shallow depths from Tudor times onwards.

In spite of huge ground disturbance in the vicinity of the Main Lode during the second part of the nineteenth century it is possible to identify features that may be interpreted as pre nineteenth century workings. A particularly well preserved openwork survives at Wheal Josiah which may correspond with New South Lode. It has been suggested that large scale ground disturbances at Wheal Maria between Morris' and Gard's Shafts and also at Wheal Fanny may be interpreted as openworks.

North of the Main Lode at Wheal Maria where later nineteenth century disturbances are less pronounced a well preserved openwork survives on Capel Tor Lode, up slope from Eastern Shaft. Similarly the valley which contains the Cat Stream flowing south from Scrub Tor and which cuts various lodes including Capel Tor and the western end of the Main Lode shows definite signs of having been worked as an alluvial or wet stream work.

South of the Main Lode in Blanchdown Wood between Wheal Maria and Wheal Frementor it is possible to find extensive evidence of lode back working. Evidence on the ground suggests the presence of at least two lodes crossing Blanchdown. These lodes have only been worked to very shallow depths and have escaped large scale nineteenth century reworking.

The various gunnises at Wheal Frementor may also be interpreted as being of some antiquity. These gunnises also appear to have escaped attention during the nineteenth century phase of working, however they were revisited during the twentieth century reworking of the mine.

To the north of Frementor lies an intriguing openwork which might be considered part of Frementor. Interestingly whilst this openwork appears to have been sunk in the vicinity of a crosscourse, shown on Brenton Symons 1848 map, the open work is on a different strike to the crosscourse and close examination reveals no sign of mineralization or indeed any identifiable geological structure.

The documentary record whilst, at best, fragmentary, is none the less illuminating, containing a number of early records relating to tin works in the vicinity.

The earliest document which may be of some significance dates from 1473 concerning lands at Woodovis (now a farm to the north of the Main Lode) and Brandonsland (a lost place). This seems to confirm the right of the Abbots of Tavistock to mining on this land, which certainly contains a number of east-west surface disturbances, marking ancient trials on lodes not reworked later.[8]

The earliest reference to a tin work which might be considered to be in the vicinity of the Main Lode dates from 1539 and relates to a tin work at "Lutchabroke" "next Hele Combe Foot",[9] Hele Combe runs east to west on a course paralleling the Main Lode, joining the Cat and subsequently the Tamar. This being the case it is arguable that Lutchabroke may correspond with the latter day Wheal Maria.

A document dated 20th September 1578 makes reference to "a tin work called the Ilande in Blanchdown within the jurisdiction of the Stannary Court of Tavistock".[10] Reference is also made to "a new work in the Ilande of Blanchdown" in a list of tin works paying doles to the Earl of Bedford dated 1583. In addition to the Ilande of Blanchedon the 1583 list also mentions a tin work at Blanchedon.[11] It is not known to what extent the area that Blanchedon covered, as the name describes an area of open down land; one of several in the vicinity. The eastern part of the Main Lode outside Grenofen Wood was certainly on open down land before the 19th century, but the references to tinworks may refer to those already described in Higher Blanchdown Wood. The Ilande can be pinpointed with some accuracy; the name being applied to a meadow by the river on the 1768 Bedford Estate Map, just east of Frementor, possibly corresponding with South Fanny.[12]

Of particular interest is a document dated 25th November 1579 referring to a pair of tin works in "Greenaven" known as the "Willys" and the "Boynabeame" which could, arguably, be workings on the course of the Main Lode.[13]

The 1583 list of tin works contains an early reference to workings in the vicinity of Frementor or "Ffremytor".[14]

In 1744 Christopher Wynne, the Duke of Bedford's agent, noted that tinners were illegally cutting timber for use in two tinworks in Blanchdown Wood.[15]

An interesting document entitled "Particulars of the late Mr Warne's Tinworks (a fourth part) in the parish of Tavistock (now in possession of Mrs Elizabeth Cake)" and dated September 1778 survives in the care of the Devon Record Office (see Note 4). The list includes references to tin works including Fremator, Fremator Combe and Higher Fremator. The document also mentions works called Hanging Cleave and Hanging Cleave Beam which may equate to Wheal Jack Thomas/ Watson's Mine in Hangingcliff Wood. Also of considerable interest are tinworks in North and South Blanch Down Woods.[16]

The author is particularly grateful to Robert Waterhouse for his assistance with this chapter.

Chapter 1 Notes

Note 1

For example Bedford Estate records refer to "Sheeperrygge Woode" tinwork in 1539.[17] It is believed that the same lode, albeit at depth, latterly developed into the George & Charlotte Mine.

Note 2

In a South West context the eighteenth century "mining interest" was well aware of the transition from tin to copper. Kalmeter in his journal of 1724-1725 notes:

> There are thus mines in Cornwall which could possibly be called copper mines, where tin and copper are found together in one and the same lode, and often also a stone of mundic with it, or that these mines were taken up an worked for tin until, when they came further and deeper down on the lode, the former diminished or disappeared, and copper came in instead.[18]

Note 3

The 1690s saw the birth of the "modern" British copper and associated brass industry. Contributing factors include: The Mines Royal Acts of 1690 and 1693 which effectively deregulated copper mining, the introduction of gunpowder into west of England mining from 1689 onwards and the development of coal fired reverbatory smelting furnaces.

Note 4

Robert Waterhouse comments that "this list has been noted by previous writers, who had however failed to realise that associated paperwork in the same bundle contained two explanations of how title to these tinworks was passed down three generations. All of these can be identified and cross-referenced to at least two tin adventurers of the mid-late 17th and early 18th centuries. The earlier of the two, Henry Nosworthy of Manaton, East Dartmoor, was an active tin speculator between c.1650 and c.1685. This would suggest that the tinworks on the list were in existence and possibly at work in that period, but may not have been active by 1778".[19]

Chapter 1 references

1. *Mining Journal* 30 July 1892
2. Devon Record Office document L1258, 1844 lease
3. *Mining Journal* 10 August 1876
4. Devon Record Office document L1258 add 8m/E11/14
5. *Mining Journal* 13 June 1863
6. *Mining Journal* 26 October 1850
7. *Mining Journal* 30 July 1892
8. Personal communication Robert Waterhouse.
9. Devon Record Office document DD1346 cited in: Greeves T. A. P., 1981, The Devon tin industry 1450-1750, Doctoral thesis, University of Exeter.
10. Devon Record Office document L1258, Bedford Estate papers.
11. Devon Record Office document L1258, Bedford Estate papers.
12. Personal communication Robert Waterhouse
13. Devon Record Office document L1258 Bedford Estate papers.
14. Devon Record Office document L1258 Bedford Estate papers.
15. Devon Record Office document L1258 Bedford Estate papers.
16. Devon Record Office document LM1258. Particulars of the late Mr Warne's Tinworks (a fourth part) in the parish of Tavistock (now in possession of Mrs Elizabeth Cake), September 1778.
17. Devon Record Office document DD1342 cited in: Greeves T. A. P, 1981, *op. cit.*
18. Brooke J., 2001
19. Personal communication, Robert Waterhouse

Chapter 2

Application and granting of the lease, 1843 – 1844

"character quite satisfactory – may make a proposal" – Christopher Haedy, July 1843.

J. H. Murchison, a well known mining commentator and speculator, writing in 1850, observed that various attempts had been made to secure the lease of the sett prior to 1844.[1] Amongst those expressing an interest were the Williams family of Scorrier: John Williams had attempted to secure the adjacent Capel Tor Sett in both 1820 and 1832.

The 1820 Capel Tor application was rejected out of hand by John Russell, the 6th Duke, on the grounds that mining would be injurious to his pheasant coverts. John Williams was informed by His Grace the Duke's agent W. G. Adam that:

> his grace has an insuperable objection to letting miners into that part of the woods if they can be excluded as they will be such dangerous enemies to the game.[2]

In 1832 a further application for Capel Tor was made by John Williams, in association with Gill and Rundle who were key figures in the Tavistock Canal Company. Interestingly W. G. Adam expressed the opinion that the 1832 application "was a trap to get...... the whole ground from the Tamar to Chip Shop wagon-lane".[3] In other words Williams and his cohorts were, in the view of the Estate, attempting to secure what became the Devon Great Consols sett. The Estate also rejected this application; this may have been because the Duke wanted "no gang of miners disturbing his pheasants".[4] However it should be noted that the Estate was not well disposed towards Gill and Rundle with whom they had

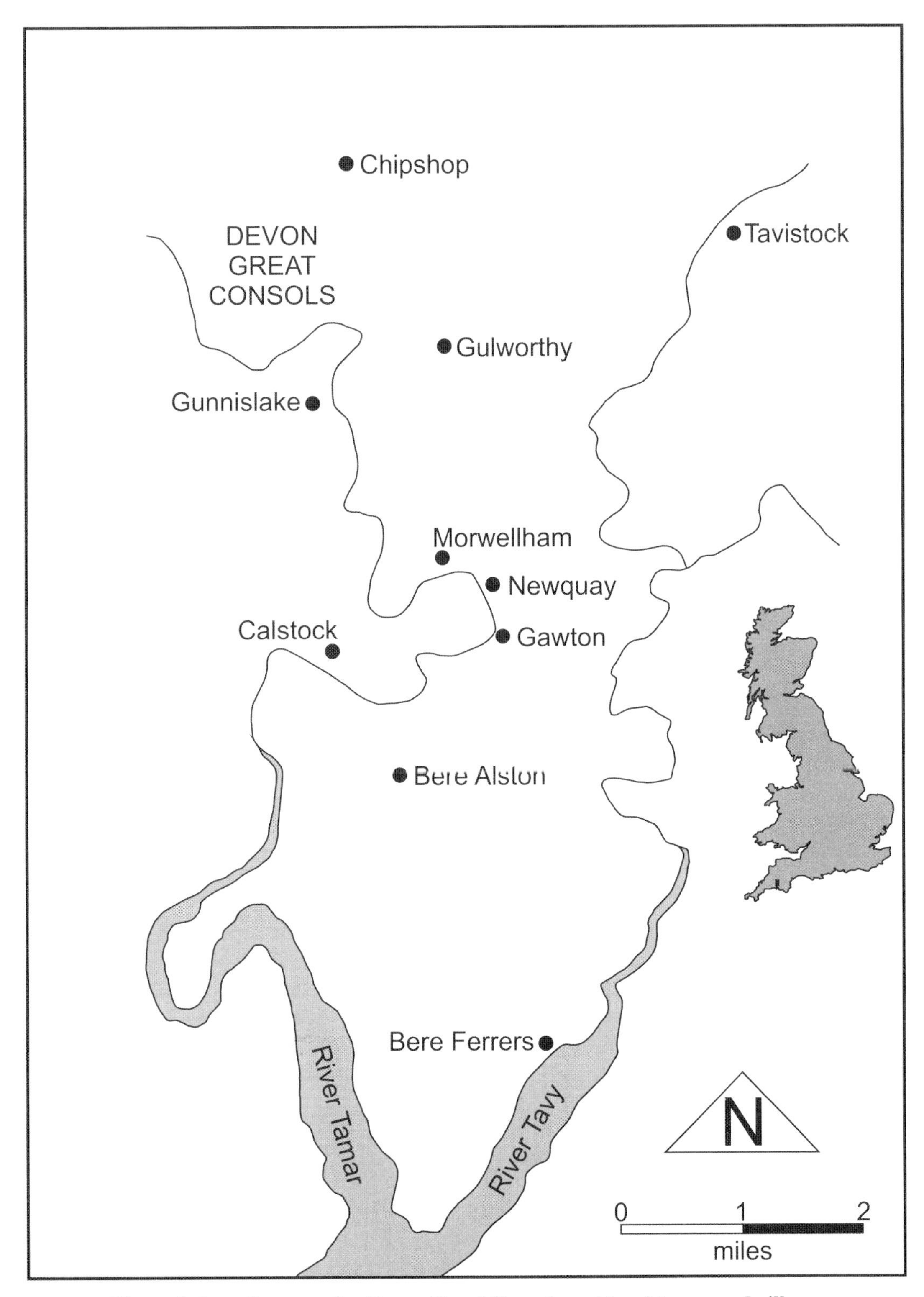

Figure 1. Location map for Devon Great Consols and local towns and villages.

had prior dealings.

With the death of the 6th Duke in 1839 the situation changed. His successor Francis Russell, the 7th Duke, seems to have been more amenable to mining, or at least the income it brought him. History would have us believe that Josiah Hugo Hitchins (see Note 1) was the main protagonist in securing the first lease for the sett, for example, writing in 1850, J. H. Murchison credits Hitchins as the motivating force behind the enterprise:

> Before the present lessees obtained it, many applications were made for the grant, all of which were refused. The matter was subsequently laid before some gentlemen in London by Mr. Josiah Hitchins, whose opinion of the undertaking was very strong.....These and similar representations induced six gentlemen to agree to subscribe among them the sum of £12,000, for the purpose of developing the mineral resources of the property; and having convinced the Duke of their ability to fully carry out their intentions in a legitimate manner he granted them a lease.[5]

This basic idea has been repeated by all subsequent commentators and on the face of it, it appears to be a perfectly acceptable interpretation. Certainly Josiah Hitchins considered himself to be the discoverer of the mine. In a letter entitled "The discovery of Devon Great Consols" to the *Mining Journal* in August 1876 Hitchins wrote:

> The highly promising character of the mineral stuff altogether more particularly the pre-eminently grand gossan (the produce of the back of the lode) found lying about here and there at the surface near some old workings, and a half and half sort of shode pit and trial shaft about 20ft deep fixed my determination the moment I caught sight thereof. I at once

Key to figure 2, opposite;

S1 New North Lode - Eastern Shaft; S2 Capel Tor Western Shaft; S3 Capel Tor Eastern Shaft; S4 Gard's Shaft (Wheal Maria); S5 Morris' Shaft (Wheal Maria); S6 Western Shaft (Wheal Fanny); S7 Eastern Shaft (Wheal Fanny); S8 Ventilating Shaft (Wheal Fanny); S9 Blackwell's Shaft; S10 Anna Maria Engine Shaft; S11 Field Shaft; S12 South Lode Shaft; S13 Richard's Shaft; S14 Hitchin's Shaft; S15 Counthouse Shaft; S16 Ages Shaft (Old and New); S17 Railway Shaft; S18 New Shaft; S19 Inclined Shaft; S20 Thomas' Shaft; S21 South Fanny Shaft; S22 Plunger Shaft; S23 Watson's Shaft.

W1 Wheal Maria Wheel; W2 South Fanny Wheel; W3 Plunger Shaf Wheel; W4 Agnes Shaft Wheel; W5/W6 Great Wheels.

----- Flat rod runs

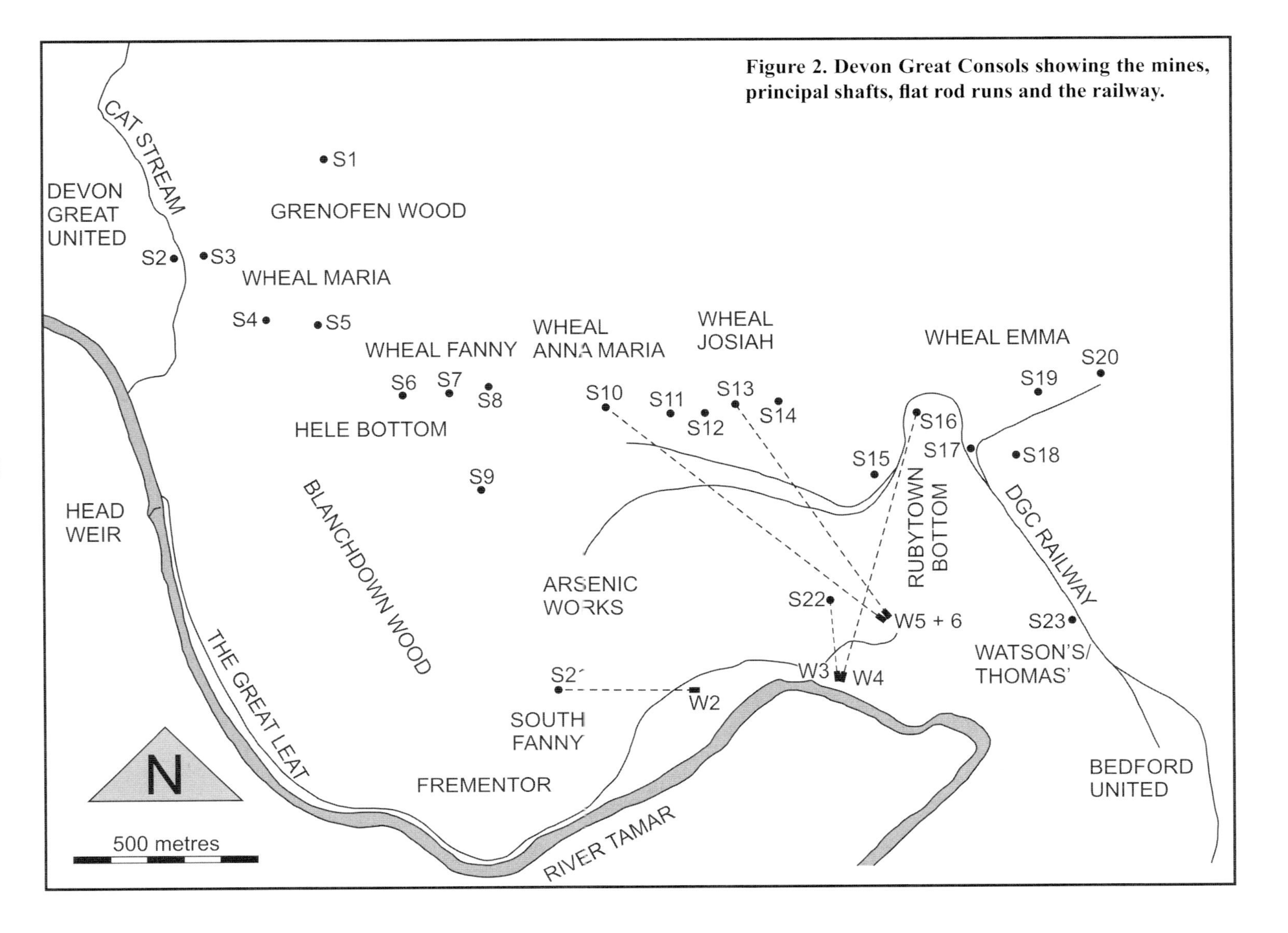

Figure 2. Devon Great Consols showing the mines, principal shafts, flat rod runs and the railway.

> set about securing the ground by lease in the usual way to authoritatively allow of the lode to be opened out; which, as is well known to the mining interest, soon proved to be one of very great productive power, being found of the value of £300 per fathom and more, at a less depth of 20 fms. The lease was obtained for the Messrs. Thomas, Morris and Gard, gentlemen of high standing in London as stock brokers, bill discounters, and wine and coal merchants ....[6]

It cannot be disputed that Hitchins was actively promoting the virtues of the sett, for example on the 29th of January 1844 he wrote to Murchison:

> With respect to this property, I have only again and again to repeat, that it is my firm conviction that it is abounding, to a dead certainty, in metalliferous deposits and resources more so by far than in any other in this great mining district. Let us but have a fair capital wherewith to bring them into operation, and it cannot fail to be rendered capable of great results.[7]

However there is no surviving contemporary evidence to suggest that Josiah Hitchins applied to the Bedford Estate for the sett and only minimal evidence to suggest that he was in any way involved in the process.

The Estate received at least two applications for the sett during 1843. One application received by The Duke's mineral agent John Hitchins (Josiah's elder brother) came from J. W Phillips, then Manager of Bedford United Mine (see Note 2):

> Tavistock July 6th 1843...Sir I beg to make an application for the sett of Wheal Frementor, the boundaries to be as follows, to extend so far south as the Bedford United Mine or sett, west as the River Tamar. North 600 fathoms and east so far as the new road leading to Chipshop (see Note 3).
>
> I have only to add that in making this application it has been at the suggestion of several highly respectable parties (see Note 4) who have a desire to embark their Capital for the trial of the lodes – I should suggest that the usual & customary grant of licence for 12 months be given which enable me to clear up the old men's workings, adits &c and to ascertain if the lodes are worth prosecuting. Awaiting the favour of your reply .....[8]

In spite of his highly respectable backers Captain Phillips' application appears to have progressed no further.

The other, ultimately successful, application came from one William Alexander Thomas of P. W. Thomas and Sons. P. W. Thomas and Sons was a successful London stockbroking business, established in 1820, with offices at 50 Threadneedle Street. The Thomases specialised in railway shares, however they were one of the few London Companies with an interest in mining shares.[9,10] Unlike Phillips, W. A. Thomas does not appear to have been known in the district, consequently he was keen to establish his bona fides. In April 1843 John Benson, the Bedford's agent at Tavistock wrote to Christopher Haedy, the Estate's agent in London outlining Thomas' credentials:

> Bedford Office, Tavistock April 1843: Mr Thomas, the applicant for a mining sett, whose letters I gave you, is one, (Wm. Alexander Thomas) of the firm of P. W. Thomas & Sons of 50 Threadneedle Street, stock brokers & Money agents – He refers to Messrs. Sanderson of 83 King William Place - but to say their firm have banked many years with Messrs. Barrett Hoare Esq., the agents to the Tavistock Bank through whom he says you may make enquiry; with the result of which he is quite sure you will be quite satisfied. He wishes you to understand he has no wish to make a company to sell shares to have the shareholders to wish they had not embarked in an unprofitable concern. His intention (with some others who will hold the mine in their own hands) is to here after come to an understanding of the best opinions of the chances before they begin. If they encourage and appearances warrant fair hopes then he says they would go on immediately. He says the more we inquire into the respectability of his firm the better pleased he will be the more certain he will feel that we shall be satisfied.[11]

The other potential adventurers were London money men including William Morris (see Note 5) and Richard Gard, both senior partners in the extremely well respected discount broking firm Sanderson & Co., William's brother Thomas Morris, a coal merchant from Camberwell with interests in South Wales and W. A. Thomas' brother John.[12]

From W. A. Thomas' perspective the process was too slow, writing to Christopher Haedy on 13th July 1843 Thomas very politely enquired as to progress. A pencilled note in Haedy's hand at the foot of Thomas' letter notes: "character quite satisfactory – may make a proposal".[13] This is confirmed in a rough note probably written by Haedy, dated 2nd August 1843 which stated that Mr Thomas "might make proposals to Mr Benson for such sett as he wishes to have" In reply Thomas wrote on the 4th August 1843 that he proposed to visit Devonshire in the next fortnight and would call on Mr Benson.[14]

Negotiations must have been ongoing throughout the remainder of 1843. Very sensibly Thomas engaged a man with local mining experience, Captain William Williams of Wheal Friendship, to act as his agent. Negotiations appear to have been fruitful; by the beginning of 1844 the basis of an agreement had been arrived at:

> Tavistock Bedford Office January 18 1844
> Wm Williams & John Benson
> Memorandum of terms of a proposed set of lands for the purpose of working for copper and other minerals, excepting manganese, in the parish of Tavistock.
>
> It is proposed by Captn. Williams on behalf of Mr. W. A. Thomas to take a set of the grounds bounding South by the Luscombe set which extends 200 fathoms south of the Marquis Lode (see Note 6) – West of the River Tamar – North by the Parish of Lamerton and East by the road leading from Morwellham to Launceston called waggon lane.
>
> The term to be twenty one years
> The dues to be 1/15th
> Doing as little damage as possible and paying for what damage done.
> The lode in Frementor and the lode in Grenofen to be each prosecuted in a mining like manner the streams passing through the set to be used for the purpose of the mine – but not to prevent their being used for other purposes on the Duke's land. In all other aspects the terms to be those that are usual in mining leases.[15]

The terms of the lease continued to be hammered out during the early part of 1844, the main issue being establishing the rate of the Lord's dues. For example in a letter dated 24th February 1844 regarding proposed dues Benson wrote:

> I have not instructed Messrs. Wing & Twining (see Note 7) respecting the lease to Mr Thomas – because I wish to send him a plan of the ground and because (author's note: John) Hitchins suggests that after a certain profit has been made the dues should be increased to 1/10th".[16]

A similar Benson letter dated 23rd March 1844 regarding proposed dues notes that:

> (author's note: John) Hitchins thinks the dues of 1/15th are too low to serve for the whole term – and thinks that after a profit of £5,000 has been realised they ought to be increased to 1/12th or even 1/10th share.[17]

On the 26th March 1844 Captain Williams wrote to Benson on the subject

> My Dear Sir, on my return from Barnstaple last night I found your favour of the 23 inst – in reference to the proposed set north of the Bedford United Mine. I received a letter from Mr Thomas about the 12 inst, he then informed me that no time should be lost on his part, to bring the business to a close as soon as possible and that Mr Thomas would take it on himself to see the business settled, since the date of that letter I have not heard from Mr Thomas. I fully expected by this time that it might be nearly settled, as Mr Thomas said he should write to Mr Benson at once on the subject, and I am rather surprised by your letter, that he is not in communication with you on the subject of Dues & C as soon as I hear from the parties will see you at once on the subject.[18]

In response to Williams' letter Benson wrote to Haedy on the 27th March:

> from a note I had from Captain Williams of Wheal Friendship yesterday evening it would appear that Mr Thomas has given no instructions to him regarding the dues, consequently I cannot proceed to Messrs. Wing and Twining instructions for preparing a set. I have everything else ready pending.[19]

Unfortunately at this critical point in the story the Benson Letters fall silent on the matter. However matters must have been mutually agreed as a lease, signed on the 26th July 1844, was granted by the Duke of Bedford to Messrs. John Thomas and William Alexander Thomas on behalf of themselves and their fellow adventurers. Whilst signed and dated 26th July 1844 the lease, which was to run for 21 years from 25th March 1844 or Lady Day.

The lease assigned to the adventurers:

> full and free liberty, licence power and authority to dig work, mine and

search for copper, copper ore, tin, tin ore, lead, lead ore and all other metals and metallic minerals (except Royal metal and manganese) ...... to dig and make such adits, shafts, pits, drifts, leats or watercourses and to make and to make such erections and buildings thereon [and] the use of all waters and watercourses arising on or running through the limits aforesaid...[20]

In return for granting the lease the Duke of Bedford was entitled to dues amounting to one fifteenth part increasing to 1/12th once a profit of £20,000 had been realised:

of all such copper and copper ore, as shall at any times during the continuance of the said term of twenty one years be dug broken raised or gotten within the limits aforesaid on the grass at the scales on the mine after the same shall have been made merchantable and fit for sale free of all costs charges and expenses.[21]

In addition to the copper dues the Duke was also entitled to one fifteenth part (again increasing to one twelfth once a profit of £20,000 had been realised) of tin, tin ore, mundic, lead and lead ore "and all other metals and metallic minerals by these presents authorised."[22]

The adventurers were obliged to work the mine seriously:

And shall and will on or before the twenty fifth day of March one thousand eight hundred and forty five commence and during the remainder of the said term continue well and effectively and without intermission to drive the deep adit leading into the workings to be carried on under these presents by four men at the least as well as by night as by day. And also shall and will work the lode at Frementor and the lode in Grenofen situate within the limits aforesaid as soon as the same shall be reached in the course of working at the least at all times thenceforward during the said term of twenty one years and drive on the same on the respective courses of such lodes....[23]

To work the mine the Thomases and their fellow adventurers set up a cost book company with 1,024 shares and a subscribed capital of £12,000 (see Note 8). It is at this stage that Josiah Hitchins would appear to have become formally involved, both as an adventurer and also as the Company's mine agent replacing Captain Williams (see Note 9).[24,25,26,27] In the first accounts of the new enterprise, in August 1844, payments were made to both Josiah Hitchins (£5 11s) and Captain Williams (£10) to cover their expenses.[28]

With the lease signed and a call made of £1 per share operations started at what was then known as "North Bedford Mine" in August 1844.[29]

The author is particularly grateful to Alasdair Neill for his assistance with this chapter.

Chapter 2 Notes

Note 1

The Hitchins "clan" can cause undue confusion to the unwary historian. Josiah Hugo Hitchins should not be confused with his brother John Hitchins who was "Toller" or mining agent to the Bedford Estate, a post which had previously been filled by their father Josiah "Jehu" Hitchins.

Note 2

The background to this application is considered in Brooke J, 1977, Who discovered Devon Great Consols?, *Memoirs of the Northern Cavern & Mine Research Society* 1977, vol. 5 pp21-22.

Note 3

J. W. Phillips diary (Cited in *Devon and Cornwall Notes & Queries* Vol. XXX, Part VII p183) puts this date several months earlier on the 6th February 1843.

Note 4

It is interesting, if ultimately futile, to speculate on who the "highly respectable parties" were, however it is worth noting that Josiah Hitchins was recorded as being the manager of Bedford United at this time![30]

Note 5

William Morris (Senior) was the father of "the" William Morris of Arts & Crafts fame. Between 1845 and 1872 the Morris family received £270,624 in dividends from Devon Great Consols, of that William Morris (Junior) earned £8,803, affording him both the income and the leisure to develop his ideas and find his vocation.[31] During the early 1870s William Morris was a somewhat ineffectual director of the Company. Digressing slightly some of William Morris' wallpapers had a significant arsenic content, although whether this was Devon Great Consols arsenic has not been definitively established.

Note 6
The extent of the proposed sett is somewhat strange in that it includes Bedford United, a sett which had already been granted.

Note 7
The Estate's solicitors.

Note 8
Cost book companies were a particular feature of mining organisation in Cornwall and Devon. The following is a description of a generic cost book company is taken verbatim from *The Jurist* Vol. XVI, Part II, of 1853:

> permission is obtained from the owner of the land to work a lode; the adventurers then hold a meeting, and decide on the number of shares into which their capital is to be divided, and the number to be allocated to each; they appoint an agent, commonly called a purser, for managing the affairs of the mine; and enter in a book, called the "cost book", the minutes of their proceedings, which are signed by all present. A licence to try for ores for some short period is then obtained, followed, if the search be promising, by a sett, that is, a lease of the minerals, or a licence to dig, or both, granted to the landowner to the purser, or to one or two of the adventurers, without any expression of trust on their part for the rest, or any other persons, for a term of years, usually twenty one, stipulating for the annual payment of some portion of the ore raised. The purser manages the works, keeps the cost book, in which he enters all the proceeds and disbursements of the mine, the names of the shareholders, together with the account for and against each, such as have been agreed to at a general meeting, and convenes those meetings by circular letters at regular intervals, commonly of two months. These general meetings review the accounts and report of the purser, and pass resolutions, either declaring dividends or authorising calls, and directing the mode of carrying on the mine. Any adventurer may relinquish his share, and with it his liabilities – at least as far as his partners are concerned – by giving notice of relinquishment in writing to the purser, and settling his account with the mine.

Note 9
The last we hear of Williams, in connection with the mine, is the correspondence with Benson at the end of March 1844. By August 1844 Hitchins was directing operations on the mine. Hitchins appears to have followed Williams into a

revolving door and come out in front of him!

Chapter 2 references

1. *Mining Journal* 26 October 1850
2. Devon Record Office document L1258SS/C letters 85
3. Devon Record Office document L1258SS/C letters 85
4. Thomas C, 1857
5. *Mining Journal* October 26 1850
6. *Mining Journal* August 10 1876
7. *Mining Journal* October 26 1850
8. Devon Record Office document L128E11/14
9. Devon Record Office document L128E11/14 Benson letter April (?) 1843
10. Harvey C. & Press J., 1990, The city and mining enterprise: The making of the Morris family fortune, Journal of the William Morris Society, 91, pp. 3-14.
11. Devon Record Office document L128E11/14 Benson letter April (?) 1843
12. Harvey C. & Press J., 1990, *op. cit.*
13. Devon Record Office document L128E11/14, W. A. Thomas letter to C. Haedy, 13th July 1843
14. Devon Record Office document L128E11/14
15. Devon Record Office document L128E11/14
16. Devon Record Office document L1258/SS/C (Letters) Mr Benson's letters, Bundle 53, 1844
17. Devon Record Office document L1258/SS/C (Letters) Mr Benson's letters, Bundle 53, 1844
18. Devon Record Office document L128E11/14
19. Devon Record Office document L1258/SS/C (Letters) Mr Benson's letters, Bundle 53, 1844
20. Devon Record Office document L1258, 1844 lease
21. Devon Record Office document L1258, 1844 lease
22. Devon Record Office document L1258, 1844 lease
23. Devon Record Office document L1258, 1844 lease
24. *Mining Journal* October 26 1850
25. *Mining Journal* July 28 1860
26. Collins J. H., 1912, Observations on the West of England Mining Region, (reprint 1988), Cornish Mining Classics, Truro.
27. Goodridge J. C., 1964, Devon Great Consols, a study of Victorian mining enterprise, Transactions of the Devonshire Association, Vol. XCVI. pp. 228-268.
28. *Devon and Cornwall Notes & Queries* Vol. XXX
29. *Mining Journal* October 26 1850
30. *Mining Journal* 4 March 1843
31. Harvey, C. & Press J., 1990, *op. cit.*

Chapter 3

The discovery of the Main Lode

"Wheal Maria is all the cry – never was there such a wonderful discovery" – Josiah Hitchins, December 1844.

Operations commenced at "North Bedford" on the 10th August 1844. Whilst Josiah Hitchins was in overall charge of operations, work on the ground was superintended by Captain James Phillips of Bedford United Mine.[1,2] Initial work was centred on an existing shaft on the western extremity of the sett subsequently known as Wheal Maria (see Note 1). The shaft was soon named Gard's Shaft after Richard Gard, one of the original shareholders.

On the first day of working Phillips recorded in his diary (see Note 2) the loan of 162lbs. of tools including six pick hilts, three shovel hilts and three shovels; loans of this nature continuing until the beginning of November 1844. Phillips also noted the names of the miners to whom the tools were issued; the list includes: J. Williams, Silas Clemo, J. Rodda, F. Warn, James Paule, George Harris and Richard Hooper. Work on clearing the old shaft progressed well, candles being issued for the first time on the 19th August.[3] Josiah Hitchins, in a report (see Note 3) dated 24th August 1844 noted that:

> We have cleared up to a depth of 14 fathoms, which we think is the bottom of the old workings, we shall, however, to a certainty, determine this very shortly. The lode altogether, from surface to bottom, is as fine a lode as can possibly be seen, and in the bottom is all of three fathoms wide, carrying an immense gossan.[4]

Whilst not carrying payable values the discovery of "an immense gossan" was highly significant (see Note 4) and no time was lost in sinking below the bottom of the old man's shaft; powder and safety fuse being issued on or around the 26th of August. Phillips noted that the shaft was being sunk by James Warn and five (unfortunately) unnamed partners (see Note 5).[5]

Whilst the centre of operations was Gard's Shaft, Hitchins did not neglect the wider exploration of the sett; John Williams, Samuel Knight, H. Kellaway, Charles Bray, and James Blanchard being engaged on costeaning.[6]

Work progressed as summer turned into autumn and Hitchins' reports from the mine became increasingly positive, on the 25th October 1844 he wrote:

> ..... our lode is apparently improving greatly in character as we are getting down deeper in the shaft and the old workings, the men having today brought in large stones of mundic, mixed with ore &c., of a very kindly nature in every respect. The men will have it that it must make one of the great concerns in the two counties; but I have heard such stories before today, and do not wish to excite any false views or expectations. Notwithstanding, I cannot conceal my feelings about the matter – I think we shall have a good concern in the Maria to a certainty.[7]

Until October 1844 Captain Phillips appears to have been in sole charge of day to day operations. However on the 8th October Captain James Richards was appointed resident Captain. Although Richards would have nominally been in charge, he and Phillips worked in conjunction with each other throughout the remainder of 1844 and into 1845.[8]

By the beginning of November Hitchins' expectations were starting to be realised. In a report dated 6th November 1844 Hitchins can hardly contain his enthusiasm:

> I have just returned from the mine, and have brought with me some stones of ore, which although not rich, are exceedingly kindly. Finer looking stuff than has been taken up today, there cannot be I assure you. The lode is four fathoms wide, and the men working in it are quite in ecstasies: they all say it cannot fail to prove in depth a grand deposit of copper ore. I cannot conceal my opinion, that I believe we shall have a wonderful lode under this great gossan, and that too at no great depth. There is copper apparently coming in more and more every fathom we sink, and it is coming in just as I like to see it, gradually and not too shallow, which will render it more

> lasting. Upon the whole, I should think that was never such a lode seen in this county. I only wish you could see it now, and the stuff coming up from the shaft; such a strong quantity of strong gossan I never saw from a lode anywhere. We shall have a great concern, depend on it, and I have no hesitation in advising you to hold as large an extent in it as possible – it will be a fortune to you in my opinion. That I cannot conceal the high spirits in with our prospects, you perceive. Really it is excusable, from the present glittering appearance of the mine. Remember also, that the lode is only one of many fine lodes we have in the Wheal Maria sett. Our shaft is down from surface 16 fathoms.[9]

Hitchins' predictions were soon realised; on the 12th November 1844, at a depth of seventeen and a half fathoms below collar Gard's Shaft cut a massive deposit of copper:

> Tavistock, Nov. 13, 1844. I told you, not long since, that we might expect something good before long, and now let me enjoy the pleasure of realising my anticipations. Last evening the lode began to assume a more decided character, although it had for many days previously been exhibiting all the characteristics, having a tendency to the successful change which I now have the happiness to announce to you. The lode (that is to say the leading part of it, composed of ore) is now four feet wide, and excellent work it is, and we are now saving some for dressing, towards our first sampling, which I hope will be a good one. The lode appears to be coming better and better every inch that it is sunk in the shaft; but of course, since the cutting or ore last evening, there cannot be enough done to develop it sufficiently to admit of anything like a correct estimate of its value. There is, however, no fear in saying, that the lode is present in the bottom of the shaft (now 17½ fathoms from the surface), it is worth £30 per fathom. Last evening the men came in with some of the ore in high spirits; and of course, this morning I went out fully prepared to make a large allowance for their sanguine representations. I am happy to say, however, that the results are equal to the report. The appearances are certainly splendid – gossan, spar, mundic, &c., accompanies the ore; there is here great excitement about the discovery. It was said last evening by some that the lode was worth £100 per fathom. Although I do not think it is worth so much, yet I will not say that it will be so before long.[10]

The scale and magnitude of the discovery was beginning to become clear; on

23rd November 1844 Josiah Hitchins wrote that:

> The North Bedford property will undoubtedly be a grand one, and Wheal Maria will no doubt cut the shine out of all with her samplings. Capt. Richards says – "You say the lode will produce 6 to 8 tons of ore per fathom". Why should it not? And much more, under so large a gossan as we have on the lode..... we have a wonderful concern in Wheal Maria.[11]

An account, dating from the 1860s, gives a picturesque impression of operations on the mine during this pioneer period:

> At the time of the discovery a rude blacksmiths shop, the walls of which were formed of gossan, was the only building on the mine, which was quite destitute of machinery. When the copper ore was cut into, the sudden influx of water drove the miners from their work and soon filled the shaft. A small water-wheel was now erected, and this with an apparatus of pumps, a whim and a capstan for some time, constituted the celebrated Wheal Maria now called the Devon Great Consols.
>
> The author well remembers descending into the mine, shortly after the great discovery...... Several miners were working abreast, on a rich vein of ore, and the extraction already made was not unlike the aisle of a cathedral church. Embosomed in a wood of oak coppice, the surface aspect of the mine was still more remarkable. So sudden and unexpected had been the discovery, and in such quantities was the ore being thrown up, that the place was crammed with ore, the richest yellow sulphuret, in appearance much more resembling heaps of gold than baser metal, and presenting a combination of gorgeous metallic wealth and sylvan beauty, the like of which will never perhaps be seen again.[12]

1844 ended on a high note; on the 26th December 1844 Hitchins was able to write:

> It would be premature, at this point of progress, to attempt to define the scale of our probable profit; but I believe there will be something very handsome for us from this concern – I hope £100,000 at least. I think that there cannot be any doubt of this. The lode is looking better than ever, coming in the shaft, the deeper it is sunk – more solid, and a greater abundance of ore, that will not only turn out more in quantity, but better in quality. The lode (taking

> the length and width of the shaft, 10ft. by 6 ft.) may be fairly reckoned as being worth at least £100 per fathom...... Wheal Maria is all the cry – never was there such a wonderful discovery.[13]

Hitchins was obviously aware that his discovery was of no little importance. However, it is arguable that the true magnitude of what was to become known as the Main Lode can only truly be appreciated in hindsight.

Chapter 3 Notes

Note 1

J. H. Murchison writing of this phase in the development of the mine refers to "Great Wheal Maria" and states that it was named after the "amiable wife of Mr. Hitchins".[14] In later usage the prefix great is absent, the mine being known simply as Wheal Maria.

Note 2

Captain James Phillips' papers were examined by the late Justin Brooke in the 1960s. Their current whereabouts, if they survive, are not known. A brief article written by Justin Brooke[15] gives a fragmentary and tantalising glimpse of their contents.

Note 3

Hitchins sent a series of reports to the mining speculator and commentator J. H. Murchison covering the discovery of the main lode. These reports were published by Murchison in an extensive article in the *Mining Journal*.[16] It is these reports that form the basis of this chapter.

Note 4

Gossan is the weathered and decomposed upper section of a lode, sometimes referred to as the "zone of oxidation". The top section of the lode is oxidised by the chemical action of percolating surface water. Typically sulphides such as chalcopyrite are converted into oxidised minerals soluble in rain water which are leached out of this upper section of the lode. Cassiterite, in contrast to the sulphide ores, is not water soluble and consequently remains present in the upper "gossanised" section of the lode, explaining the presence of tin works at shallow depths. The significance of the gossan is that it overlies what is termed the "zone of secondary enrichment". The zone of secondary enrichment is hugely important in that it here that minerals leached from the zone of oxidation are re-deposited.

Note 5

Writing in 1914 Moses Bawden records that Josiah Hitchins employed three brothers by the name of Clemo on this work.[17] Bawden's article needs to be treated with some caution. Writing of the discovery of the Main Lode he states that it was discovered in 1845 not 1844. Having said that, Bawden's ongoing involvement with the mine means that this article is of value. Regarding the three Clemo brothers Bawden notes that:

> William Clemo....became mine captain, and in 1891 was appointed manager, and kept that position up to the time of his death on the 3rd September 1900, after being on service at the mines for over 55 years. Of the two brothers who came with him..... one served fifty years and the other fifty five years, both of them in responsible positions as pit men.[18]

Chapter 3 references

1. *Devon and Cornwall Notes & Queries* Vol. XXX,
2. *Mining Journal* 26 October 1850
3. *Devon and Cornwall Notes & Queries* Vol. XXX
4. *Mining Journal* 26 October 1850
5. *Devon and Cornwall Notes & Queries* Vol. XXX
6. *Devon and Cornwall Notes & Queries* Vol. XXX
7. *Mining Journal* 26 October 1850
8. *Devon and Cornwall Notes & Queries* Vol. XXX
9. *Mining Journal* 26 October 1850
10. *Mining Journal* 26 October 1850
11. *Mining Journal* 26 October 1850
12. Chowen G, 1863, Some Accounts of the rise and progress of mining in Devonshire: from the time of the Phoenicians to the present, Thomas S. Chave.
13. *Mining Journal* 26 October 1850
14. *Mining Journal* 26 October 1850
15. *Devon and Cornwall Notes & Queries* Vol. XXX
16. *Mining Journal* 26 October 1850
17. Bawden M., 1914, Mines and mining in the Tavistock district, Transactions of the Devonshire Association, Vol. XLVI. pp. 256 - 264.
18. Bawden, M., 1914, *op. cit.*

Chapter 4

Exploration and development 1845-1849

"The Devonshire Great Consolidated Copper Mining Company"

1845

In January 1845 the original cost book company was reconstituted as a joint stock company, becoming fully registered on 25th March 1844. The new joint stock company was known as the "Devonshire Great Consolidated Copper Mining Company".[1] A memorandum, dated 28th January 1845, appended to a copy of the 1844 lease, notes that the Thomases, as the original signatories; "covenanted and agreed" that the "liberties licences and authorities" of the 1844 lease were granted to the new company.[2]

The joint stock company looked very similar to the earlier cost book company: It was divided into 1,024 shares. It had the six shareholders: Josiah Hitchins (144 shares), Richard Gard (288 shares), William Thomas (144 shares), John Thomas (144 shares) William Morris (272 shares) and Thomas Morris (32 shares).[3] However the principal difference between a cost book company and a joint stock company was that in a cost book company control of management of the mine was directly exercised by the whole body of adventurers, whereas in a joint stock company control was delegated to a representative board of directors.[4]

The first Annual General Meeting was held on the 5th May 1845 at the company's office in Barge Yard Chambers, Bucklesbury in the City of London. That meeting elected the first board of directors comprising R. S. Gard (chairman), Thomas Morris and W. A. Thomas; William Morris and John Thomas being appointed as auditors.[5]

The first annual report of the 5th May 1845 notes that prior to full registration as a joint stock company:

> all matters relating to the mines were conducted under the direction of the Lessees whose account of disbursements for the concern having been approved of was assumed by the Company and paid accordingly.[6]

In practical terms this meant the call of £1 made by whilst the mine operated under a cost book was repaid; the adventurers had made sure that they recouped their initial outlay! 1845 could not have started better for the Company. From the first sale of ore (382 tons, 1 cwt, 2qrs.) in January up to April 1st ore sales had realised £19,157 6s 9d and by April 1st 1845 a profit of £17,940 13s had been made.[7]

The process of selling the ore to the smelters was quite involved, being a two stage process, the first stage being sampling which took place at Morwellham and ticketing which took place at Truro three weeks after sampling.[8,9] Writing in 1850, J. H. Murchison has left us a characteristically detailed picture of the process as practiced during this period:

> The samplings of ore take place on the last Friday in every month. This having occurred during my stay in Tavistock, I willingly accepted the kind offer of Mr. Morris to drive me to Morwellham Quay on the River Tamar, to witness this operation. The quay is about four miles from Tavistock, and five miles from the mines. New Quay being about half a mile down the river. The ore is brought down in carts.... and put in a well paved yard; that from different mines being kept separate. It is then divided into six doles, or equal weighted parts, two of which are again divided across the middle, and the sample taken from the centre of the dole. The samples are taken and separately mixed by several persons, on a square iron plate, fixed in the middle of the floor, and set upon sand, and upon the iron plate the ore is riddled, and ground to powder, by men with "bucking irons", after which it is again well mixed, and samples taken and put into small bags, one for each agent of the purchasers and for the mines. These are taken away to be assayed, in order to ascertain the price to be paid at the sale or "ticketing". This Company have their samples assayed by an assayer at Truro, and another at Tavistock...... When bought, a mark is put upon the doles; that for instance, of Williams, Foster & Co., being a brick and a turf.....

> On the ticketing days a dinner is given by the mines having ore for sale, which is attended by their agents, and those of the smelters. Before dinner, tickets, upon which are written offers for the different lots, are handed by each of the latter to the chairman (see Note 1), who reads them aloud, and the highest bidders obtain them; but if two or more offer the same price for any lot, it is divided equally between them; and if it is not intended by an agent to purchase any particular parcel, so low a price is offered as to preclude the possibility of his obtaining it.[10]

The Bedford Estate was delighted with progress being made; writing from the office in Tavistock on 10th January 1845 John Benson was almost eulogistic:

> I yesterday went to the new mine opened by Thomas & Co. and certainly it is one of the most wonderful things ever known..... such has been their good luck that with an expenditure of I should think not more than £500 or £600 they are likely to sample 100 tons of rich ..... ore this month which is supposed to be worth about £10 per fathom ......[11]

Benson went on to contrast the management and operation of Devon Great Consols and Bedford United with local endeavour:

> Both these mines you know are worked, by London Companies and certainly they set about their works more systematically than any of the people here have done – Gill & Co. and the others who have dabbled here never go on without they are near a prospect of pay & they have not been made to work so closely as they might (see Note 2).[12]

The Main Lode was certainly living up to expectations; the operations at Wheal Maria were proving highly satisfactory. Gard's Shaft had been sunk to a depth of 28 fathoms and the lode varied in width from nine to eighteen feet wide.[13] In addition to developing Wheal Maria, Hitchins, in his role as agent, had been busy proving the lode beyond Wheal Maria, to the east of the Great Crosscourse. By the time of the first A.G.M. he had "very recently opened the lode 650 fathoms from Wheal Maria, which from the distance will hereafter be designated as a distinct mine, and has been, in accordance with the wishes of the agent named Wheal Josiah."[14]

To work the two mines efficiently Hitchins recommended that the Company should enter in to a contract for the purchase of two steam engines, one for Wheal

Maria, one for Wheal Josiah.[15] Both engines were designed by William West and were built by the Perran Foundry.[16]

1846

At the second A.G.M., held on the 5th May 1846, at Barge Yard Chambers, the number of directors was increased from three to five, William Morris and John Thomas joining Richard Gard, William Thomas and Thomas Morris. The role of auditors was filled by Joseph Deane Browne of Jermyn Street and Charles Fox of Perran near Truro.

The twelve months up to the 31st January 1846 had been profitable indeed: ore sales totalled £116,068 14s whilst costs totalled £42,445 17s 11d, leaving a profit of £73,622 17s 1d. On the back of such large profits six dividends had been declared, a total of £72,704 having been divided amongst the shareholders, amounting to £71 per share.[17] This prosperity had a phenomenal impact on the Company's share value, £1 shares changing hands for as much as £800.[18]

The prosperity of the mine was the more remarkable in that all the ore had been raised from Wheal Maria.[19] By April 1846 Wheal Maria was a hub of activity, a 24′ x 4′ waterwheel was in use for pumping. Two other wheels, one for hoisting and one for crushing, were on the mine and nearly ready for erection. Additionally a 40-inch steam engine, costing £1,250, had been delivered, although had not yet erected. Ancillary buildings such as a count house, stables and workshops for engineers, blacksmiths, carpenters, sawyers, shaft and timber men had been constructed. The deepest point in the mine was Gard's Shaft which was sinking below the 40 fathom level (see Note 3), the 40 being the deepest level in the mine. To the east of Gard's Shaft lay Morris' Shaft which had reached a few fathoms below the 28 fathom level before work had been suspended due to "quickness of water". To obviate this, the Company intended to erect the 40″ engine at the shaft.[20] Morris' Shaft was sunk on a major fault or crosscourse, presumably for ease of sinking, however this would certainly have contributed to problems with water. The crosscourse, known the Great Crosscourse (see Note 4) is one of the major geological structures in the district, heaving the Main Lode approximately eighty fathoms south,[21] it also marks the boundary between Wheal Maria and Wheal Fanny.

Named after Josiah Hitchins' infant daughter, Wheal Fanny lies on the eastern side of the Great Crosscourse. At the end of April 1846 it was very much a new mine. Eastern Engine Shaft had been sunk to a depth of 23 fathoms below surface and the only level was the 15 fathom. At surface a "first rate water engine with powerful crushing apparatus had been erected".[22]

Between Wheal Fanny and Wheal Josiah lies Wheal Anna Maria; like Wheal Fanny very much a new mine. By 31st April 1846 work had started on sinking Engine Shaft which had reached a depth of eleven fathoms. At surface the lode had been "laid open..... for 100 fathoms" exposing the "splendid character of gossan".[23] Wheal Anna Maria was named after "Her Grace the Duchess of Bedford".[24]

Work at Wheal Josiah had progressed further than at either Wheal Fanny or Anna Maria. By the 31st April work had started on sinking two shafts, sixty fathoms from each other. The easternmost shaft, Hitchins', had reached a depth of thirty five fathoms whilst the eastern shaft, Richards', had reached 28 fathoms. A steam engine had been erected on the mine and was pumping from both shafts; the engine, including bob, rods etc., was valued at £1,400 (see Note 5). Work had also started on driving the 35 fathom level from Hitchins' Shaft. At this depth the Main Lode looked promising although payable ore had not yet been encountered, the lode comprising of gossan with good stones of ore.[25]

In addition to ongoing development on the Main Lode work was also being carried out elsewhere on the sett. It will be recalled that the lease specified that the Company were obliged to work "the lode at Frementor". Frementor lies in the south west of the sett and, unlike the other lodes in the set, lies in granite. In accordance with the lease and adit level was being driven at Frementor in April 1845. Work was also underway driving an adit at Wheal Jack Thomas in the south east corner or the sett (see Note 6).[26]

As early as December 1845 the Company was considering constructing a railway to link Wheal Maria with the quays down at Morwellham. Plans solidified during the spring of 1846 however the Company and the Estate could not agree terms and by July 1846, the plan was formally abandoned by the Company directors (see Note 7).[27] Plans to construct a railway between the mine and Morwellham would not resurface until the late 1850s.

1847

The 3rd A.G.M. was held on the 3rd May 1847 and, as usual was held at the London office in Barge Yard. Richard Gard was in the chair and the board of directors remained unchanged. In recompense for their past year's service the directors were paid £200. Josiah Hitchins as agent was paid 6d in the pound on all profits made. It is instructive to contrast the directors pay with that of the men developing the mine and extracting the ore: the average monthly earnings of tutworkers was £3 2s 7d whilst that of tributers was slightly higher at £4 15s 3d per month.[28]

During the year ending 30th April 1847, 15,709 tons of ore had been sold realising £103,899 7s 1d; however the year had been an expensive one for the mine, expenditure being described as "enormous"; in consequence "only" £25,600 was divided between the shareholders. This is not surprising given the huge amount of development work and exploration being undertaken along the length of the Main Lode. A major item in the mines expenditure was timber, during the year ending April 1847 £5,426 12s 10d was spent on this item alone.[29]

At Wheal Maria the 40-inch engine on Morris' Shaft had been completed and was actively keeping the water in check. The 40″ was also pumping from the nearby Gard's Shaft via a 60-fathom run of flat rods thus replacing the 24′ by 4′ waterwheel. As well as pumping from the two Wheal Maria shafts the engine was also pumping the two engine shafts at Wheal Fanny via two, 100 fathoms long, flat rod runs.[30]

In addition to the 40-inch steam engine on Morris' Shaft the two water wheels mentioned in the 1846 report were progressing well. By the end of April 1847 the 40-foot "grinding wheel" (see Note 8) was in service and was capable with dealing with ten tons per hour. An extensive ore floor capable of holding five hundred tons and a "house for weighing machines" had also been constructed. The second wheel, reported as being 40 feet in diameter, was, by April 30th 1847, reported to be within a week of completion (see Note 9). The intention was to use it for hoisting from Gard's and Morris' Shafts, replacing horse whim haulage which was costing the mine £50 a month.[31]

Underground Gard's Shaft had been sunk down to a depth of 60 fathoms. The bottom level was the 50 fathom where the ore ground was proving of poorer quality than that in the shallower levels. The 40 fathom level and those above it had all been driven east as far as the Great Crosscourse. Thanks to the 40-inch engine, work had restarted on Morris' Shaft which was down to a depth of fifty fathoms.[32]

In the year ending 30th April 1847, 251 fathoms, 5 feet, 6 inches had been driven at a cost of £852 4s 5d.[33]

At Wheal Fanny embankments had been constructed to form reservoirs for "the important objects of keeping back, and economising the surface water; for application for the more important operations of grinding and dressing; and for all the other more subordinate purposes".[34]

Two shafts were being sunk: Eastern and Western Engine Shafts, the two shafts being connected on the 15 fathom level. By 30th April 1847, Eastern Shaft had been sunk below the 25 fathom level, the intention being to start driving the 35 fathom level east and west. As has been previously mentioned both shafts were being pumped by the 40-inch engine on Morris' Shaft via long flat rod runs.[35]

During the year ending 30th April 1847, 245 fathoms, 3 feet, 9 inches had been driven in Wheal Fanny at a cost of £1,231 0s 9d.[36] At Anna Maria an extensive dressing floor had been constructed by the end of April 1847. This floor not only served Anna Maria but also Wheal Josiah where there was a distinct lack of surface water.[37]

Anna Maria Engine Shaft had reached the 30 fathom level where a plat and bob were being constructed with a view to sinking deeper. The shaft was being pumped by the steam engine at Wheal Josiah via a one hundred and seventy fathom long flat rod run. The 30 fathom level had been driven both east and west of the shaft however no payable ore had been found, the lode containing a high proportion of mundic which, at this stage in the mine's history, was not profitable to recover.[38]

In addition to development work from Engine Shaft an adit upwards of 150 fathoms long had been driven with the intention to "ease the underground water not only in this concern but also in Wheal Josiah, rather than under expectations of discovering ore ground".[39] This adit became known as the Blanchdown Adit (see Note 10).

At Wheal Josiah, by the end of April 1847, Hitchins' Engine Shaft was down to the 50 fathom level, whilst Richards' Engine Shaft was seven fathoms below the 50. The 40 and 50 fathom levels were being driven from both shafts with the intention of communicating on the 50. As Wheal Josiah went deeper the lode was improving, with good payable ore being encountered on the 50 fathom level.[40]

In year ending 30th April 1847, 438 fathoms, 2 inches had been driven at a cost of £1,654 16s 2d.[41] At Wheal John Thomas and Frementor work continued driving their respective adits.[42]

1848

The 1848 Annual General Meeting, held on the 1st May, saw some significant changes to the management structure of the Company. Firstly William Morris had died, his place on the board of directors being taken by Francis Morris who had been elected to the board at an E.G.M. held on the 23rd December 1847. Secondly, and more significantly for the day to day running of the mine, Thomas Morris was appointed resident director. Morris also took over the day to day financial management of the mines from Josiah Hitchins. Hitchins continued to look after operations on the mine, retaining the role of "principal superintendant of the mining department".[43]

During the year preceding the 1848 A.G.M., 14,413 tons of ore had been sold, realising £101,916 13s 8d.[44] At Wheal Maria it was noted that a 50′ x 4′ hoisting wheel had been completed resulting in a saving of £60 per month in comparison with the horse whims previously in use. In addition to hoisting the new wheel was also pumping Gard's and Morris' Shafts. This cannot have been an ideal arrangement, however it freed up Morris' engine which was being used to pump both Wheal Fanny and Anna Maria, the Anna Maria flat rod run being two hundred fathoms in length.[45]

Dressing facilities at Wheal Maria had been extended by the end of April 1848; additional floors had been constructed to cope with halvans and slimes dressing. This was an important development in that it marks the beginning of the mines very successful policy of dressing lower grade ores.[46]

By the end of April Morris' Shaft was down to the 70 fathom level where a plat was being cut for a cistern and, "lodgement of stuff", the intention being to sink a further ten fathoms and to drive an 80 fathom level. Unfortunately the lode was not proving rich at depth, the lode as encountered on 60 fathom level being unproductive. Gard's Shaft had also been sunk down to the 60 fathom level with no better results. At the time of the 1848 A.G.M. there was no intention to sink Gard's any deeper owing to the hard nature of the ground in comparison with that at Morris'. Whilst Wheal Maria was proving disappointing at depth the shallower parts of the lode were much more satisfactory, the 28 fathom level, for example, had opened up, some very promising "tribute ground".[47]

At Wheal Fanny the "hawling machine" had been completed and was in service. To keep the wheel supplied with water additional reservoirs and associated leats had been constructed.[48] In line with developments elsewhere on the mine, the Wheal Fanny dressing floors had been expanded.

Nine men were engaged on the sinking of Wheal Fanny Eastern Engine Shaft which had reached a depth of 48 fathoms. Western Engine Shaft was not quite so advanced; by the end of April 1848 sinking was going on below the 35 fathom level. The 25 fathom level was proving as rich as any ground thus encountered anywhere on the Main Lode. A winze sunk to a depth of five fathoms below the 25 confirmed that the ground remain rich. Indeed the ground remained rich down to the 45 fathom level, the lowest level in the mine. The 35 fathom level was being driven from both shafts with the intention of establishing a connection, presumably to improve ventilation.[49]

The Anna Maria dressing floors were rapidly becoming the most extensive on the mine. The floors were capable of dealing with not only the Anna Maria ore but also the Wheal Josiah ore, a 210 fathom long self acting incline having been built to transport Josiah ore to the floors.[50] In addition to the Anna Maria main dressing floor a subsidiary floor which could hold 500 tons of dressed ore had been constructed.

Using Morris' Engine to pump Anna Maria was very much a temporary measure. In April 1848 a new steam engine was being erected at Anna Maria. The Company had great expectations of the new engine which would be.[51]

> of sufficient power, for the purposes of pumping and effectually keeping under control the water of this concern, and much of that at Wheal Josiah, as intended, by flat rods to Richards Shaft. This engine also possess the power which will be applied to the hawling of all the stuff from this mine, and to the grinding of all the ore to be raised therefrom, as well as Wheal Josiah.[52]

By the end of April 1848 Anna Maria Engine Shaft had reached a depth of 50 fathoms below surface, at which depth the lode, in contrast to shallower levels, was beginning to prove rich. The 50 fathom level was being driven both east and west of the shaft, twelve men being engaged on this work.[53]

At Wheal Josiah adequate surface arrangements were largely in place by April

1848. Both Hitchins' and Richards' Shafts had reached 70 fathoms and good ore was being raised from above the 50 fathom level and below.[54]

During the year a major discovery had been made at Wheal Josiah: South Lode. A whim and other necessary machinery had been erected to facilitate the sinking of a shaft. By the end of April 1848 a shaft had been sunk to a depth of fourteen fathoms. At this depth the lode at this point was dipping to the north; it was about five feet wide and composed mainly of "capel", with a little ore (see Note 11).[55]

South Lode was not the only major discovery during the year to the end of April 1848. The Main Lode had been proved towards the eastern end of the sett, where 60 fathoms of promising ground had been opened up. This eastern extension of the Main Lode was named Wheal Emma "in compliment to the lady, the relict of our late lamented director William Morris Esq".[56]

Work also continued at Frementor, the adit having reached a length of one hundred and twenty fathoms and had intersected several lodes, only one of which had been driven on. The 1848 Annual Report noted that Frementor would be developed at some "future and no doubt distant period." One gets a strong impression that Company was not committed to Frementor which, in comparison to the Main Lode, was insignificant and they were only pursuing development work to comply with the stipulations of their lease.[57]

By early June 1848 the *Devonport Journal* reported that due to the low price of copper "a large number of men" had been discharged from Devon Great Consols.[58]

The discovery of the Main Lode sparked what may be termed a mining boom in the Tamar Valley, setts in the vicinity rapidly being taken up. The ground which naturally excited most interest lay to the east of the Devon Great Consols sett which should contain the eastern extension of the Main Lode. During the latter part of the 1840s the Bedford Estate had received a number of applications for the sett, including one from the Williams family. The potential of the Eastern Ground had not escaped the attention of the Company who had, by September 1848 had made an application for a sett. Mr Thomas argued that the Company had the best claim to the sett because the potential of the eastern ground had only been made apparent by the work already carried out by the Company. Thomas also argued that the Eastern ground lay higher that their sett and consequently the Devon Great Consols workings would unwater the new ground, reducing the cost of working it. Whilst accepting Thomas' arguments and the justice of the Company's claim the Estate did not grant them, or indeed anyone else, a sett.[58]

Christopher Haedy, on behalf of the Estate, felt that the Eastern Ground should be kept in reserve "prolonging employment of people working in the Maria &c. Mines when they are worked out, and not to let the whole be worked out together at the risk of having the population the two setts would draw around them, or give employment to, suddenly thrown out of work, a burden to the district.[69] Haedy suggested that the "Maria Mines" were not close enough to the Eastern ground to prove the mining prospects of the ground; if the ground should prove very rich the Haedy felt that the Estate should have first option of working it. The low price of copper was also cited as a reason for not granting the sett.[61] There the matter of the Eastern Ground rested until 1856-1857 when the Company again made an application for the grant of a sett.

1849

The 5th A.G.M. was held at Barge Yard on the 7th May 1849. Richard Sommers Gard, Francis Morris, Thomas Morris, John Thomas and W. A. Thomas were in post as directors, with W. A. Thomas acting as chairman. The directors each received £250 for their previous year's endeavours. It was announced that 2,167 tons of ore had been raised during the year and dividends of £30,720 had been divided between the shareholders.

The meeting itself was a significant one. Prior to 1849 a mixture of water and steam power was employed on the mine, however as the mine developed the increasing cost of coal was becoming prohibitive. To that end the decision was taken to adopt water power where practical:[62]

> The increasing depth of shafts and the extension of the levels of the mine have demanded a corresponding power of machinery. It has been the anxious desire of the directors to adopt that which at the same time would be most effectual and most economical. With this view, they have consulted other engineers as well as those of the Company, and after careful deliberation they have decided on adopting water – power in preference to steam power, it having been satisfactorily proved by competent parties that a great annual saving would be effected thereby at a comparatively trifling increase of outlay.[63]

Whilst "other engineers" may well have been consulted the decision to adopt water power was very much a reflection of Hitchins' thinking at the time. This would have been heavily influenced by his experience at Bedford United; particularly the extension of the Tavistock Canal leat from the Impham Valley to the mine in 1841-1842 and the erection of powerful wheels on both Marquis and North Lodes.[64,65]

As usual the Annual Report contained Hitchins' report on progress at the mine, the 1849 report being dated 4th May. At Wheal Maria, Morris' Shaft was three fathoms below the 80. Unfortunately the lode was proving poor at this depth: drives on both the 60 and 80 fathom levels had shown the lode to be "without character equal to induce further prosecution at these levels, being composed chiefly of coarse and uncongenial spar and caple, intermixed with mundick only". In the light of this the plan was to sink the shaft down to a depth of 100 fathoms to ascertain whether the lode improved with depth. As Wheal Maria went deeper the quality of the lode was decreasing; the real riches of the mine lying in the shallow levels. At Gard's Shaft the ground above the 50 was still proving rich, the average tribute being paid being 5s 2d in the pound.[66]

Wheal Fanny had not reached the same depth as Wheal Maria: Eastern Engine Shaft was three fathoms below the 55 and was being sunk by nine men who were making slow progress due to the hardness of the ground. Western Engine Shaft had reached a similar depth, having just reached a depth of fifty five fathoms. The miners were encountering good ore particularly in the western sections of the mine. However the ore ground on the 50 fathom level, in the vicinity of Eastern Shaft was not as promising.[67]

At Anna Maria the agent's house had been finished. The steam engine mentioned as under construction in April 1848 had also been completed and was pumping both Anna Maria and Wheal Josiah, however due to the amount of water the engine was, hardly surprisingly, unable to drive a crusher as well. This meant that the ore which should have been crushed at the Anna Maria dressing floor had to be transported to the floor at Wheal Fanny for grinding which must have been both inconvenient and expensive. Having said that, this rather clumsy arrangement was transitional in nature and arose as a consequence of the new policy to employ water power. To this end a large waterwheel was under construction at Blanchdown which was intended to pump both Anna Maria and Josiah. The construction of this new wheel meant that the engine at Wheal Josiah, which had formerly been employed as a pumping engine, was now in use as a whim, hauling ore from both Anna Maria and Josiah. Hitchins noted that this represented a monthly saving in the region of £70. (5th Annual Report, 1849) By May 1849 Anna Maria Engine Shaft was five fathoms below the 60 fathom level. The lode on the 60 fathom level was unproductive however the foot of the shaft had cut good ore ground; an augury of things to come.[68]

The shafts at Wheal Josiah were both sinking below the 80 fathom level by May 1849, the 80 connecting both shafts. Richards' Shaft was five fathoms below the 80 in tolerably good ore ground and Hitchins' was three fathoms below and in good ground.[69] In addition to work on the Main Lode at Wheal Josiah a crosscut was being driven south from the 60 fathom level to intersect South Lode. No progress appears to have been made at South Lode Shaft which remained at a depth of fourteen fathoms.

Work had commenced at Wheal Emma by this point. A count house had been constructed as had smiths' and carpenters' workshops. Work had also started on sinking an engine shaft, which had reached a depth of twenty two fathoms by May 1849, although it had yet to cut the lode.[70]

Since Hitchins' previous annual report at the end of April 1848 good progress had been made driving the Frementor adit with upwards of seventy fathoms having been driven, bringing the total length of the adit to 190 fathoms. Limited work had also been undertaken on lode although this had since been suspended.[71]

Hitchins' plan to harness the power of the River Tamar to drive a series of extremely powerful waterwheels was well under way. An agreement had been reached with the Duchy of Cornwall to abstract water from the River Tamar for an annual rent of £250.[72,73] By November 1849 the first phase of the work had been completed, albeit at huge expense.[74]

A huge leat, known as the Great Leat, two miles long and up to eighteen feet wide had been constructed from a weir at Latchley down to Blanchdown. The weir itself predated the Great Leat, having been constructed by South Wheal Maria Mine on the Cornish bank of the Tamar. In addition to the "Great Leat" a second leat 1,600 fathoms long was constructed to recycle water from the mines and deliver it to the wheel at Blanchdown.[75,76] A sketch map of September 1850 shows this leat running from Wheal Maria to Blanchdown.[77]

At Blanchdown the leats fed a massively powerful wheel, constructed by Messrs. Nicholls, Williams and Co. of Tavistock, forty feet in diameter with a twelve foot breast, turning at between four and a half and five revolutions a minute, developing in the region of 140 horsepower.[78,79,80,81] It had an oak axle five feet in diameter running in fifteen inch diameter journals. The wheel had 112 buckets, each "formed of two deal boards, whose ends rest in sockets formed in the cast iron rings or shroudings".[82] The arms of the wheel were also deal. In consequence of the great weight and huge torque of the wheel the arms were arranged in four series. The wheel's feed was interesting, being fed at two points. The main supply being brought in about ten feet above centre in a ten foot wide launder; this supply was supplemented as required by an "occasional auxiliary stream of mine water on top of the wheel".[83] The weight of the cranks and connecting rods were balanced by two balance bobs placed behind the wheel.[84]

The wheel was connected by 399½ fathoms of flat rods to pitwork at Wheal Anna Maria Engine Shaft, Richards' Shaft and Hitchins' Shaft. The flat rods, constructed of three and a quarter inch round iron bar, were "very strong....., working upon rollers, on the top of very substantially erected supports, and of which there were upwards of 130".[85,86,87,88,89,90]

Chapter 4 Notes

Note 1

The Chair was taken by the mine offering the largest quantity of ore for sale, usually Devon Great Consols.

Note 2

Gill & Co. was a local company who were heavily involved in Tamar valley metal mining, particularly via their involvement with the Tavistock Canal of which they were major shareholders. In the course of their business Gill & Co. developed a reputation for both parsimony and a certain acuteness in their dealings.

Note 3

In Wheal Maria the depths of levels were measured from the collar of Gard's Shaft.[91]

Note 4

Some early references refer to the Great Crosscourse as Morris' Crosscourse (see also Appendix 2 "Geology and Lodes").

Note 5

The engine erected on Wheal Josiah was, in the context of the other Devon Great

Consols engines, rather exotic being a Sims 16″/30″ compound.[92] Barton,[93] whilst agreeing that the Wheal Josiah engine was a 16″/30″ compound incorrectly suggests that it was a horizontal whim manufactured by the Perran Foundry and was erected by William West in 1844.

Note 6

This lode lies about a third of a mile south of the Main Lode and north of Bedford United North Lode. Initially worked as Wheal Jack Thomas or Wheal Thomas it was worked in the latter part of the nineteenth century as Watson's Mine after Peter Watson who became the company chairman and managing director in the late 1870s.

Note 7

The story of the 1846 railway proposal is comprehensively documented in Dickinson.[94]

Note 8

Presumably a roller crusher.

Note 9

The 1847 annual report notes that this wheel was forty feet in diameter; this may be an error as the 1848 annual report records this wheel as being fifty feet in diameter by four feet breast.

Note 10

The Blanchdown Adit was driven as a crosscut. The adit cuts New South Lode, South Lode and the Main Lode. New South Lode is encountered ninety five fathoms from the portal where it cuts the 60 fathom level. South Lode is cut at one hundred and five fathoms on the 60 and Main Lode is cut at one hundred and thirty five fathoms also on the 60, thirty two fathoms east of Hitchins' Shaft. In Wheal Josiah and Wheal Emma adit level is the 60 fathom level, whilst in Wheal Anna Maria this becomes the 40, in Wheal Fanny the 35 and in Wheal Maria the 28 Fathom Level.[95]

Note 11

References to South Lode are apt to be confusing (at least the current author finds them so!). There appear to be two South Lodes and differentiating between the two can be problematical. Most important is the South Lode which branches off from Wheal Anna Maria, runs parallel to the Main Lode and rejoins the Main

Lode in Wheal Josiah. A lode equating to this is clearly shown on Symons' plan of the Tavistock mining district of 1848.

The second South Lode, not shown on Symons' 1848 map, appears to be a continuation of the Jack Thomas/Watson's mine lode. The 1867 lode map makes reference to a South Lode Shaft sunk on this lode. Likewise certain references from the 1850s mentioning South Lode certainly refer to this lode. However the 1884 Ordnance Survey map and a plan of this section of the mine included in the abandoned mines record allude to this section of the mine as South Fanny and South Wheal Fanny respectively.

To avoid confusion, the 1848 South Lode associated with the Main Lode is referred to as South Lode whilst the second South Lode is referred to as South Lode (South Fanny).

Chapter 4 references

1. *Mining Journal* 26 October 1850
2. Devon Record Office document L1258, 1844 lease
3. Goodridge J. C., 1964, Devon Great Consols, a study of Victorian mining enterprise, Transactions of the Devonshire Association, Vol. XCVI. pp. 228-268.
4. *The Jurist* Vol. XVI , Part II, 1853
5. 1st Annual report 1845
6. 1st Annual report 1845
7. *Mining Journal* 26 October 1850
8. *Mining Journal* 26 October 1850
9. *Tavistock Gazette* 2-23 December
10. *Mining Journal* 26 October 1850
11. Devon Record Office document L1258, Benson letters, Bundle 54
12. Devon Record Office document L1258, Benson letters, Bundle 54
13. *Mining Journal* 26 October 1850
14. 1st Annual Report 1845
15. 1st Annual Report 1845
16. Browne, W., 1848. *The Cornish Engine Reporter* No.12, January 1848, St. Austell.
17. 2nd Annual Report, 1846
18. *Mining Journal* 26 October 1850
19. *Mining Journal* 26 October 1850
20. 2nd Annual Report, 1846
21. Phillips J. A., 1884, *A Treatise on Ore Deposits*, Macmillan & Co.
22. 2nd Annual Report 1846

23. 2nd Annual Report 1846
24. *Mining Journal* 26 October 1850
25. 2nd Annual Report, 1846
26. 2nd Annual Report, 1846
27. Dickinson M, 1985, The Duke's men, the Wheal Maria people and others, the story of the Devon Great Consols "ghost railways", Tamar Journal, Vol. 7, pp. 31-40.
28. 3rd Annual Report 1847
29. 3rd Annual Report 1847
30. 3rd Annual Report 1847
31. 3rd Annual Report 1847
32. 3rd Annual Report 1847
33. 3rd Annual Report 1847
34. 3rd Annual Report 1847
35. 3rd Annual Report 1847
36. 3rd Annual Report 1847
37. 3rd Annual Report 1847
38. 3rd Annual Report 1847
39. 3rd Annual Report 1847
40. 3rd Annual Report 1847
41. 3rd Annual Report 1847
42. 3rd Annual Report 1847
43. 4th Annual report, 1848
44. *Mining Journal* 26 October 1850
45. 4th Annual report, 1848
46. 4th Annual report, 1848
47. 4th Annual report, 1848
48. 4th Annual report, 1848
49. 4th Annual report, 1848
50. 4th Annual report, 1848
51. 4th Annual report, 1848
52. 4th Annual report, 1848
53. 4th Annual report, 1848
54. 4th Annual report, 1848
55. 4th Annual report, 1848
56. 4th Annual report, 1848
57. 4th Annual report, 1848
58. *Devonport Journal* 8 June 1848
59. Devon Record Office document L1258 add 8m/E11/15

60. Devon Record Office document L1258 add 8m/E11/15
61. Devon Record Office document L1258 add 8m/E11/15
62. 5th Annual Report, 1849
63. 5th Annual Report, 1849
64. Waterhouse R. E. in prep.
65. Watson J. Y., 1843, *Compendium of British Mining*, London.
66. 5th Annual Report, 1849
67. 5th Annual Report, 1849
68. 5th Annual Report, 1849
69. 5th Annual Report, 1849
70. 5th Annual Report, 1849
71. 5th Annual Report, 1849
72. *Mining Journal* 12 May 1849
73. *Mining Journal* 26 October 1850
74. 6th Annual Report 1850
75. *Mining Journal* 26 October 1850
76. *Mining Journal* 7 July 1860
77. *Mining Journal* 2 November 1850
78. Andre G. G., 1878, *A Descriptive Treatise on Mining Machinery, Tools and other Appliances used in Mining,* Vol. 2, E. & F. N. Spon.
79. *Mining Journal* 18 May 1850
80. *Mining Journal* 29 June 1850
81. *Mining Journal* 26 October 1850
82. Andre G. G. 1878, *op. cit.*
83. *Mining Journal* 7 July 1850
84. Andre G. G. 1878, *op. cit.*
85. *Mining Journal* 29 June 1850
86. *Mining Journal* 10 May 1851
87. *Tavistock Gazette* 2 December 1864
88. *Tavistock Gazette* 9 December 1864
89. *Tavistock Gazette* 16 December 1864
90. *Tavistock Gazette* 23 December 1864
91. Dines H. G, 1956, *The Metalliferous Mining Region of South West England*, H.M.S.O.
92. *Mining Journal* 27 July 1860
93. Barton D. B., 1966, *The Cornish Beam Engine*, D. Bradford Barton Ltd.
94. Dickinson M, 1985, The Duke's men, the Wheal Maria people and others, the story of the Devon Great Consols "ghost railways", Tamar Journal, Vol. 7, pp. 31-40.
95. Dines H. G., 1956, *op. cit.*

Chapter 5

Consolidation and growth 1850 – 1859

"One Could almost imagine that all the copper in the neighbourhood was concentrated in this spot" – Mining Journal 1858

As the new decade began, the mine was in a wholly admirable position: the Main Lode had been proved along its length, ongoing exploration and development work confirming and indeed surpassing initial expectations. The 1850s would prove to be the most profitable in the mine's history. Copper prices were booming and no one was in a better position to meet that demand than Devon Great Consols. During the 1850s the mine's management could have become complacent, content to ride the prosperity of the copper boom; however and much to their credit, they constantly strove to increase returns and reduce costs. To this end large capital projects were embarked upon throughout the decade notably the ongoing development of the water power system, the expansion of dressing facilities to process lower grade copper ores and the construction of a railway linking the mine to the quays at Morwellham. The Directors also had the vision to look beyond the boundaries of their current sett, finally securing a lease for the "Eastern Ground" from the Estate in 1857.

1850

It is somewhat ironic that such a successful decade should start on a less than positive note Writing of a visit to the mine in January 1850 J. H. Murchison recalls, in outraged tones, a strike by child pickers employed on the mine's dressing floors:

> I had scarcely arrived at the mines, on the first day of my first visit (in January last), than about 200 of these youngsters of both sexes came clamouring to

> the counting – house, complaining that their wages had been "cut", and that they could not live on 10s. 6d. a week. Of course, their demand for higher wages was not acceded to, upon which the whole number, influenced no doubt by one or two obstinate characters, left the mine and went home. Next morning they returned; but they were very properly told that they were not wanted, much to their surprise and disappointment, as they thought they were indispensable. It appears that, previous to the rise in provisions in 1847, the wages of the young men were 1s. 9d., and the young women 10d. a day; but which were then advanced to 2s. and 1s.; at that time the sack of wheat was 25s. to 30s. About a year ago their wages were again reduced to 1s. 9d. and 10d.; the present cause of complaint was the notice of a further reduction to 1s. 8d. and 9d. – the price of a sack of wheat being now only 10s. 6d. If these healthy and well fed young people were aware of the earnings of the poor needle women and others in London, who have scarcely more in a week than the former have in a day to live on, with food and lodging, at the same time much larger their complaints would assuredly be exchanged into expressions of thankfulness. Some reflections of this kind had perhaps crossed their minds, or perhaps they felt that "strikes" could only bring punishments on themselves, for on the second day they again presented themselves and offered to resume work on any terms, they were, in fact, engaged at a penny a day less than they had previously been offered, and than was given to strangers and those who had not misconducted themselves.[1]

Although minor in itself, the January 1850 strike may be seen as the precursor of the more significant labour unrest that the mine would experience in later decades.

The 6th A.G.M. of the Company was held on the 6th May 1850 at Barge Chambers. The directors remained the same as the previous year. In terms of dividends the previous year had been a poor one by Devon Great Consols standards with only £3,580 divided amongst the shareholders. The reason for this, the Directors explained, was the very heavy expenditure involved in executing the first phase of the water power scheme, to the 1st March 1850 the project had cost the Company £5,800.[2]

Whilst the water power scheme was the major project that year work also continued apace on the rest of the mine: At Wheal Maria a waterwheel was erected to drive twelve Brunton frames, two of which had been completed by

May 1850. Underground, Morris' Shaft had reached a depth of 99 fathoms. Both the 50 and 60 fathom levels were still unproductive whilst the 40 and 28 were producing good ore. During the year ending 4th May 1850, 241 fathoms, 2 feet, 2 inches had been driven in Wheal Maria.[3]

The plant on the dressing floors at Wheal Fanny had been increased by the addition of nine heads of stamps which were driven by the crusher wheel. Both Eastern and Western Engine Shafts had reached the 65 fathom level, at which point the lode, which measured five feet wide, was proving unproductive. A third shaft 65 fathoms to the east of Eastern Shaft, subsequently known as Ventilating Shaft, had been started to improve ventilation in this section of the mine. By the start of May 1850 this shaft had reached a depth of six fathoms. Whilst the lode was unproductive on the 65 it was proving productive on the 45, the 35 and 15 fathom levels. During the year up to May 1850, 276 fathoms, 6 inches had been driven in the mine.[4]

The commissioning of the "Great Wheel" meant that the Anna Maria engine was no longer required to pump both Anna Maria and Josiah. The engine was converted to a whim for hoisting on Anna Maria Engine Shaft. Engine Shaft had, by May 1850, been sunk six fathoms below the 70 fathom level and was, at last, opening up good ground, twelve men being engaged on this work. During the year only 85 fathoms, 4 feet, 11 inches had been driven at Anna Maria reflecting the poor nature of the lode on and above the 60 fathom level.[5]

Remaining at Wheal Anna Maria the company was in the process of erecting "a powerful crushing and stamping machine" which "will return 40 tons of stamps ore per month from the "halvans".[6] Prior to this date it had not been considered worth dressing second grade ore or "halvans". This was largely due to the incredibly high quality ore being raised in the first years of the mine's operation. The decision to start treating poorer grade ore would lead to the development, under the guiding hand of Captain Isaac Richards, of a sophisticated ore dressing process which would help keep the mine profitable in the hard days of the late 1860s and into the 1870s. Arguably the move to recover lower grade ore was as important and far sighted as the move to water power.[7]

Richards' Shaft at Wheal Josiah was the deepest point on the Devon Great Consols sett. By May 1850 it was two fathoms below the 103 fathom level, at which point the lode was productive. Richards' Shaft and Hitchins' Shaft had been connected on the 90 fathom level. The 90 was productive as was the 80,

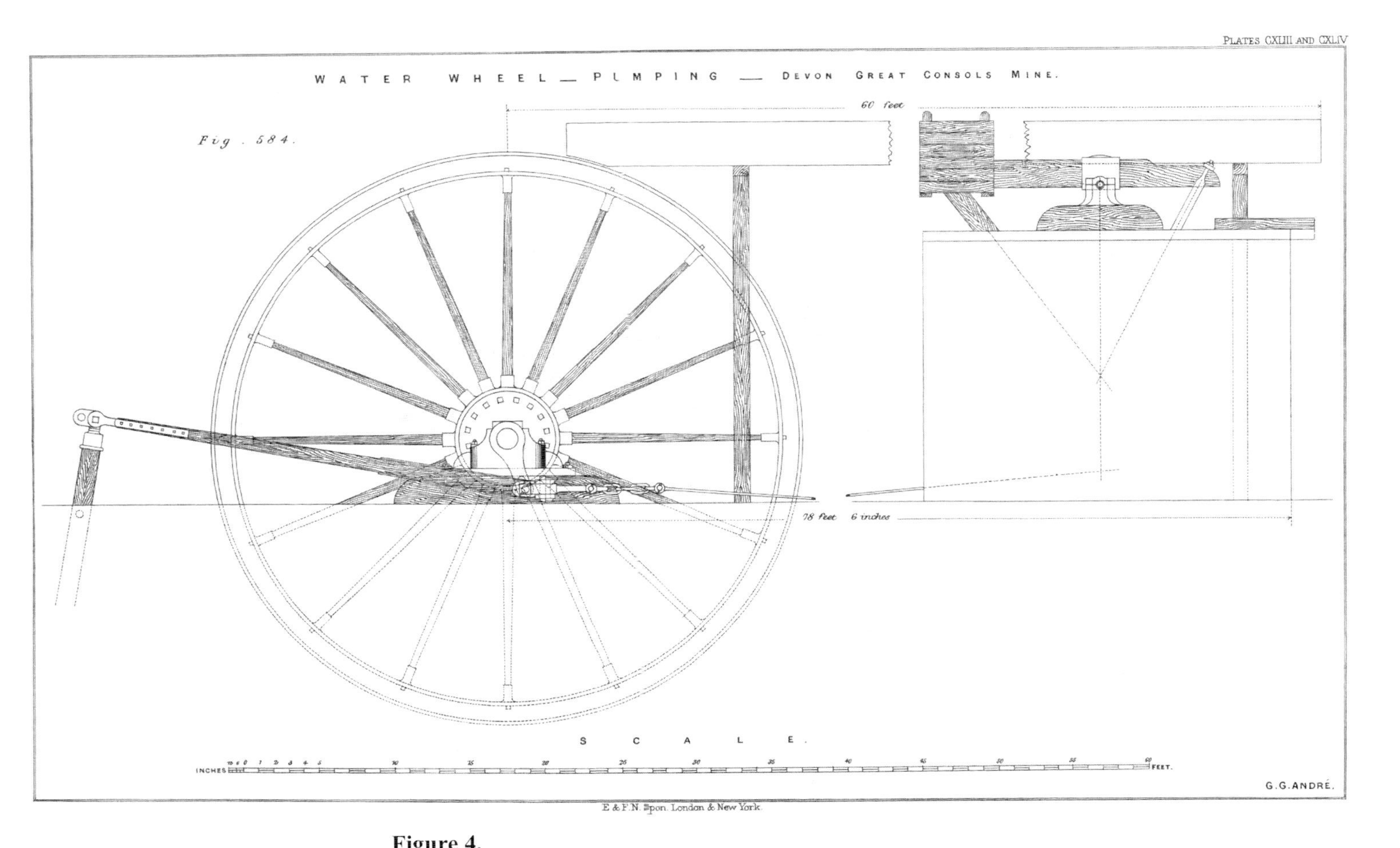

Figure 4.
Side elevation of 40′ x 12′ "Great Wheel" at Blanchdown.

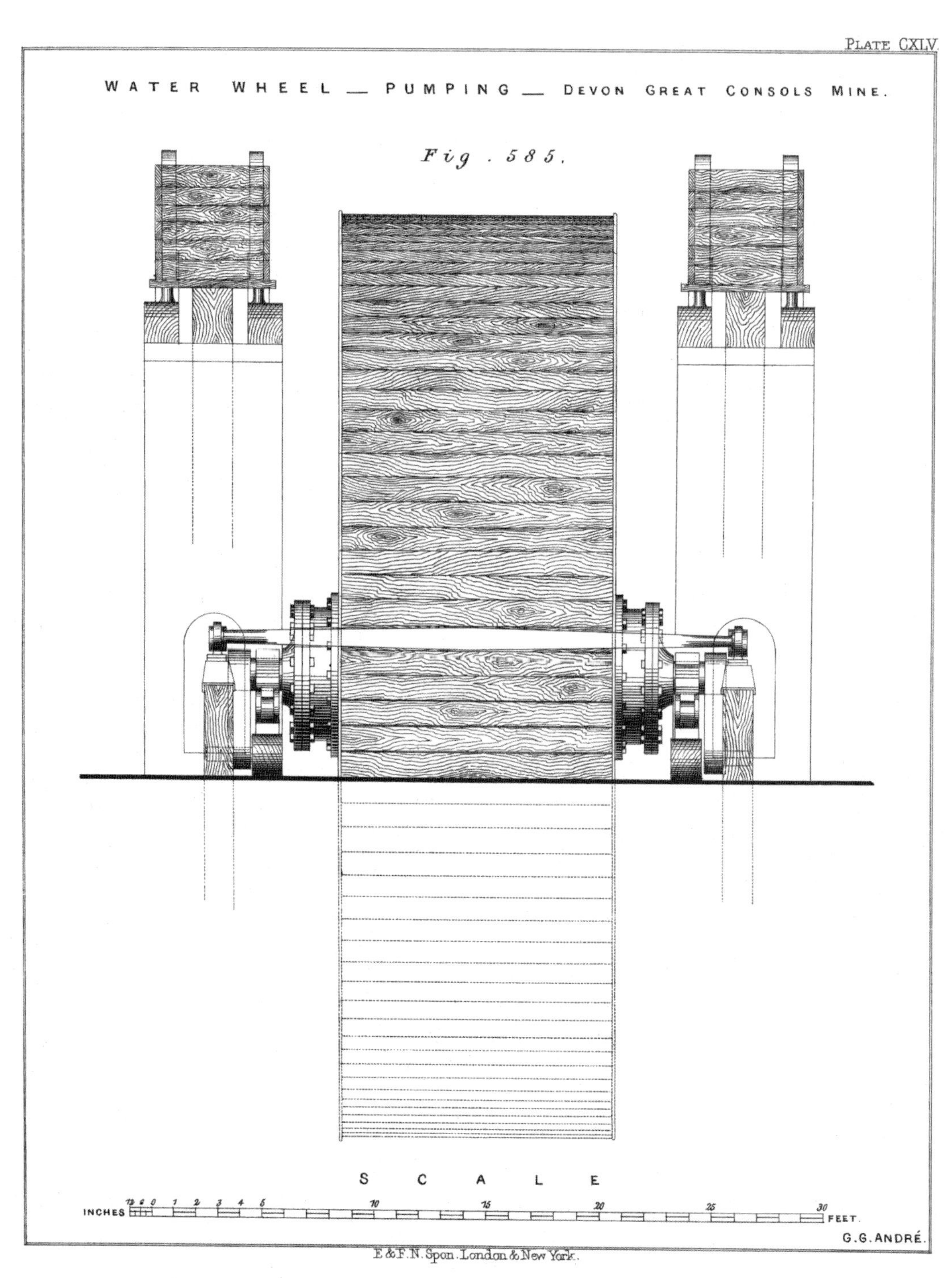

Figure 5. End elevation of 40′ x 12′ "Great Wheel" at Blanchdown.

however the lode on both the 70 and 60 fathom levels was poor. On the 60 fathom level the crosscut had been driven 72 fathoms at which point it intersected South Lode. Unfortunately at this depth the lode was only two feet wide and poor. A second crosscut had been driven north from the 50 fathom level and after only eighteen feet a lode five feet wide was intersected. During the year 375 fathoms, 3 feet, and 2 inches had been driven in Wheal Josiah.[8]

Wheal Emma was still very much at the early development stage: a 40″ engine had been completed and a second engine, a whim, was in the course of erection. Thomas' Engine Shaft had reached a depth of 55 fathoms and whilst the lode had been cut it was "not of sufficient inducement, as yet, to drive on its course". This is reflected in fact that only 21 fathoms and 1 inch had been driven during the year ending May 1850.[9]

Attention had also been paid to Wheal Thomas during the year. A 36-foot diameter wheel had been erected to handle both hoisting and pumping, the intention to try the mine below adit. To improve ventilation work had started on sinking a shaft ninety fathoms from the adit portal.[10] Little progress had been made at Frementor "due to the hard nature of the ground, 73 fathoms, 4 feet and 5 inches having been driven up to the year ending 4th May 1850.

In a year of significant developments a key event at the 1850 A.G.M. was the announcement of the resignation of Josiah Hitchins:

> One of the last acts of the Directors has been to receive and accept Mr Josiah H. Hitchins' resignation of the office of Superintending Engineer and Mine Surveyor. They do not expect now that the mines are in a regular course of working, the interest of the shareholders will materially suffer by that gentleman's retirement from active duty, especially as the directors have nominated him, agreeably to his request, the Consulting Engineer and Mine Surveyor to the Company, in which capacity he will be at all times ready to advise them in cases of difficulty, or on matters of importance.[11]

Hitchins' resignation effectively marked the end of his direct involvement with the mine, his appointment as "Consulting Engineer and Mine Surveyor" being purely honorary and nominal. Why Hitchins resigned is unclear, however H. C. Salmon, writing in 1860, commented that:

> He had such an interest in Devon Great Consols as should have ensured

> him an ample fortune, but his sanguine and generous temperament, which induced him to trust men who were worthy of neither, led him into a sea of speculation which it was beyond the power of any single individual to control or escape from.[12]

Whatever the reasons for his resignation and, indeed, whatever his failings, Josiah Hitchins was fundamental to the early development of the mine. In the role of Superintending Engineer and Mine Surveyor Hitchins oversaw the exploration, development and exploitation of the Main Lode from Wheal Maria in the west to Wheal Emma in the east. Hitchins realised the needs of long term development outweighed those of short term expediency: The development of the mine's infrastructure initiated by Hitchins laid the foundations for the mine's future prosperity. His most significant achievement was the move to water power; which resulted in the development of one of the most sophisticated and complex water power systems seen in British metalliferous mining.

During 1850 J. H. Murchison paid a number of visits to the mine. One such visit he devoted "to viewing the underground works..... accompanied and guided by Captain James Richards":

> The first thing to be done was to change all our own clothes, and put on miners dresses, composed principally of flannel, with duck coats and trowsers, lined with serge. It is very warm underground; but scarcely possible to emerge again dry. We also had to provide ourselves with strong hats, of the "wide-awake" shape, and a few tallow candles, which being lit and embedded in clay, we stuck on our hats while proceeding up and down the ladders, but carried in our hands when going along the levels. Thus decked, we first proceeded down Hitchins' Shaft by ladders, which were each 30 feet long, as far as the 50 fathom level; the shaft being perpendicular for the same distance, and the ladders inclining in every thirty feet nearly from side to side (six feet). The staves were in some cases made of wood, but principally of iron; and a few were ten inches apart, but the greater proportion were twelve inches. For myself, I found the former far more convenient; but the men did not appear to think so. I think, however, there were not a sufficient number of the ten inch apart staves to enable them to judge. From the 50 fathom level downwards the ladders went on the course of the lode, which dipped twenty inches in the fathom. We descended to the 60 fathom level direct, which is 130 fathoms in length; to the end of which we went. No men were at work here until the winze was sunk to the 70, for

Figure 6. Contemporay photograph of one of the locomotives built by Gilkes, Wilson of the Tees Iron Works in Middlesbrough.

Figure 7. Recently discovered panorama of the western end of the mine. The large engine

Figure 8. Devon Great Consols Band, led by Captain Cock, the agent at Wheal Maria.

house on the left stands on Watson's Shaft at Devon Great United.

Figure 9. The Wheal Josiah Counthouse; the posed group appears to be the mine's agents.

Figure 10. Image of Morwellham reputed to date from the late 1860s. The Great Dock, constructed in 1857-1858 is centre left. Prominent on the quay are the doles of copper ore awaiting shipment to South Wales. Also visible is the DGC railway: the incline is seen running down through the woods in the centre of the picture (the shorter incline to the right is the Tavistock Canal Incline).

the purpose of ventilation. The main workings are all ventilated by winzes, and the crosscuts by air machines; for the most part, I found the air very good indeed. We then went down by a winze to the 70 fathom level, and found a fine course of ore all the way between the two levels. We then went direct to the 90 fathom level, and found a good course of ore in the end of the driveage, which was twenty fathoms from the shaft. We then returned to the 80, and back through the level to Richards' Shaft; 50 fathoms. From this we descended to the sump, then about 98 fathoms (588 feet) from the surface. Here we found a good deal of water; and after breaking a few specimens of ore, we re-ascended to the 90, and went westward to the end; 28 fathoms; here the lode was looking well. On reaching the 80 fathom level, we proceeded eastward 85 fathoms, and found an improvement had just taken place. In going along the levels, owing to the quantity of attle in some places, we had to creep on our hands and knees, and could then scarcely squeeze through. From the 80, at Richards' Shaft, we ascended to the surface. At Wheal Fanny, we went down to the 25 fathom level; but this mine was not so dry as the other; the lode was thirty feet wide, very rich the whole breadth, and mixed with a good deal of spar. It was timbered with six pieces of timber over each other, and each piece was about one foot square, and yet they were broken a little. Our proceedings underground occupied between four and five hours.[13]

1851

The 1851 Annual Report paints a picture of a Company confident in both its performance and prospects. Since the previous Annual General Meeting the second of the "Great Wheels" had been completed costing £3,366 7s 6d up to 31st January 1851. The Directors had to report that expenditure on the Great Wheels had exceeded estimates. Having said that, the subsequent history would more than vindicate the Hitchins' strategy. So proud were the company of the Great Wheels that models of them were sent up to the Great Exhibition as was a subscription of 25 guineas. The Company had also purchased the weir at the head of the Great Leat from South Maria mine; in doing so they had secured their water supply. Staying on the subject of water the directors had, during the year, defended a law suit regarding pollution of the River Tamar. To prevent the issue arising in future the Company bought the rights to the fishery.[14]

With the departure of Josiah Hitchins the role of Thomas Morris, as resident director, became more important. In recognition of the growing importance of the resident director, the Company decided to construct a residence, known as

Abbotsfield, for Morris at Tavistock.[15]

Turning to the mine, Captain James Richards gave a comprehensive report on progress. During the year ending May 1851 the average number of men employed on tutwork was two hundred and seven, at an average monthly wage of £3 7s 3d. During the same period an average of 169 men were employed on tribute earning an average monthly wage of £3 18s 6d.[16]

At surface at Wheal Maria the Brunton frames had been completed. Two additional waterwheels had also been erected for ore dressing purposes, one of twelve feet and one of four feet. Underground 168 fathoms, 6 inches had been driven. Morris' Shaft had reached a depth of 100 fathoms below surface. A crosscut was being driven from the foot of the shaft to intersect the lode. The intention was to make the crosscut large enough to carry flat rods from the shaft, allowing sinking to continue on the lode, a cheaper option than sinking a vertical, or near vertical shaft and crosscutting to the lode. Crosscuts had also been driven south from the 28 and 40 fathom levels to intersect a minor lode known as Woolridge's Lode. The 28 and 40 crosscuts were extended south of Woolridge's Lode, however nothing of interest was intersected.[17]

At Wheal Fanny driveage totalled 269 fathoms, 4 feet, 5 inches. Work was progressing on the 55 and 45 fathom levels, with good ore on the 45. Ventilating Shaft had reached the 45 whilst Western Engine Shaft was down to the 75 fathom level where the lode was found to be poor.[18]

The dressing floors at Anna Maria continued to develop reflecting the decision to expand the processing of halvans. By May 1851 work on the crushing and grinding machine had been completed. Underground Anna Maria Engine Shaft was five fathoms below the 80 fathom level in good ground, as was the 80 itself. During the year 155 fathoms, 6 inches had been driven in Anna Maria.[19]

The Company intended to erect a steam whim between Anna Maria and Wheal Josiah to speed up hoisting.[20] In the year preceding Richards' 1851 report 520 fathoms, 2 feet, 8 inches had been driven in Wheal Josiah, at that time the focus of development of the mine. Richards' Shaft was the deepest on the mine having reached a depth of nine fathoms below the 115 level. The shaft was being sunk through good ground by twelve men, to prove the Main Lode at depth. In this vicinity the 103 fathom level was proving productive. Field Shaft had reached the 103 fathom level and Hitchins' Shaft was seven fathoms below the 103.[21]

Development was continuing apace at Wheal Emma. The steam whim had been completed. To drive pumps in the Inclined Shaft workings, flat rods had been run from the engine at Thomas' Shaft to Inclined Shaft. Sinking was ongoing in Thomas' Shaft which had reached a depth of nine and a half fathoms below the 60 by May 1851. Nine men were engaged in sinking the shaft, the intention being to continue sinking down to a depth of 72 fathoms and drive a level at that depth. An adit was being driven in from the Rubbytown Valley, which, by May 1851, had reached a length of 32 fathoms, a further drive of 35 fathoms being required to intersect Inclined Shaft. It was reported that the character of the lode was improving on the 60 with good quality ore in places, albeit with an abundance of mundic. Work was also progressing at the Inclined Shaft which had reached the 32. Twelve men were engaged on sinking, the intention being to intersect the 60.[22]

At Wheal Thomas a flat rod run had been completed from the waterwheel providing power for pumping and hoisting in Engine Shaft and a capstan and shears had been erected on the shaft. Engine Shaft had communicated with the adit and had reached a depth of fourteen fathoms below the 40. The Adit had reached a distance of thirty five fathoms beyond the shaft.[23] Driving the Frementor Adit was proving to be slow due to hard ground.

1852

The eighth Annual General Meeting was held on the 3rd May 1852 at the Company's London offices at Barge Yard. The Directors informed the meeting that the mine was experiencing rising costs due to the increased depth of the mine and increased ore dressing costs. The increase in cost associated with depth is fairly self explanatory: increased depth meant increased pumping and hoisting costs. The increase in ore dressing costs is interesting as it represents a significant change in emphasis on the mine. During the earliest phase of production at the mine poorer grade ores were ignored in favour of the abundant top quality "crop" ore. The best quality ore required minimal dressing mainly consisting of hand reduction and picking. The lesser quality ore or "halvans" required much more extensive dressing to render it marketable, hence the ongoing expansion of the mine's ore dressing facilities, particularly on the floors at Anna Maria.[24]

In addition to the development of facilities to process the increasing tonnages of halvans work also continued on the water power scheme. At the 1852 Annual General Meeting the directors announced that £4,000 had been budgeted to erect another large pumping wheel, known as the "Plunger Wheel", to pump water for ore dressing and condensing from the River Tamar to Wheal Josiah. Whilst not as imposing as the Great Wheels this new wheel must have been an impressive sight, being 30′ x 16′ and driving eighteen inch pitwork. A fourth big pumping wheel had been erected and was engaged in pumping from South Lode (South Fanny).[25]

During the year ending May 1852 the mine sold 18,946 tons 8 cwts of copper ore, realising £110,379 2s 11d. Whilst copper ore was by far and away the main product of the mine it was not the only one: The 1852 accounts contain a very interesting item relating to the sale of "mundic" which realised £214 10s 8d.[26] Whilst insignificant in itself this item appears to be the first recorded sale of "mundic" or arsenopyrite from the mine. What makes this item significant is the huge role that arsenic would play in the mine's fortunes from the late 1860s onwards.

During the course of the year ending May 1852 a monthly average of two hundred and twelve men had been employed on tutwork compared with 176 men employed on tribute. The average monthly wage of the tutworkers was £3 7s, whilst the tributers were, on average, earning £4 14s 7d a month.[27]

Also of note in the 1852 Annual Report was the decision to set up a "casting foundry" on the mine. This was located at Wheal Maria and contained a "powerful hammer" and a "superior lathe"; the machinery being water powered.[28] At the time Devon Great Consols was believed to be unique amongst British metal mines in establishing its own foundry. H. C. Salmon, writing in 1860, attributed this to the fact that the number of shareholders was small and did not include local founders and merchants who would have insisted on supplying the mine with their own goods. Salmon felt that having a foundry on site was an advantage in that it would save time and would allow the mine to monitor quality directly.[29]

At Wheal Maria underground development work was slowing down with only 55 fathoms, 5 feet and 4 inches being driven in the year up to May 1852. Gard's Shaft had been sunk below the 60, however progress was proving slow due to the hardness of the ground. Morris' Shaft had reached the 100 where the lode was proving poor.[30]

Progress was better at Wheal Fanny where 154 fathoms 3 feet had been driven during the preceding year. The most productive ground proved to be in the vicinity of Western Engine Shaft. By May 1852 the shaft had reached a depth of nine fathoms below the 75, the 75 cutting good ground. The ground around Eastern Engine Shaft was not so good, the lode proving unproductive on the 55. Ventilating Shaft was seven fathoms below the 55 in good ground, however sinking had been suspended due to the amount of water being encountered.[31]

In early 1852 Anna Maria was a hive of activity: A 24-inch horizontal whim engine had been erected to work both Anna Maria Engine Shaft and Field Shaft. The former whim engine was employed driving a powerful crusher for reducing halvans, sizing apparatus, eight jigging machines and four round buddles with tossing and packing gear. This plant was capable of dealing with fifty tons a day. In addition two small waterwheels had been erected to work "our improved patent frames" for slime dressing.[32]

Underground at Anna Maria 171 fathoms, 2 feet, 4 inches had been driven up to May 1852. Engine Shaft had reached a depth of between twelve and thirteen fathoms below the 95 were the lode was twenty four feet wide and rich. Drives on both the 80 and 70 fathom levels were cutting good ground.[33] By November 1852 the lode on the 95 west was being described as "a most magnificent course of ore, worth 40 tons per fathom".[34]

Wheal Josiah continued to be the focus of underground activity on the mine with 429 fathoms, 4 feet, 5 inches of driving being recorded in the year up to May 1852. Richards' Shaft had reached the 130 level, where the lode was noted as being reasonable. It was also noted that the lode on the 115 was largely productive whilst the 103 was good. Hitchins' Shaft was down to the 115 level where the lode was twelve feet wide and good. This must have been something of a relief as the lode on the 90 had been disordered by a crosscourse. Field Shaft had also reached the 115, where the lode was only two feet wide; the Company were preparing to sink deeper using six men.[35]

Two hundred and fifteen fathoms, two feet, five inches had been driven at Wheal Emma during the year. Thomas' Engine Shaft had reached the 87 where the lode was described as promising. Inclined Shaft had reached the 63 fathom level. The 47 fathom level at Inclined Shaft had been connected to Thomas' Shaft.

Work had been stepped up on South Lode (South Fanny), in order to give it a fair trial. The Engine Shaft had been sunk to a depth of thirty fathoms from surface by May 1852, the intention being, as has already been noted, to sink deeper. A 30′ x 8′ water wheel had been erected for both pumping and hauling. An adit had been driven for a distance of 69 fathoms, 20 fathoms from engine shaft.[36]

By May 1852 Wheal Thomas had reached a depth of ten fathoms below the 52. At this depth the lode was five feet wide with occasional stones of ore.[37] Between January and December 1852 20,886 tons 14 cwts., 3 qrs. were sold realising £138,728 5s 9d. During the same period 418 tons, 12 cwt. 2qrs. of mundic were sold realising £450 14s 11d.[38]

The dressing of copper ores

Given the increasing importance being placed on recovering lower grade copper ores it is worth departing from the chronological structure of this narrative and examine at in some length the processes as developed on the Anna Maria floors by Captain Isaac Richards, giving as it does some insight into what can be considered "best practice" as far as mid-nineteenth century copper dressing is concerned.

The first stage in the process was to divide the vein stuff into two distinct classes: Crop ore and halvans. Simple hand dressing was all that was required to separate the crop from the halvans, the ore being spalled, cobbed, picked, riddled and jigged. Both the crop and the halvans were divided into two different sizes, the larger known as rough and the smaller known as smalls. The crop smalls (less than ½ inch) were of a marketable size, whilst the rough crop ore (greater than ½ inch) was sent through the crusher to reduce it to a marketable size. That done the crop ore was ready for market.[39,40]

Figure 11. Bal maidens picking copper ore.

However for the halvans the process was only just beginning. As with the rough crop ore the halvans were also sent through the crusher.[41,42] The crushers employed at Devon Great Consols are what have been termed "Cornish Rolls" more commonly referred to as roller crushers. Devon Great Consols employed a variety of roller crushers, a typical example consisted of two crushing rolls, two foot ten inches in diameter, with seven shallow flutes on their surface. One roller was twenty two inches wide, the other twenty four inches wide. The rolls were fixed to strong longitudinal beams which were integral to the walls of the crusher house. The bearing in which the rollers are housed moved horizontally, and a weighted beam kept the rolls in tension. Ore was passed into the rolls via a hopper. Having passed through the rolls the ore fell into a rotating, cylindrical sieve or trommel. Ore which passed through the sieve was ready for the next stage in the process, ore which was too large to pass through the mesh was automatically returned to the feed hopper by a raff wheel and sent through the rolls again. Crop ore had to pass a ½ inch mesh sieve whilst the halvans had to pass a 1/8 inch mesh. In 1878 the cost of crushing one tone of ore was 3¾d.[43,44,45,46,47]

A stream of water carried the crushed ore to a series of sizing sieves or trommels. This apparatus consisted of a series of four inclined cylindrical sieves, each being six feet long and two feet in diameter at one end and eighteen inches diameter at the other. The mesh sizes of the sieves in use in 1857 (mesh size appears to have varied over time see Philips & Darlington 1857, *Mining Journal* 28 July 1860, *Tavistock Gazette* 2-23 December 1864, Darlington 1878) were 1/20 inch (first two sieves), 1/12 inch, and 1/16 inch. The sieves rotated at a rate of 20 revolutions per minute. [48,49]

The coarsest material which was also the richest passed through the larger mesh sections and was mechanically jigged and "tyed" until it was marketable.[50] The finer material which passed through the other two sieves was carried by a stream of water into a separating cistern. The separating cistern was designed by a Captain T. B. Wilken of Wendron. This device separated the finer material into two classes: slimes and roughs.[51]

From the separating cistern the roughs passed at once into a round buddle. The roughs were buddled several times to separate the good ore from the waste. Having been buddled the ore was then jigged and tyed to remove further waste material. Treating the roughs in the round buddles, jiggers and tyes concentrated the ore content from 1½ per cent to 10 per cent.[52] The slimes from the Wilken separating cistern passed into a slime sizing cistern, a device invented by Captain Richards. Using water under pressure the slime sizing cistern separated the feed into rough and fine slimes.[53,54]

The rough slimes were then treated on Zenner's rotating buddle.[55] Zenner's buddle was a modified version of a type of slime buddle first erected at Clausthal in the Harz. The main difference between the round buddle and Zenner's buddle was that Zenner's had a conical rotating table with static brushes or rakes whilst the round had a static floor and rotating rakes or brushes.[56] Zenner's buddle concentrated the feed into three parts: The best which were "tozed" clean, seconds which were buddled again and then tozed, and finally thirds which were buddled two or three further times and then tozed.[57]

The fine slimes from the slime sizing cistern were delivered by a launder into a slime pit. The slime pits at Devon Great Consols were designed by Captain Richards and can more properly be described as slime dressers. The slime pit consisted of an inverted cone. The slimes in solution run slowly down the sides of the cone, the heavier and more valuable material being deposited at the bottom of the cone. This richer material escapes through a launder at the bottom of the slime pit, regulated by a plug.[58,59,60]

The launder from the bottom of the slime pit fed the concentrated slimes onto Brunton's frames, an apparatus for dressing very fine slimes. Before the feed reached the frame it was disintegrated by "devils and tormentors" and passed through a 1/60 inch trommel.[61] The Brunton comprised of a frame twelve feet long by three foot four inches wide over which ran an endless belt of coarse, painted canvas, strengthened and kept level by elm laths a few inches apart. The belt was inclined at an angle of about one in six, and was kept moving upwards by means of a small water wheel. The feed was introduced about four feet from the top of the rotating band where it was sprayed with water. The water carried the waste over the bottom end of the frame whilst the heavier ores were carried over the top end and into a wagon cistern, full of water, which served the dual purpose of collecting the concentrate and washing the belt.[62,63,64,65]

The wagon cistern was removed as required and was emptied into a packing kieve. At Devon Great Consols kieving was done automatically; mechanical hammers, driven by a small water wheel, beat on the sides of the kieve (a wooden tub), expelling water and separating the ore "according to their several forms and densities". The waste from this process was either rejected or returned to the Brunton, whilst the good concentrate was ready for sale.[66]

In 1857 it was calculated that the old floors could process from forty to fifty tons of "stuff" during a nine hour shift. This was achieved at a cost of twelve shillings per ton of dressed ore, or at a tribute of four shillings in the pound.[67]

1853

The ninth Annual General Meeting of the company was held on May 2nd at the Company's office at Barge Yard. It was noted that the Estate had provided twenty workers cottages at Wheal Maria and whilst this was to be welcomed the Directors felt that at least one hundred were required. Staying on the subject of housing it was also noted that Abbotsfield House, for the Resident Director, would be completed during the course of the month.[68]

As ever Captain James Richards' report contained monthly averages for both tutwork and tribute. During the year up to May 1853 and average of 216 men had been employed on tutwork at an average of £3 7s 2d a month. For the same period the mine employed an average of 196 tributers at an average of £6 9s 10d a month. This staggering increase in the cost of tributers was accounted for thus:

> owing to the prosperous state of mining generally, and the consequent demand for labour it has been found difficult to obtain good miners at the ordinary state of wages.[69]

Whilst the mining boom being experienced in the Tamar Valley would have been a significant contributing factor to the rapid inflation of tributers earnings other factors are also important. A major factor would have been the isolation of the mine from major settlements such as Tavistock and Bere Alston. The Company's Directors certainly recognised this as a significant issue and it explains their ongoing attempts to persuade the Bedford Estate to provide more workers housing. The nature of the mines themselves also needs to be taken into consideration: By 1853 Wheal Anna Maria, Wheal Josiah and Wheal Emma were over 100 fathoms deep and the only way to reach the bottom was on ladders. Time spent ladder climbing was both tiring and unpaid and must have had an impact on the price of tribute bargains.

At Wheal Maria the foundry was proving a success and, consequently, was to be expanded. Underground 88 fathoms, 1 foot, 11 inches had been driven in Wheal Maria up to May 1853. Gard's Shaft had reached nineteen fathoms below the 60 where the ground had improved for sinking. On the 28 fathom level a cross cut had been driven north to intersect a minor lode known as North Lode.[70]

Driveage in Wheal Fanny totalled 130 fathoms, 1 foot. From Western Engine

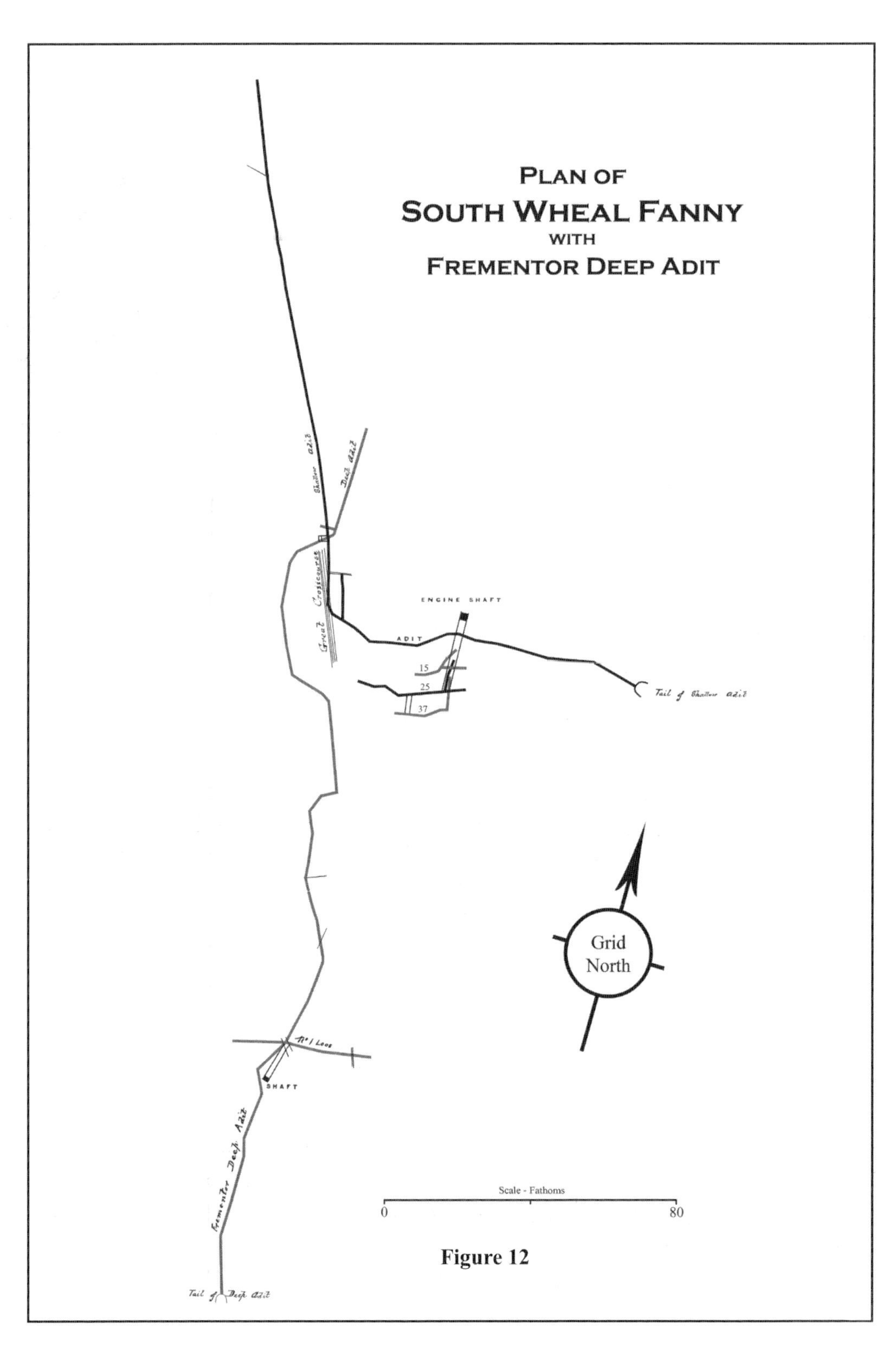
PLAN OF
SOUTH WHEAL FANNY
WITH
FREMENTOR DEEP ADIT
Shallow adit
Deep Adit
Great Crosscourse
ENGINE SHAFT
ADIT
15
25
37
Tail of Shallow adit
Grid
North
No 1 Lode
SHAFT
Frementor Deep Adit
Scale - Fathoms
0
80
Tail of Deep adit

Figure 12

Shaft drives had been pushed on the 90, 75 and 65 fathom levels. Eastern Engine Shaft was, by May 1853, twelve fathoms below the 65 where a very rich caunter lode, eighteen inches wide, had been intersected. The problems with water in Ventilating Shaft having been overcome, the shaft had been communicated with the 65.[71]

During 1853 the Anna Maria dressing floors were working flat out. The crusher, sizing apparatus, jigging machines and buddles completed and were capable of handling fifty tons a day. The dressing floors had been extended to cope with the accumulated halvans. So great was the amount that the crusher had to work both day and night to deal with them.[72]

To speed up haulage, Anna Maria Engine Shaft was fitted with guide rails for skip haulage, superseding the less efficient kibble haulage. In January 1853 the Engine Shaft had reached the 110 which was in magnificent ground (see Note 1).[73] By May 1853 Engine Shaft was three fathoms below the 110. Above the 110 the 95 had been communicated with Field Shaft and the 70 west was being driven towards Wheal Fanny to improve ventilation.[74] A new line of flat rods was constructed from the main line of the "large water wheel" to drive pumps in Field Shaft.

The high level of activity at Wheal Anna Maria was matched by that at Wheal Josiah. A new line of flat rods had been constructed from the mainline of the "Richards' Wheel" flat rod run to pump Hitchins' Shaft. A new 25′ x 2′ waterwheel had also been erected by May 1853 pumping the New Engine Shaft which had been sunk 200 fathoms east of Hitchins' Engine Shaft. The shears and capstan at Richards' Shaft were transferred to Inclined Shaft at Wheal Emma and were replace with a more substantial set capable of dealing with the depth that Richards' was reaching. To power the Wheal Josiah saw mill a 32' x 3' waterwheel was under construction, which Richards estimated would be completed by July 1853. To meet the increased demand for water on this section of the mine a large storage pond had been constructed to hold all the water raised at night and at other times when it was not immediately required.[75]

By May 1853 Richards', Field's and Hitchins' shafts had all reached the 144 fathom level, both Richards' Shaft was seven fathoms below the 144 whilst Hitchins' Shaft was seven fathoms below the 115. On the 144 in the vicinity of Richards' Shaft the lode was described as being eight feet wide comprising "caple, an abundance of mundic, fluor, prian and a little ore". At Field Shaft the lode on the 144 was reported to be four feet wide and of good quality. The lode at Hitchins' was slightly better being eight feet wide comprising of good ore. In addition to the 144 drives had been made on the 130 at Richards' and the 115, 80 and 70 fathom levels at Hitchins'. Total drives during the year ending May 1853 totalled 308 fathoms, 1 foot, 11 inches.[76]

In common with both Wheal Anna Maria and Wheal Josiah, Wheal Emma had also gone deeper than 100 fathoms; Thomas' Shaft being below the 100 fathom level. At this depth the lode was 14 feet wide with good stones of ore. During the year work concentrated on the 100 and 87 levels. During the year ending May 1853 only 56 fathoms, 4 feet, 9 inches had been driven in Wheal Emma.[77]

At Wheal Thomas work to open up and prove the lode continued. Engine Shaft was down nine fathoms below the 64, the lode being five feet wide and comprising of hard caple, mundic, fluor, prian and "a little ore of good quality". Drives had been made on the 30 and 40 fathom levels. Additionally, New Engine Shaft had been sunk on lode and had reached a depth of 27 fathoms by May 1853. To pump New Engine Shaft a line of flat rods was extended from the "old line". For hoisting a new horse whim with "poppet heads" was erected. In total the year's driveage at Wheal Thomas was 59 fathoms, 3 feet, 8 inches.[78]

The Frementor Adit continued to be driven, the year's driveage rate being 27 fathoms, 3 feet, 10 inches by May 1853. During the course of the year the adit had cut two minor lodes. The Company announced the intention to communicate the Frementor Adit with South Lode (South Fanny) at depth, effectively becoming the South Fanny Deep Adit.[79]

Exploratory and development work was well underway on South Lode (South Fanny) by May 1853 with sixty eight fathoms, eleven inches having been driven during the preceding year. Driving on the 15 showed the lode to be only two feet wide with occasional stones of ore. The adit had reached a length of 90 fathoms and was, at the beginning of May 1853, thirty six fathoms west of the shaft. The adit had intersected "Morris' Crosscourse" (see Note 2) and, in May 1853, was being driven through it.[80]

On August 12th 1853 the South West of England experienced an earthquake, the effects of which seriously alarmed those underground at the time:

> Mr Ambrose Barrat, of Tavistock, was in Wheal Fanny mine....., in the

> fifteen fathom level, sitting with five other persons, and heard a rumbling noise like thunder, ran out to look at the weather, which was clear, and returned, and remarked the men and ground trembling, sat down, but felt as if raised from his seat, with a tremulous motion of the mine, which lasted about a minute and a half. He subsequently descended to a lower level, and found the workmen so much frightened at the idea that the mine was falling in, that they had moved to another part of the level. In Wheal Maria....., in one of the levels men were working in 100 fathoms to the north of the shaft, and there the concussion was so great, that they thought that the shaft had fallen in, and ran to see what was amiss.[81]

Between January and December 1853 the mine sold 24,378 tons, 19 cwts., 2 qrs. of copper ore, realising £143,562 1s 4d. During the same period the mine also sold by private contract 1,055 tons, 17 cwts., 3qrs. of mundic, realising £1,045 8s 3d.[82]

1854

Some idea of the scale of Devon Great Consols activity during the mining boom can gained from *Devonport Journal* report from March 1854:

> The sale of ore by the Devon Great Consols last week was the largest monthly sale by one company ever made in Cornwall, and realised nearly £10,000 profit for the month.[83]

At the 10th Annual General Meeting of the Company held at Barge Yard on the 1st May 1854 the Directors commented on the high copper prices then prevailing and the fact that this allowed them to work lower grade ores at a profit. A consequence of this was a backlog of halvans awaiting dressing. In response the Company proposed to erect "a double grinding steam engine" to deal with the halvans whilst the price was high.

The booming copper prices had to be balanced against the increasing scarcity of skilled miners, which, the Directors felt, was a consequence of the lack of housing in close proximity to the mine. Due to the shortage of skilled men the company decided to abandon "less profitable and merely experimental operations, and devote the entire disposable force to the development of the more essential parts of the mines".[84]

In terms of the mine infrastructure the Directors were pleased to report that the "Great Plunger Lift" supplying water to Wheal Josiah had proved a great success. Likewise the foundry was deemed to have been successful and had been considerably expanded in the preceding year to a point where it could deal with the "whole running work of the mines". As part of the expansion the blast furnace had been upgraded giving a seventy per cent saving in fuel. The company felt that having their own foundry saved the company £1,500 a year. The Directors were also able to report that the saw mill at Wheal Josiah was in operation.[85] As was customary the meeting agreed to continue the education grant.[86]

Captain Richards' report, dated 29th April 1854 noted that a monthly average of 207 tutworkers had been employed at an average monthly wage of £3 13s 6d, compared with 180 tributers at an average of £6 3s 9d, a sum which Richards considered too high but, he conceded, reflected the scarcity of skilled labour.[87]

During the year to the end of April 1854 driveage in Wheal Maria amounted to 71 fathoms, 2 feet and 3 inches. Gard's Shaft had reached a point four fathoms below the 80. At this depth, it was felt, the lode was looking better and the decision was taken to keep sinking.[88]

Wheal Fanny had joined Anna Maria, Josiah and Emma in breaking the 100 fathom barrier. By the end of April 1854 Western Engine Shaft had reached a depth of fifteen fathoms below the 105. The lode at depth appears to have been disappointing: On the 90 the lode contained spots of ore, whilst on the 105 occasional stones of ore were noted. Eastern Engine Shaft was four fathoms below the 80 sinking in hard ground. At Ventilating Shaft the 55 was proving unproductive. In total 111 fathoms, 2 feet, 7 inches had been driven during the year.[89]

The large output from Wheal Anna Maria and Wheal Josiah meant that the Anna Maria dressing floors had expanded again during the year ending April 1854. As part of this expansion a small waterwheel was erected on the floors to drive round buddles and riddles. Whilst physical space on the Anna Maria floors was a problem an equally pressing problem was the large and increasing amounts of water required for dressing the halvans. In an attempt to meet this ever growing demand, water was diverted from the waterwheel driven crusher which was replaced with a new "grinder" attached to a steam engine.[90] This ran counter to the policy of replacing steam with water power and was obviously driven by expediency. This situation does highlight one of the major problems at Devon Great Consols, namely limited water supply at the level of the dressing floors.

What water there was on the mine was inconveniently located at the bottom of the site and required pumping to get it up to dressing floor level as in the case of the "Great Plunger Lift".

By the end of April 1854 Anna Maria Engine Shaft was six feet below the 124 fathom level. The 124 was in the process of being driven towards the lode. During the preceding year both the 95 and the 110 had continued very productive. During this period length of drives in Anna Maria totalled 170 fathoms, 1 foot, 3 inches.[91]

At Wheal Josiah, guides for skip hauling had been installed in Field Shaft. Richards' Shaft had reached a depth of four fathoms below the 158 fathom level; the lode on the 158 being spotted with good ore. At Hitchins' Shaft, which was down to the 130, the lode was poor at depth, the 130, 115, 90 and 80 fathom levels being reported as unproductive. At the eastern end of Wheal Josiah Agnes Shaft was five fathoms below the 80. Both the 70 and 80 fathom levels at Agnes Shaft were in good ore ground. The Company intended to connect the 80 west at Agnes Shaft with the corresponding level at Hitchins' Shaft to improve ventilation in this section of the mine. Agnes Shaft was being pumped by a flat rod run coming from the shaft bob on Inclined Shaft at Wheal Emma. During the year to the end of April 1854, 359 fathoms, 5 feet and 6 inches had been driven in Wheal Josiah.[92]

By the end of April 1854 Thomas' Shaft at Wheal Emma was down to the 115 fathom level whilst Inclined Shaft was twelve fathoms below the 75. One hundred and forty fathoms, one foot, two inches had been driven during the year.[93] At Wheal Thomas Eastern Engine Shaft had communicated with the 30 fathom level. To the end of April 1854, 84 fathoms, 34 feet 6 inches had been driven during the preceding year.

Progress at Frementor to the end of April 1854 was limited, only five fathoms, four feet, four inches having been driven that year.[94] South Lode (South Fanny) was gradually improving as exploration went deeper, good stones of ore being reported on the 25 fathom level. By the end of April 1854 South Lode Engine Shaft had reached a depth of twelve fathoms below the 25. During the year driveage totalled 76 fathoms, 1 inch.

1855

The 1855 Annual General Meeting was held at the company office at 77 Gresham House, Old Broad Street. The move of the Company office to a "better address" may well reflect the ongoing and increasing prosperity of the Company. At the meeting the Directors announced that the Estate had erected 24 cottages at Mill Hill, which, given the scale of the housing problem must have seemed like a drop in the ocean.[95]

Captain Richards' report, dated May 5th 1855, was presented at the A.G.M. As usual Richards provided average monthly figures for the number of workers employed on the mine. On average 180 tutworkers had been employed earning, on average, £3 12s 2d per month compared with £5 9s 10d earned by the tributers whose numbers averaged 152 men.[96] Gard's Shaft at Wheal Maria had reached the 95, however at that depth hard "spar-caple" had been encountered and, in consequence Richards recommended that development work be discontinued.

Like Wheal Maria, Wheal Fanny was also proving less good at depth. By May 1855 Western Engine Shaft was seven and a half fathoms below the 120, the ground on the 120 being composed of caple, mundic and quartz. However at shallower depth Wheal Fanny was in good ground: Richards reporting the 65 was being driven from Eastern Engine Shaft, likewise good ground was being encountered on and below the 55 in the vicinity of Ventilating Shaft. During the year the 55 had communicated with the 70 fathom level in Anna Maria.[97]

At Anna Maria steam capstans had been installed on both Engine Shaft and Field Shaft. The new halvans dressing floors, it was noted, were in a forward state of completion. A "steam grinder" supplied by Nicholls, Matthews, at a cost of £990, had been erected on the Anna Maria dressing floors. The new "steam grinder", which had been completed and tested a few days before the A.G.M, was installed to speed up the processing of halvans. The "steam grinder" comprised a thirty inch steam engine driving two grinders or roller crushers capable of processing fifty tons of halvans a day. To provide water for the halvans dressing a new pond was constructed fed from the Great Leat by two columns of seven inch pumps. Richards also noted that a 20′ x 2′ waterwheel had been erected to drive both sizing apparatus and round buddles. Underground Anna Maria was going deep, Engine Shaft having reached the 137 level by May 1855; this would prove to be the deepest level in this part of the mine. Unlike Wheals Maria and Fanny, Anna Maria was proving rich at depth, drives on the 124, 110, 95 and 80 fathom levels opening up good ground.[98]

Steam capstans had also been installed at Wheal Josiah to serve Richards' and Hitchins' Shafts. A thirty foot waterwheel and "hauling machine" had been erected at Agnes Engine Shaft.[99]

Field Shaft had reached the 141 fathom level and was reported to be in good ground which contained some fluorspar. The 131 at Field Shaft had communicated with the 110 at Anna Maria. Richards' Shaft was now down to the 175 level with drives on the 144 and 130. Richards' and Field Shaft had communicated on the 130. By May 1855 Hitchins' Shaft was five fathoms below the 130. A winze was being sunk below the 130, however progress was hindered by the amount of water encountered. To ameliorate this pumps connected to the main rods in Hitchins' Shaft were installed in the winze.[100]

Sinking at Agnes Engine Shaft was proving extremely difficult and dangerous below the 103 where the ground was proving increasingly soft. So bad was the problem that the Company intended to bypass the loose ground by sinking a new shaft to the south which would intersect the lode at the 130 fathom level.[101]

At Wheal Emma, Thomas' Engine Shaft had reached the 130 from where a crosscut had been driven in a southerly direction to intersect the northern part of the lode. Above the 130 driving had progressed on the 115, 100 and 87 fathom levels. Inclined Shaft was a fathom below the 87 with dives on the 87, 63 and 32 fathom levels.[102] Progress at Wheal Thomas was limited to drives on the 30 and 25 fathom levels.

Developments at Wheal Frementor were more interesting than they had been for a number of years. During the year up to May 1855 Richards reported that the Frementor Adit had communicated with South Lode (South Fanny). The connection with South Lode meant a drastic improvement in ventilation, pumps which were used for conveying air into the adit becoming redundant for that purpose.[103] Apart from the connection with the Frementor adit work on South Lode was limited to drives on the 37 fathom level.

1856

The twelfth Annual General Meeting held in May 1856. Of some significance was the fact that Richard Gard did not stand for re-election as a director of the company.[104]

James Richards, as usual, presented a comprehensive report on the progress of the mines, the 1856 report being dated 5th May. Richards informed the shareholders that during the year up to May they had employed a monthly average of 257 tutworkers and 80 tributers the tutworkers receiving £3 12s 8d a month, the tributers £5 10s 6d, highlighting the premium that skilled miners were able to demand at this time.[105]

Whilst some development work had been undertaken at Wheal Maria it had not met with success. A crosscut had been driven north for seventeen fathoms from Gard's Shaft on the 60. The crosscut had intersected North Lode, unfortunately this lode proved to comprise of hard "caple-quartz" and consequently Richards recommended that work be abandoned at this point. A second crosscut, known as John's crosscut, had been driven south from the 28 fathom level but had cut nothing of importance.[106]

Development work was also ongoing at Wheal Fanny. Western Engine Shaft had reached the 135 from which a crosscut was being driven to the lode. A further crosscut had been driven from the 45 fathom level. At Eastern Engine Shaft drives were being pushed on the 80 and the 65, the 65 had proving unproductive. Above the 65 a series of winzes had been sunk from the 55.[107]

To deal with the vast quantities of halvans the new dressing floor at Wheal Anna Maria was rapidly expanding. Richards in his May 1856 report noted that a twenty foot diameter water wheel had been erected to drive sizing machinery and buddles whilst six other wheels ranging in size from four feet to twenty feet had been installed to drive jigging machines. The shareholders were informed that the new floors were in "full course of working". Underground development was proceeding apace with drives from Anna Maria Engine Shaft on the 80, 95, 110 and 124 Levels. At Field Shaft winzes were being sunk below the 130 in good ground, whilst below that the 141 level was being opened up. To speed up both development work and the extraction of ore at Field Shaft a "hauling machine" had been installed driven by the Anna Maria crusher engine.[108]

At Wheal Josiah, Richards' Shaft had reached the 175, Richards reporting that "good stones of ore" we being found, whilst Hitchins' was down to the 144 in unproductive ground.[109]

At Agnes New Shaft extremely rapid progress had been made, the shaft having been sunk to a depth of eight fathoms below the 103, twelve men being engaged on sinking. A 30' x 10' water wheel taking water from the Great Leat had been erected to drive the pumps in Agnes New Shaft via 400 fathoms of flat rods. At Agnes Old Shaft, Richards reported work on the 60, 80, 90 and 103 fathom levels, the 90 having communicated with the 90 east of Hitchins' Shaft.[110]

Thomas' Engine Shaft at Wheal Emma had reached the 130 from which a crosscut driven south had cut the lode which, according to Richards report, contained good stones of ore. Above the 130 Richards' comments on drives on the 100 and 87 fathom levels. A crosscut from Inclined Shaft on the 100 fathom level had also cut the lode which looked promising. Richards also notes drives on the 75, 63, 47 and 32 fathom levels.[111]

Richards' May 1856 report dealt briefly with subsidiary workings: Wheal Thomas was proving unproductive on the 30 fathom level. Likewise there was little to report from South Lode except for a drive on the 25. At Frementor the adit had reached a length of 349 fathoms without encountering anything of importance.[112]

Lease negotiations and the 1857 leases.

As early as 1854 the Company had been looking to extend the term of their lease.[113] However it was not until May 1856 that serious negotiations began, when, on the 20th May 1856, John and W. A. Thomas made a formal application to the Duke. This application revisited the 1848 application for a lease for the Eastern Ground, extending as far east as the River Lumburn. Additionally the company wanted to extend the term of their original lease which had, at that time, nine years to run. The Company argued that the extension of the term would allow it "to ensure that gradual taking away of the ore discovered, thereby preventing waste alike injurious to the Lord and the Adventurers".[114] In effect this was a veiled threat, if the Estate did not renew the lease the Company would "pick the eyes" of the mine, that is to say extract as much ore as possible during the remainder of their lease without regard to future development (see Note 3). The Company wanted the lease for both the Eastern Ground and the existing sett to run for a period of thirty years. They also wanted their dues reduced from 1/12th to 1/15th.[115]

On May 24th 1856 John Benson wrote to Christopher Haedy from the Tavistock Office strongly recommending that the Thomas' application should be rejected without hesitation.[116] Haedy accepted Benson's judgement and wrote to Thomas on May 27th 1856:

> The Duke of Bedford having considered your application together with the opinion of his agents upon it, has desired me to inform you that your application is one with which he does not feel that it is proper that we should comply.[117]

In early June 1856 W. A. Thomas submitted an amended proposal to Haedy. This proposed that a lease be granted on the existing sett and the Eastern sett at 1/15th dues. In return the Company would undertake to develop the mine in gradual and cautious manner would agree to not exceeding specified tonnage per annum. Thomas argued that the Company were legitimate miners, not speculators and as such they supported the principle of a uniform Royalty, that is to say 1/15th. Thomas made the point that higher dues meant that the speculator could outbid the legitimate miner. The Company offered to pay the Estate "£4 or £5,000" if granted 1/15th dues, this being the sum which they saw as the difference between 1/12th and 1/15th dues.[118]

Haedy wrote to Benson on the 10th June 1856 regarding Thomas' amended proposals. From the tone of the letter Haedy was far from impressed. The first point he makes is that a reduction in dues from 1/12th to 1/15th was not possible. He also felt that the £4,000 or £5,000 compensation offered by Thomas did not even begin to meet the difference between 1/12th and 1/15th dues over the thirty year period of the lease as requested by Thomas. Haedy suggested that a fine of "at least £20,000" ought to be levied at the existing dues and regardless of the dimensions of the sett granted.[119] On 12th June Haedy wrote to Thomas informing him that he would consider the application over the summer in consultation with Benson and John Hitchins.[120]

Haedy was as good as his word and on the 23rd October he wrote to W. A. Thomas:

> Dear Sir, I beg to inform you that your application for an extension of the term in the set of the Devon Great Consols Mines & for an enlargement of the boundary of it has again been taken into consideration by the Duke of Bedford's Agents, and that we do not consider it proper at present to recommend his grace to comply with it.[121]

Towards the end of November 1856 Thomas Morris visited the Bedford Estate Office at Tavistock to pay the dues. Whilst there he had a conversation with John Benson and John Hitchins about a proposal to build a railway from the mines to the quays at Morwellham. Writing on the 25th November 1856 Benson informed Haedy of Thomas Morris' proposal; Benson had no objections to the railway *per se*, however he implied that it was the thin end of the wedge, the wedge being the extension of the lease and the granting of a lease for the Eastern ground. Thomas Morris informed Benson that a fresh proposition would be made for "a grant of

21 years from the present time of the ground comprised in the present set, of the ground Eastward as far as Lumburn and also for the Railway". Benson advised Morris that it would be useless to make an application if that application included the request for a reduction in dues.[122]

On the 26th November 1856 Haedy wrote to Benson regarding the discussion with Morris. With regard to the railway; Haedy felt that there would be no reason to decline provided that there was a sufficient inducement for the Duke. The Estate was also prepared to look favourably on an application for the Eastern Ground with the proviso that they could not accept less than 1/12th dues. Likewise the Estate was prepared to grant an extension to the existing lease. In light of the fact that the Company had made a net profit of half a million pounds Haedy opined that "I think to ask a fine of "£20,000 down for an addition of 21 years to the existing set would be a reasonable fine to ask".[123]

W. A. Thomas submitted a formal application to Benson on January 9th 1857 comprising of three parts:

> First – A Lease of so much land as will be required to build a Railway for the purpose of carrying Ores and materials to and from the Mines and the Morwellham and New Quays.
>
> Second – A Lease of a Sett of Lands for Mining purposes - situate to the East of Devon Great Consols, and more particularly described in the application made by this company on the 20th May 1856
>
> Thirdly – And a renewal of the Lease of the Lands now occupied by the Company and known as Devon Great Consols."[124]

The question of dues was still a sticking point, Thomas still arguing for a reduction. On the 20th January 1857 Thomas Morris wrote to Benson regarding the new application. By way of a carrot Morris suggested that the railway could be extended from Morwellham to "The New Quay". However the crux of the letter was the Estate's terms regarding both the "fine" and dues. By this time the Company must have been aware of the Estates thinking regarding the £20,000 "fine". Morris commenting that

> With reference to a fine for an extension of the lease you certainly labour under a misunderstanding in supposing that we ever anticipated anything

of the kind.

> The sum you alluded to was mentioned only in consideration of the dues being reduced from 1/12th to 1/15th. We cannot expect to make the same profit during the subsequent Grant as the present owing to the increased depth of the mines.[125]

Morris continued to argue the Company's point in a letter to Benson dated 26th January 1857. In it he acknowledges that the mine had unquestionably made large profits but, he argued, Cornish copper mines were paying 1/15th and in many instances 1/18th thus he considered 1/12th to be excessive.[126]

In spite of the Company's arguments the Estate remained firm on the subject of dues, by February 1857 the Company had come to accept this. On February 14th 1857 W. A. Thomas wrote to Benson that the "dues on royalty being the same as in their present lease".[127] Likewise the company had accepted the principal of the fine but felt that £20,000 was excessive given the potential cost of the railway and the fact that the mine was looking poor in depth. In his letter of 14th February W. A. Thomas suggested that £10,000 was ample for a 21 year extension of the existing lease and a 29 year lease on the Eastern Ground.[128]

Unfortunately, from the Company's point of view, the Estate was negotiating from a very strong position and was not prepared to make concessions. On March 2nd 1857 Haedy and W. A. Thomas had a meeting in London at which Thomas agreed to pay the £20,000 fine (see Note 4). It was agreed that the fine could be paid in four installments: £5,000 on signing the agreement, £5,000 in September 1857, £5,000 in March 1858 and £5,000 in September 1858.[129] In a masterpiece of understatement Haedy commented that:

> I feel quite satisfied with this conclusion of the negotiation, and the fine and 1/12th dues being reserved to the Duke.[130]

The negotiation process resulted in the Estate granting the Company two leases: one for the newly negotiated sett, the other for the railway.

The first lease was preceded by the surrender of the 1844 lease, the surrender document being dated the 2nd November 1857; this was signed for the Company by the Thomases.[131] The surrender of the 1844 lease was followed on the 4th November 1857 by the lease for the "set in lands for mining in the Parishes of Tavistock & Lamerton and County of Devon"; again this was signed by the Thomases on behalf of the Company. The lease was to run for twenty nine years from the 25th March 1857.[132] The lease document notes that the lease was granted in consideration of the surrender of the 1844 lease and:

> in consideration of the sum of twenty thousand pounds paid or agreed to be paid to the said Francis Duke of Bedford by the said John Thomas & William Alexander Thomas and their fellow adventurers in the manner and at the time hereinafter mentioned that is to say ten thousand pounds upon the execution of these presents and the remaining sum of ten thousand pounds by two equal installments of five thousand pounds each on the twenty fifth day of March one thousand eight hundred and fifty eight and the twenty ninth day of September one thousand eight hundred and fifty eight.[133]

The new sett included the ground covered by the 1844 lease plus the eastern ground which extended as far east as the Lumburn Valley. As agreed the Duke was to receive 1/12th dues on any ores raised.[134]

The second lease "of lands in the Parish of Tavistock in the County of Devon and liberty to construct a Railway thereon" was granted to the Thomas' by the Duke of Bedford on the 4th November 1857; the lease running for a period of twenty nine years from the 25th March 1857. The lease granted the Company the "liberty to make, maintain, manage work and use a railway or railways and other conveniences". In return for the granting of this lease the Company agreed to pay the Estate the very reasonable sum of £15 a year. This sum was to be paid in equal proportions on 25th March, 24th June, 29th September, 25th December, first payment to be made on 25th December 1857. The Company was also required to carry "ores, goods and merchandises" for the Duke and his tenants at a cost not exceeding one shilling a mile (see Note 5).[135]

1857

In terms of the mining enterprise 1857 started well: At Anna Maria Engine Shaft the 124 west, the 110 west and the 95 west were all reported as being in good or promising ground, the lode in the 110 being "worth full £100 per fathom". Whilst not as rich as the 110 at Anna Maria the ground around both Field Shaft and Agnes Shaft were proving productive.[136]

The thirteenth Annual General Meeting of the company was held on Tuesday, 12th May 1857. As usual the preceding year had been profitable for the company a profit of £30,080 8s 6d having been made; the directors taking great pleasure in being able to announce a dividend of £63 a share. The dividend would have been larger but the directors had felt it prudent to "commence a reservation" in anticipation of the fine that the Duke was intending to levy in return for assigning the new lease.[137]

1858

By the beginning to 1858 the "fine" exacted by the Duke had become common knowledge, the *Mining Journal* commenting:

> The immense quantities of ores now being raised in the different mines are a sufficient reason why such a large amount of money was paid on the renewal of the lease; and one could almost imagine that all the copper in the neighbourhood was concentrated in this spot.[138]

In February 1858 it was noted that the 130 on South Lode at Wheal Josiah was cutting a good course of ore.[139] Agnes Shaft was sinking on lode below the 130 and was cutting rich ore.[140] At Wheal Emma a crosscut on the 112 at Inclined Shaft was being driven in a northerly direction through promising ground. In the back of the 63 Bray's rise and Rowe's rise were cutting "a most magnificent course of ore" yielding 18 tons of ore with a value of £216 per fathom. The 87 east of Thomas' Engine Shaft was also proving particularly rich. It was felt that the increasing productivity of Wheal Emma indicated that the newly acquired Eastern Ground would be equally productive.[141,142]

At the May 1858 Annual General Meeting the directors announced that during the year ending 1st May 1858 the company had sold 3,663 tons of ore less than in the previous year. The directors explained that this was not due to a falling off in the productivity of the mine, but rather due to stagnation in the copper trade. W. A. Thomas explained that it was not thought advisable to "press upon a falling market an excessive quality of ore". The accounts presented at the meeting show that to the 1st May 1858 £1,966 9s 9d had been spent on the railway whilst £26 16s 10d had been spent on the new dock at Morwellham.[143]

The railway and quay at Morwellham

The railway, as authorised by the 1857 lease, was to start in the vicinity of Agnes Shaft at Wheal Josiah. The map appended to the 1857 Mining Lease shows that there was already a railway in situ from Wheal Anna Maria to the start point as stipulated in the Railway Lease.[144,145]

Work on the railway was well in hand by the beginning of June 1858; the formation having been completed as far as the incline plane and the permanent way completed "for a considerable distance.[146] The permanent way was worthy of note in that it reflected, then, current broad gauge practice, comprising of thirty nine pounds per yard bridge rail laid on longitudinal timber baulks albeit laid to 4′ 8½″ standard gauge, as opposed to Brunel's 7′ ¼″.[147]

Work on the railway progressed rapidly during the summer and autumn of 1858, the line opening for traffic towards the end of November of that year.[148]

The railway commenced at Wheal Anna Maria with branches serving both Wheal Josiah and Wheal Emma; neither Wheal Fanny nor Wheal Maria were connected by rail (see Note 6). The railway headed east from Anna Maria, crossing the Rubbytown Valley, between Wheal Josiah and Emma on a substantial "horseshoe" embankment before heading south east past Wheal Emma and Wheal Jack Thomas. The line continued in a southerly direction, past Bedford United, to which a branch was laid, Bedford United ore being carried at 6d per ton.[149] Crossing under the Tavistock-Gunnislake road the line followed the contours along the top of the valley to a point above the port of Morwellham, to which it was connected by an incline plane.

Between Anna Maria and the incline head the line, trains consisting of eight or ten wagons hauled by a locomotive,[150] took twenty minutes to complete the journey between the mine and the incline head.[151] The company accounts for the period February to March 1858 contain an item for a locomotive engine at a cost of £500,[152] probably having been supplied by Messrs. Hopkins Gilkes. By the May 1859 Annual General Meeting the railway was coming into its own and the directors felt that they would soon require a second locomotive (see Note 7).[153]

The incline ran for approximately half a mile to the quays at Morwellham. A 22-inch stationary steam engine, constructed in the Company's foundry at Wheal Maria, was sited at the head of the incline, wagons being hauled up and lowered down the incline attached to a four inch wire rope.[154,155,156] The incline must have been partially self acting, the weight of the descending wagons, at the least, assisting the engine in hauling the ascending wagons.

At Morwellham the incline entered a tunnel under the newly constructed Estate cottages before debouching onto the quays. On the quay the railway split into a number of sidings, three of which ran on elevated trestles allowing the wagons to empty their loads directly onto the quay, the wagons being bottom dischargers (see Note 8).[157]

Construction of the railway, including the cost of "plant" came to £12,595, although this figure did not include any portion of the £20,000 "fine". That the construction of the railway was a shrewd move there is little doubt: prior to the construction of the railway cartage to the Tamar quays was five shillings per ton, the cost of transport via the railway to Morwellham was one shilling per ton; in 1860 this represented a saving of £518 per month.[158]

The new quay and dock at Morwellham were constructed at the same time as the railway by the Company under contract to the Estate at a reputed cost of £5,000.[159,160] As the mine's gateway to the world the company's dock and quay at Morwellham developed into a hive of activity:

> The ore floors of the company at Morwellham are several acres in extent, and have a dock in the middle capable of accommodating six vessels of about 300 tons burden. Ships of that size, however, rarely come up the river, and the average burden would be more nearly 200 tons. The trucks run out over the floors from the incline on staging, so constructed that when the bottoms of the trucks - which slide backwards and forwards - are drawn out, their contents fall through upon the floor, which is tiled. Every truckload of ore when sent from the mine is ticketed with its quality, and this ticket guides the men in unloading. Each kind being deposited by it self, the heaps are turned over and thoroughly mixed as a preliminary to subsequent operations. For sale the ore is divided into "parcels" and subdivided into "doles" - a work which is effected in a very peculiar manner and with singular rapidity by the men employed on the floors. Fourteen men are usually engaged in the operation of dividing, twelve being occupied in pairs, carrying the ore in handbarrows, one as a "wiper," and one as a "striker." The parcels of ore are the lots in which it is offered for sale, and the quantity in each is generally kept within 100 tons. The doles are the heaps into which the parcels are divided, there being commonly six. They are formed in the following way: - The carriers fill their barrows from a heap of thoroughly mixed ore - the contents of each being struck level with the edges by the wiper - and then turn them out at one spot, the striker giving the bottoms of the barrows as they are reversed a blow with a mallet, so that every particle of ore is knocked out. The carriers' next round is deposited at another point, and so they proceed until the whole of the doles

> have been commenced, when they go in succession to each as before until the entire parcel has been distributed. A barrow load is weighed at intervals, and as it is known how many go to a dole the weight of the parcel can be estimated with considerable accuracy. The men engaged in this work labour very hard, each barrow containing two hundred weights; but they obtain some relief by taking turns to wipe and strike. The ore is now ready for "sampling" by the various smelting companies.....[161]

1859

Devon Great Consols entered the last year of the 1850s in a very strong and prosperous state, the situation being neatly summarised in the *Mining Journal*:

> on the whole, the present state and future prospects of this great concern, coupled with the concessions and advantages obtained by the renewal of the lease, &c., and the improvements which have been introduced, appear to warrant a long continuance of the great prosperity which has attended it from the beginning.[162]

Whilst the company was undoubtedly in a fine condition, hairline cracks were beginning to appear. At the 1859 A.G.M., held on Tuesday, 24th May, the directors informed the meeting that the copper price was falling, the best quality "fine copper" realising £8 15s 5d less than it had the previous year. In real terms this meant a £12,000 drop in the mine's income during the year. It was noted that the copper standard had further fallen in May, the directors felt that if the situation continued they would restrict the amount of ore they released onto the market and they hoped that the shareholders would accept a temporary reduction in their dividend rather than receive an inferior price for their ore. Captain James Richards' annual report, which whilst typically positive, could not hide the fact that some of the levels were not proving as rich as hoped.[163,164] With the benefit of hindsight this might be seen as an early indication of future exhaustion of the mine's copper reserves. However it would be unfair to focus too much on these problems; at the time they were less portents of doom as opposed to pinpricks in an otherwise bright and prosperous situation.

On the positive side the shareholders' attention was drawn to the railway and dock at Morwellham which were proving their worth. More exciting was the potential that was about to be unlocked in the newly acquired Eastern Ground. Captain Richards was able to report that work had started on driving an adit from

Mill Hill. The Mill Hill adit had intersected the lode which consisted of capel, quartz and mundic.[165,166]

By Mid June 1859 it was being reported that lode in the Mill Hill adit was twenty feet wide and "exceedingly promising". It was noted that the north part of the lode in particular was composed of "gossan of very rich description".[167,168]

Whilst the Eastern Ground was very much of the future work continued apace on the original sett. For example the Main Lode at Wheal Josiah was being explored at depth: At Richards' Shaft the 200 fathom level crosscut had intersected the lode which was composed of "capel, mundic and fine stones of ore". Work on the drive was slowed by water which was noted as being "very quick and quite warm".[169]

In spite of the declining copper standard Devon Great Consols was still a stellar copper producer. On the 25th of November the mine sold 1,835 tons of copper ore which realised £9,722 11s 9d. The sale included ore from Wheal Maria, Wheal Fanny, Wheal Anna Maria and Wheal Anna Maria.[170]

Chapter 5 Notes
Note 1
In early 1853 the *Devonport Journal* carried a number of reports regarding the 110 level at Anna Maria:

> The report from the mine states that the 110-fathom level at Anna Maria is in 13 feet; there is no sign of a north wall, and it is magnificent course of ore, worth 14 tons per fathom.[171]
>
> From the mine the advices are very flattering, at Anna Maria cross-cut north in the 110 level, the lode has been cut into, and is 17 ft. wide, worth 10 tons per fathom. There is also a considerable improvement in various workings at Wheal Josiah".[172]
>
> In the 110 cross-cut north, the main part of the lode has been cut through, and proves to be 24 ft. wide, a most magnificent course of ore, worth 30 tons, or £210 per fathom".[173]

Note 2
Probably the Great Crosscourse.

Note 3
This was no idle threat. At Consolidated Mines in the late 1830s John Taylor had tried and failed to negotiate a new lease on the mines which he had worked incredibly successfully since 1818. The Mineral Lords, casting a greedy eye on the profits made by Taylor, were anxious take a much larger share, insisting on a one third share in the Company in return for a renewal. This was totally unacceptable to Taylor who, having failed to renew the lease on reasonable terms, felt that he had no option but systematically strip as much of the payable ore ground within the mine before his lease expired. Consolidated was taken over by Messrs. Williams in July 1840 and whilst they continued to work the mines for another 18 years' production and profits rapidly fell off.[174,175]

In 1863 the *Mining Journal* noted that:

> The leases.....expiring, these mines were ruthlessly torn from his (Taylor's) hands by parties who, not content to receive their portion of the regular "golden eggs" determined to seize the "goose" and have the whole: When lo! on opening the "goose" the "golden eggs" had vanished.[176]
>
> Everyone knows these great mines were irretrievably ruined and the entire district damaged beyond reparation by a policy as remarkable for its short-sightedness as for its intense selfishness; a policy marked by its disregard of every other consideration than the gratification of domineering self will.[177]

Note 4
The £20,000 "fine" paid to the Duke by the Company would set an awkward precedent. When in the early 1880s a similar "fine" was requested from the Dolcoath Mine adventurers by Mr Basset, the precedent of Devon Great Consols was cited in the discussions:

> The mining world has been paralysed recently at the action taken by Mr. Basset, in an application to renew the lease of Dolcoath, the richest tin mine in Cornwall. The present lease does not expire for five years, and to put the matter very tersely, it was intimated that a lease for 21 years would be granted to take effect from the present time, instead of commencing at the end of the five years and only upon payment of a fine of £40,000, and on condition that a shaft was sunk which would cost another £40,000, and other expensive works undertaken. The neighbourhood of Camborne where Dolcoath is situated has been in a state of great excitement, consequent

on this extraordinary demand, which is declared to be unparalleled in the history of mining. At a special meeting on Tuesday, the matter was simplified to this extent, that the 21 years lease was to take effect from the expiration of the present term, the sinking of the shaft was abandoned, and the fine of £40,000 was to be paid only out of the profits, one quarter of them being set aside for that purpose. A committee, of whom Captain Andrews, of Tavistock, is one, was appointed, to confer with Mr. Basset and his advisers on the subject, with a view to further simplify the matter. In the course of the numerous discussions which have taken place relative to this matter, it has been stated that a parallel case has been found in Devon Great Consols, where a fine of £20,000 was imposed by the Duke of Bedford when he renewed the lease in 1857, when a new lease was granted, the enormous sum of £555,000 had been paid in dividends on a called up capital of £1,024. In 1857 it was believed that the vast deposit of ore was going eastward, and the shareholders applied for a lease of the eastern ground, and it was no secret that numerous applications were made to the Bedford Office by independent parties who were willing to give any sum for working this eastern ground. The Duke of Bedford of that time, however, felt that the shareholders of Devon Great Consols were justly entitled to a prior claim, and negotiations were opened for a new lease, the result being that a lease of 29 years was granted (which allowed for eight of the unexpired term) and the sum of £20,000 was paid as a premium for granting the eastern ground. From that time to 1872 when dividends were suspended, the further sum of £630,000 was divided among the shareholders, making the almost fabulous sum of £1,185,000 given in dividends in twenty-eight years, on an outlay of £1,024.[178]

Note 5

Whilst the new mine and railway leases are dated the 4th November 1857 it is interesting to note that they had not been signed at the time of the May 1858 A.G.M when an E.G.M. was convened to confirm the arrangements made with the Estate.[179] This time lag does not appear to be unusual, the company and the estate being happy to work, in the short term, under less formal agreements, the details being ratified by formal leases. A parallel example being the 1874 lease, which is discussed in the following chapter.

Note 6

The lack of a rail connection to Fanny and Maria might be explained by the intervening ridge which was crossed by two narrow gauge inclines during the 1920s reworking.

Note 7

Early evidence regarding locomotives on the DGC railway is contradictory. The company owned a single locomotive at the time of opening. This might have been built by Nicholls Williams; D. B. Barton suggests that they supplied such a locomotive in 1856 which "was put into service on the various lengths of railway track laid on the mines to connect the main shafts with the dressing floors".[180] The *Tavistock Gazette* of June 3rd 1859 noted that an order for a second locomotive was about to be made. *Mining Journal* of July 28th 1860 notes that "The work is done by three locomotives, two of which have 12-inch cylinders, and one 7 inch cylinders, each weighing 12 tons 2 cwts without water, or 15 tons running". If this is the case it seems odd that *Mining Journal* of May 16th 1863 in reporting the 1863 Company A.G.M should note: "In order to provide against accidents or delay, a second locomotive has been purchased of Messrs. Gilkes, Wilson, & Co., of Middlesbrough, the first one having proved itself to be an admirable and effective engine in all respects". The *Tavistock Gazette* recorded in 1864 that the "line is worked by two locomotives".[181] J. H. Trounson (unpublished letter to Frank Booker 18th November 1976) suggests that there were two Gilkes, Wilson locos and that they were named "Billy" and "Jack".

Note 8

Little detail survives regarding rolling stock on the railway. All that is definitely known regarding the ore wagons is that they were bottom dischargers. However it may be instructive to look at the ore wagons employed on the Liskeard & Caradon Railway, particularly given DGC's link with Nicholls Williams:

> The earliest reference to wagons in the minutes is the order in 1855 of two copper ore wagons from Nicholls Williams. These were quoted at £95 per ton and in the event cost £90 each thus weighing about 19cwt apiece. The ore wagons were of the chaldron type with tapering sides and a bottom discharge door in the floor permitting them to discharge straight into the canal barges or onto the ore floors from trestles above the wharves at Moorswater.[182]

Writing in 1914 Moses Bawden records that there were 35 ore and timber wagons on the railway.[183]

Chapter 5 references

1. *Mining Journal* 26 October 1850
2. 6th Annual Report 1850
3. 6th Annual Report 1850
4. 6th Annual Report 1850
5. 6th Annual Report 1850
6. 6th Annual Report 1850
7. 6th Annual Report 1850
8. 6th Annual Report 1850
9. 6th Annual Report 1850
10. 6th Annual Report 1850
11. 6th Annual Report, 1850
12. *Mining Journal* 28 July 1860
13. *Mining Journal* 26 October 1850
14. 7th Annual Report, 1851
15. 7th Annual Report, 1851
16. 7th Annual Report, 1851
17. 7th Annual Report, 1851
18. 7th Annual Report, 1851
19. 7th Annual Report, 1851
20. 7th Annual Report, 1851
21. 7th Annual Report, 1851
22. 7th Annual Report, 1851
23. 7th Annual Report, 1851
24. 8th Annual Report, 1852
25. 8th Annual Report, 1852
26. 8th Annual Report, 1852
27. 8th Annual Report, 1852
28. 8th Annual Report, 1852
29. *Mining Journal* 28 July 1860
30. 8th Annual Report, 1852
31. 8th Annual Report, 1852
32. 8th Annual Report, 1852
33. 8th Annual Report, 1852
34. *Devonport Journal* 18 November 1852
35. 8th Annual Report, 1852
36. 8th Annual Report, 1852
37. 8th Annual Report, 1852

38. 9th Annual Report, 1853
39. Darlington J. & Phillips J. A. 1857, *Records of Mining and Metallurgy,* E. & F. N. Spon.
40. *Mining Journal* 28 July 1860
41. Darlington J. & Phillips J. A. 1857, *op. cit.*
42. *Mining Journal* 28 July 1860
43. *Mining Journal* 28 July 1860
44. *Tavistock Gazette* 2-23 December 1864
45. Darlington J., 1878, *On the Dressing of Ores* (reprint 2002), extracted from Ure's dictionary of arts, manufactures and mines, Dragonwheel Books.
46. Andre G. G., 1878, *A Descriptive Treatise on Mining Machinery, Tools and other Appliances used in Mining,* Vol. 2, E. & F. N. Spon.
47. Davies E. H., 1902, *Machinery for Metalliferous Mines*, Crosby Lockwood.
48. Darlington J. & Phillips J. A. 1857, *op. cit.*
49. Darlington, J., 1878, *op. cit.*
50. *Mining Journal* 28 July 1860
51. *Mining Journal* 28 July 1860
52. Darlington J. & Phillips J. A. 1857, *op. cit.*
53. *Mining Journal* 28 July 1860
54. Darlington J. & Phillips J. A. 1857, *op. cit.*
55. *Mining Journal* 28 July 1860
56. Darlington, J., 1878, *op. cit.*
57. *Mining Journal* 28 July 1860
58. Darlington J. & Phillips J. A. 1857, *op. cit.*
59. *Mining Journal* 28 July 1860
60. Darlington, J., 1878, *op. cit.*
61. Darlington, J., 1878, *op. cit.*
62. Darlington J. & Phillips J. A. 1857, *op. cit.*
63. *Mining Journal* 28 July 1860
64. Darlington, J., 1878, *op. cit.*
65. Davies E. H., 1902, *op. cit.*
66. Darlington J. & Phillips J. A. 1857, *op. cit.*
67. Darlington J. & Phillips J. A. 1857, *op. cit.*
68. 9th Annual Report, 1853
69. 9th Annual Report, 1853
70. 9th Annual Report, 1853
71. 9th Annual Report, 1853
72. 9th Annual Report, 1853
73. *Devonport Journal* 20 January 1853
74. 9th Annual Report, 1853

75. 9th Annual Report, 1853
76. 9th Annual Report, 1853
77. 9th Annual Report, 1853
78. 9th Annual Report, 1853
79. 9th Annual Report, 1853
80. 9th Annual Report, 1853
81. *Pharmaceutical Journal and Transactions of the Royal Pharmaceutical Society of Great Britain* 1852-1853 Vol. 12, page 308
82. 10th Annual Report, 1854
83. *Devonport Journal* 9 March 1854
84. 10th Annual Report, 1854
85. 10th Annual Report, 1854
86. 10th Annual Report, 1854
87. 10th Annual Report, 1854
88. 10th Annual Report, 1854
89. 10th Annual Report, 1854
90. 10th Annual Report, 1854
91. 10th Annual Report, 1854
92. 10th Annual Report, 1854
93. 10th Annual Report, 1854
94. 10th Annual Report, 1854
95. 11th Annual Report, 1855
96. 11th Annual Report, 1855
97. 11th Annual Report, 1855
98. 11th Annual Report, 1855
99. 11th Annual Report, 1855
100. 11th Annual Report, 1855
101. 11th Annual Report, 1855
102. 11th Annual Report, 1855
103. 11th Annual Report, 1855
104. 12th Annual Report, 1856
105. 12th Annual Report, 1856
106. 12th Annual Report, 1856
107. 12th Annual Report, 1856
108. 12th Annual Report, 1856
109. 12th Annual Report, 1856
110. 12th Annual Report, 1856
111. 12th Annual Report, 1856

112. 12th Annual Report, 1856
113. Devon Record Office document L1258 add 8/m/E11/15 Devon Record Office document L1258 add 8m/E11/15
114. Devon Record Office document L1258 add 8m/E11/15
115. Devon Record Office document L1258 add 8m/E11/15
116. Devon Record Office document L1258 add 8m/E11/15
117. Devon Record Office document L1258 add 8m/E11/15
118. Devon Record Office document L1258 add 8m/E11/15
119. Devon Record Office document L1258 add 8m/E11/15
120. Devon Record Office document L1258 add 8m/E11/15
121. Devon Record Office document L1258 add 8m/E11/15
122. Devon Record Office document L1258 add 8m/E11/15
123. Devon Record Office document L1258 add 8m/E11/15
124. Devon Record Office document L1258 add 8m/E11/15
125. Devon Record Office document L1258 add 8m/E11/15
126. Devon Record Office document L1258 add 8m/E11/15
127. Devon Record Office document L1258 add 8m/E11/15
128. Devon Record Office document L1258 add 8m/E11/15
129. Devon Record Office document L1258 add 8m/E11/15
130. Devon Record Office L1258 Surrender Lease 1857
131. Devon Record Office L1258 Mining Lease 1857
132. Devon Record Office L1258 Mining Lease 1857
133. Devon Record Office L1258 Mining Lease 1857
134. Devon Record Office L1258 1857 Railway Lease
135. *Mining Journal* 3 January 1857
136. *Mining Journal* 16 May 1857
137. *Mining Journal* 27 February 1858
138. *Mining Journal* 6 February 1858
139. *Mining Journal* 27 February 1858
140. *Mining Journal* 6 February 1858
141. *Mining Journal* 27 February 1858
142. *Mining Journal* 29 May 1858
143. Devon Record Office L1258 1857 Railway Lease
144. Devon Record Office L1258 1857 Mining Lease
145. *Tavistock Gazette* 4 June 1858
146. *Mining Journal* 28 July 1860
147. *Mining Journal* 1 January 1859
148. *Mining Journal* 28 July 1860

149. *Mining Journal* 28 July 1860
150. *Tavistock Gazette* 2-23 December 1864
151. *Mining Journal* 28 May 1859
152. *Mining Journal* 28 May 1859
153. *Mining Journal* 28 May 1859
154. *Tavistock Gazette* 3 June 1859
155. *Mining Journal* 28 July 1860
156. *Tavistock Gazette* 2-23 December 1864
157. *Mining Journal* 28 July 1860
158. *Mining Journal* 28 July 1860
159. *Tavistock Gazette* 2-23 December 1864
160. *Tavistock Gazette* 2-23 December 1864
161. *Mining Journal* 1 January 1859
162. *Mining Journal* 28 May 1859
163. *Tavistock Gazette* 3 June 1859
164. *Mining Journal* 28 May 1859
165. *Tavistock Gazette* 3 June 1859
166. *Tavistock Gazette* 17 June 1859
167. *Tavistock Gazette* 24 June 1859
168. *Tavistock Gazette* 26 August 1859
169. *Devonport Journal* 2 December 1859
170. *Devonport Journal* 20 January 1853
171. *Devonport Journal* 3 February 1853
172. *Devonport Journal* 17 February 1853
173. Barton D. B. 1966, *The Redruth and Chasewater Railway 1824-1915.* 2nd ed., Bradford Barton, Truro.
174. Burt R., 1977, *John Taylor, mining entrepreneur and engineer 1779-1863.*
175. *Mining Journal* 19 September 1863 cited: in Burt R., 1977 *op. cit.*
176. *Mining Journal* 14 March 1863 cited in: Barton D. B., 1966 (reference below)
177. *Tavistock Gazette* February 16 1883
178. *Mining Journal* May 29 1858
179. Barton D. B., 1964, *A Historical Survey of the Mines and Mineral Railways of East Cornwall and West Devon*, D. Bradford Barton Ltd.
180. *Tavistock Gazette* December 1864
181. Messenger M., 2001, *Caradon and Looe – The canal, railways and mines.* Twelveheads Press.
182. Bawden M., 1914, Mines and mining in the Tavistock district, *Transactions of the Devonshire Association*, Vol. XLVI. pp. 256 - 264.

Chapter 6

The 1860s: Maturity and change, from copper to arsenic

"We can walk three miles underneath" – James Richards, May 1863

Devon Great Consols entered the 1860s in a very strong position. The Company could approach its twenty first birthday, due to be celebrated in 1865, firm in the knowledge that Devon Great Consols was the finest mine in the country. However as the decade progressed clouds began to appear on the horizon, copper prices began to fall and unrest manifested itself amongst the workforce.

1860

In July 1860 H. C. Salmon, a regular *Mining Journal* correspondent, compiled an extensive review of the mine, providing a fascinating snap shot of the mine at the start of what was to be a challenging decade:

Salmon lists the management of the mines as of July 1860:

> The Chairman of the Board is Directors is Mr. W. A. Thomas, of Threadneedle – Street, London, and the Resident Director is Mr Thomas Morris, of Abbotsfield, Tavistock. The present management of the mines is as follows: Capt. James Richards, principal managing agent, which post he has occupied for the last ten years, since the resignation of Mr. Hitchins; Capt. Wm. Clemo, principal underground agent. Underground agents at different mines – Capt. Henry Cock, at Wheal Maria; Capt. Henry George at Anna Maria; Capt. James Blunt, at Wheal Josiah; and Capts. Wm. Woolcock and Henry Rodda, at Wheal Emma and Wheal Thomas. Captain Richards is the dressing and surface agent, and Capt. John Clemo the captain superintending the entire timber and pitwork of the mines. Capt.

> Phillip Richards is the dialler, sometimes assisted by his brother, Captain Joseph Richards. The principal accountant is Mr. Henry Youren, and the clerks Mr. Thomas Youren and Mr W. H. Barret. There are two engineers – Wm. Matthews of Tavistock, for the steam machinery, and Mr Nathanial Smith, of Wheal Friendship, for the hydraulic machinery. All reside on the mines, except the engineers, the two junior clerks, and the dialler.[1]

At the time of Salmon's article both Gard's and Morris' Shafts were down to the 100, however the lower levels were no longer at work, the ground below the 50 proving poor. Wheal Maria was drained to the 50 via pitwork in Gard's Shaft driven by the 50′ x 4′ waterwheel.[2]

Wheal Fanny had reached a depth of 135 fathoms, the deepest shaft being Western Shaft. As with Wheal Maria, Wheal Fanny was proving poor at depth, little good ore ground being found below the 55. In consequence the mine was only being kept drained to the 65, the pumps in Western shaft being driven via a flat rod run by the 50′ x 4′ waterwheel at Wheal Maria.[3]

In December 1860 Josiah Paull, the Estate's mineral agent, noted that exploratory work at depth had stopped in both Wheal Maria and Wheal Fanny although tributers continued to extract large quantities of good ore.[4]

Further east the lode, as we have seen, was improving at depth. In July 1860 Salmon noted that the lode in Wheal Anna Maria proved extremely productive right down to the 137. Salmon comments that the lode "made exceedingly large" singling out Cox's stope on the 95 where the lode was seven fathoms wide and comprised of solid ore.[5] South Lode, which branched off from the Main Lode thirty fathoms west of Fields Shaft in Anna Maria, was proving particularly rich albeit very much in the early stages of exploration and development.

Moving further east, Richards' Shaft at Wheal Josiah had reached a depth of 200 fathoms whilst both Hitchins' and Agnes New Shaft had reached the 170. Whilst productive the ground was not a good as that encountered in Wheal Anna Maria.[6] Salmon was pleased to report that Wheal Emma was starting to prove rich, although £30,000 to £40,000 had been spent on development before ore was reached. In Reddicliff's stope the lode was noted as being twenty feet and very rich.

Salmon is fairly dismissive of both Wheal Thomas and Wheal Frementor,

Figure 13.

Section on South Lode

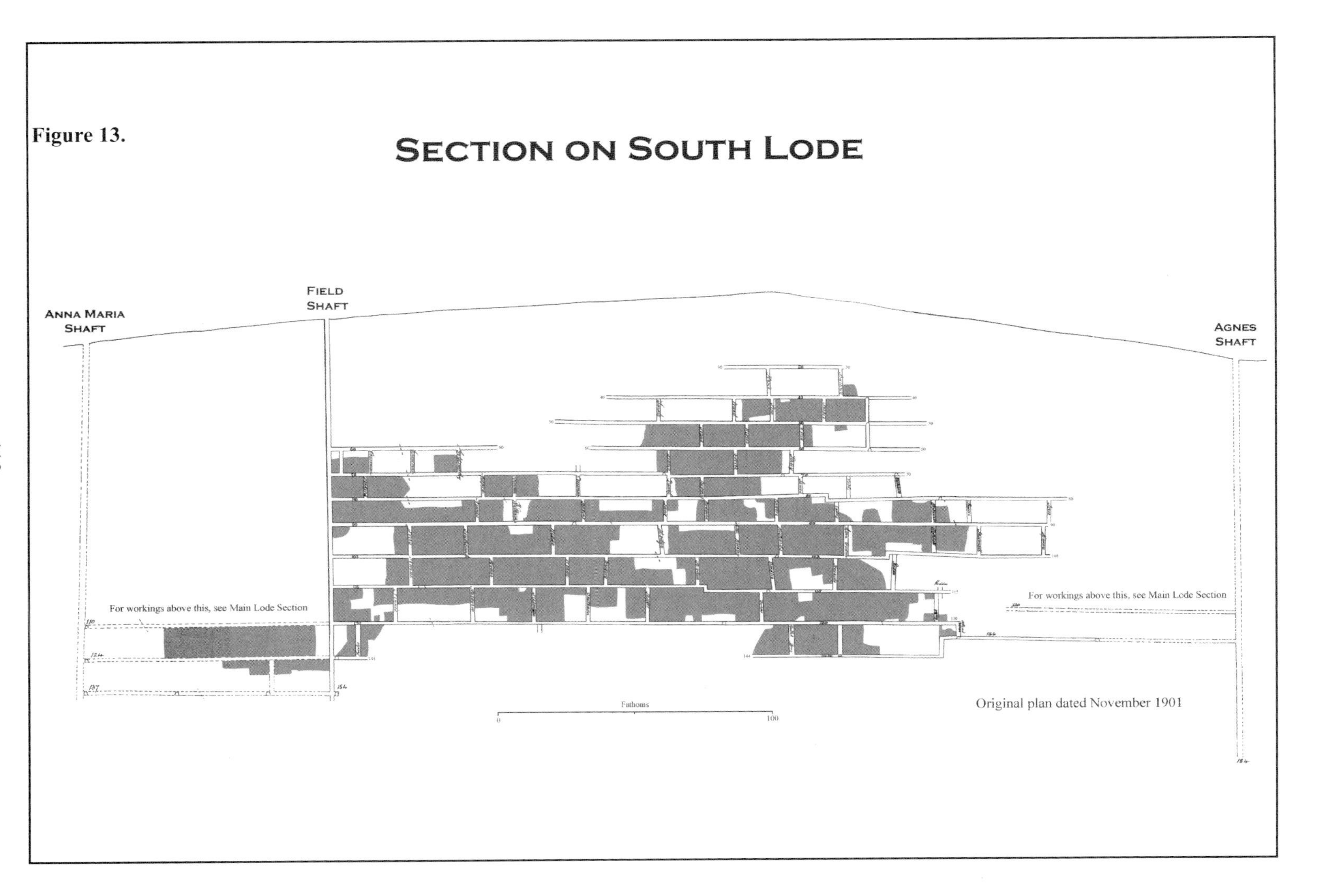

commenting that Wheal Thomas had "about paid cost". In December 1860 Josiah Paull observed that there were no tributers working in Wheal Thomas but that there was still ore ground to be taken away.[7]

In connection with Frementor, Salmon notes that a Dr Wagstaffe of Kensington was working a granite quarry in the vicinity; Salmon dismisses Wagstaffe's endeavours with the caustic comment; "It is to be sincerely hoped that Dr Wagstaffe may be more successful in this than in his previous mineral speculations; but this is very doubtful".[8]

Salmon's July 1860 article paints a picture of a mine entering a self assured maturity: Hitchins' integrated waterpower system (see Note 1) had come into its own, there were good reserves of ore, the halvans floors were greatly improving recovery rates, the railway was completed reducing transport costs significantly and the signing of the 1857 lease had secured both the Eastern Ground and extended the lease period.

By the start of the 1860s the extent of the Main Lode, as granted in the 1844 lease, was well established. However there was still a great deal of potential ore ground to be developed both in the newly granted Eastern Ground and also on South Lode, both of which were, in 1860, largely unknown quantities. Writing in the *Mining Journal* at this time Captain James Richards makes a passing reference to further lodes lying south of South Lode.[9]

Although South Lode had been discovered as early as 1848 only a limited amount of work had been undertaken during the 1850s, mainly due to the demands of opening up the Main Lode. However, as Salmon noted, South Lode was proving rich by 1860, albeit it not yet fully developed. By December 1860 the workings on South Lode had been extended westward into Anna Maria where the lode had been encountered on the 70 west of Engine Shaft.[10]

Work had also started on developing the Eastern Ground. An adit was being driven westwards from Mill Hill as rapidly as possible, the intention being to connect with the 60 fathom level at Wheal Emma.[11] By July 1860 the adit had been driven for 90 fathoms to a point where it had met a considerable crosscourse.[12]

On 17th November 1860 Josiah Paull noted that the mine appeared "to be doing a great deal towards returning the halvans as well as the arsenical mundic".[13]

1861

Writing in February 1861 James Richards summarised the current state of development of South Lode which was being developed from Field Shaft. The lode had been opened up as far down as the 90 fathom level in good ore ground, this is confirmed by Josiah Paull who notes that the 90 was producing six tons of ore per fathom. A crosscut was being driven on the 103 east of Field Shaft to intersect South Lode. Richards also expressed the intention to drive a crosscut north from Richards' Shaft, which was currently twelve fathoms below the 200, in order to prove South Lode at depth.[14,15]

On 9th February 1861 Paull commented that the shallow adit, which lay midway between Mill Hill and Wheal Emma's boundary, had cut "a very fine gossan". It was hoped that this was would "lead to discoveries of Copper ore as was the case in the old mine".[16] By late February 1861 Captain Richards was able to report that both deep and shallow adits progressing well.[17]

Elsewhere on the mine steady progress was being made. As already noted Richards' Shaft was twelve fathoms below the 200. Agnes Shaft was down to the 180. At Wheal Emma, Inclined Shaft which was in "regular course of sinking" was twelve fathoms below the 150.[18]

In his April 1861 report Josiah Paull commented that ground cut on the 212 where was said to be "favourable for ore", although at the time of his inspection water was up to the 200 fathom level and he was unable to examine the 212 himself. By mid June Paull was able to inspect the 212. Unfortunately the lode did not appear to be living up to earlier expectation, Paull noting that it was "hard and poor".[19]

On the 14th May 1861 Francis Russell the 7th Duke of Bedford died at the age of 73; he had held the title since October 20th 1839. He was succeeded by his son, William Russell, the 8th Duke, a title which he held until his death on May 27th 1872.

The main issue discussed at the 1861 Annual General Meeting, held on 15th May 1861, was an award for damages against the Company. Lord Fortescue and a Mr Willesford were awarded £878 18s 4d for "damage done to land by the weir".[20]

On May 31st it was noted that South Lode was showing signs of considerable improvement, the 90 east of Field Shaft was yielding fourteen or fifteen tons of ore per fathom. Writing on 9th November 1861 Josiah Paull reported that "the different backs on the South Lode are opening out large quantities of ore".[21] Evidently South Lode was becoming increasingly important. Paull's final report for 1861, dated December 1861, observed that "These mines have been worked with great success during the present year".[22]

1862

In his January 1862 report Paull noted the tribute pitches in Wheal Maria continued to yield good ore. Wheal Fanny, on the other hand, was not proving as productive as previously, however Paull comments that the agents held out hopes for the ground above the 45.[23]

Various points were being developed in Anna Maria in January 1862 including the 40 east of Engine Shaft on the Main Lode. Jeffery's crosscut was being driven south from the 80 fathom level although work had been suspended by mid July. On South Lode the 60, 70, 80, 103, 115 and the 130 fathom levels were being driven east of Field's Shaft. On the 90 a crosscut was being driven south of South Lode with the intention of cutting a portion of the lode encountered on the 80.[24]

In Wheal Josiah a winze was being sunk below the 200 fathom level in January 1862. East of Hitchins' Shaft the 50 was being driven on the Main Lode which, Paull notes, was unproductive. As at Anna Maria the South Lode was receiving considerable attention. At the time of Paull's January 1862 report work on the 60 had be suspended until a winze had been sunk to communicate with the 70 fathom level. A crosscut was being driven south from the 60 along a crosscourse to intersect the Blanchdown Adit which was being driven north towards Wheal Josiah. At the end of May two men were driving the crosscut south and four men driving the adit north, the two ends being within eighty fathoms of each other.[25]

Development continued at depth in Wheal Josiah: In July 1862, for example, the 212 was being driven both east and west of Richards' Shaft, six men in each end. The lode on the 212 was recorded as being regular and large but also hard and poor. However by the latter part of November things were beginning to improve, small stones of good quality ore were being encountered in the 212 east of Richards' Shaft.[26]

With regard to Wheal Emma, Paull commented in his January 1862 report that "there is nothing doing in the deep levels from the incline shaft". The men who had been working in these deep levels had been transferred to shallow works, the intention being to drive a shallow adit between Inclined Shaft and Thomas' Shaft.

By the end of July the shallow adit had been driven 80 fathoms from Rubbytown Bottom and was said to be following a "promising lode". By 31st March 1862 a new whim engine was being erected at Inclined Shaft. Paull hoped that the erection of the new engine would mean that the deep levels would be brought back into production. By the end of May both the 137 and 150 levels were being driven east from Inclined Shaft; one assumes that the long term intention was to prove the Eastern Ground at depth.[27]

The 60 fathom level east of Thomas' had been driven into the Eastern Ground and was, in January 1861, under the shallow adit. A rise was being put up from the 60 to the shallow adit to prove the lode and also to facilitate ventilation. On July 31st Paull noted that the deep adit continued to be driven west towards Wheal Emma but that the lode was "very poor".

By the end of March a "very fine course of ore" had been cut on the 50 fathom level west of Hitchins' Shaft on South Lode; Paull commenting that it was "the best I have seen on the South Lode". This part of the lode is recorded as being worth fifteen tons of ore or £120 per fathom. South Lode continued to prove rich throughout the year, for example on 1st September it was noted that "the different backs on the South Lode continue to look well and large sales of ore a being made".[28]

In his 30th September 1862 report Paull noted that the Company intended to resume work at both Wheal Thomas and Frementor ("Fremator"). At Wheal Thomas the intention was to drive the Deep Adit east of Engine Shaft under "a long piece of promising ground" whilst at Frementor "some trials are to be made". By the end of November the Frementor Deep Adit had been cleared and repaired. The company intended to recommence work on "a promising lode" sixty fathoms north of the portal. In addition crosscuts were to be driven from the South Fanny workings (see Note 2).[29]

The Company had always taken the education of children of its employees seriously, making a significant donation to local schools every year. In 1862 the company took their philanthropy a stage further by establishing a school on the mine. The 1862 Annual General Meeting being informed that the school was attended of upwards of 60 children.[30] In 1863 C. Twitte, reporting to Lord Kinnaird, observed that:

A school is held on the mines for all the children of the men under 10 years of age. The charge per child of one penny per week.[31]

Writing slightly later, in 1868 Gilson Martin, the Estate's agent, noted that:

> The Devon Great Consols company has always contributed liberally to all schools in the district, it has a school upon its own premises at Wheal Josiah which is maintained entirely by the Company and the children's pence. I believe this is chiefly intended for children too small to walk to the Gulworthy School, during this year there has been an average attendance of 45, 17 of whom were agents children and 28 mining labourers.[32]

1863

During the period 1862 to 1864 a Royal Commission under the chairmanship of Lord Kinnaird made exhaustive enquiries into the conditions prevailing in British metal mines, reporting on 4th July 1864. On 25th May 1863, C. Twitte Esq. presented his "report on the Devonshire Mines" to Lord Kinnaird. Not surprisingly a significant portion of Mr Twitte's report dealt with Devon Great Consols:

Twitte noted that in April 1863 the following were employed on the mine:

Underground	
Agents	10
Pit and timber men	11
Tutwork men	331
Tributers	13
Trammers, fillers and landers	79
Labourers	7
Boys	19
Total	**473**

Surface	
Agents and Engineer	9
Smiths	20
Carpenters and sawyers	18
Masons	13
Engine men	15

Labourers	128
Foundry	15
Railway	22
Total	240

Dressing pans (query should read pares? – author)

Men	117
Boys	181
Girls	199
Total	497
Total	1210

Twitte also provided an extensive outline of activity on the mine. At Wheal Maria the most westerly shaft was Castle's Shaft which was twenty three fathoms deep and was used as a ventilation shaft. It was a relatively small shaft being nine feet by five feet. Both Gard's and Morris' Shafts were slightly larger being ten feet by six feet. Gard's, which was used for both pumping and hoisting had reached a depth of one hundred fathoms. Morris' had reached the 100 fathom level. Only one end was being driven, a crosscut from the 23 fathom level from Morris' Shaft. Twitte note that Wheal Maria was the only part of the mine in which tributers were at work.

Moving east to Wheal Fanny, Twitte observed that Western Engine Shaft was twelve by eight feet in size and had reached a depth of 135 fathoms. It was perpendicular "to the 18, and then underlays 2 feet in a fathom". The shaft was used for both pumping and hoisting, hoisting being done in kibbles. Eastern Engine Shaft was down to a depth of 80 fathoms, the first fifteen fathoms were vertical, below that it was sunk on lode. Only used for hoisting, Eastern Engine Shaft was slightly smaller than Western Engine Shaft at ten feet by six feet. The final Wheal Fanny Shaft was Ventilating Shaft which was sixty five fathoms deep and eight by four feet in size.

Only two ends were being driven in Wheal Fanny: A crosscut from the 25 fathom level. It was noted that this drive was well ventilated as "air pipes are placed to carry the air to the end". The second drive was also a cross cut; it was driven from the 55 to "intersect a lode 10 fathoms south", presumably this refers to Woolridge's Lode.

Anna Maria Engine Shaft was twelve feet by seven feet and was used for both pumping and hoisting. In contrast to Wheal Maria skips were in use in the shaft; Twitte reporting that two skips were in use drawn by a seven inch flat hemp rope. The shaft had reached the 137, the top thirty fathoms were perpendicular, the remainder being sunk on lode. Field Shaft had reached the 154 and was used for hoisting, a single skip being employed in the ten by six foot shaft.

There were four development ends in Anna Maria; Jeffries' crosscut which was being driven south from the 80 (see Note 3) by two men. The drive was currently sixty fathoms long and was ventilated by air pipes. The 95 was being driven ninety fathoms west of Engine Shaft as was the forty fathom level, one hundred fathoms west of the shaft. Finally the 80 was being driven forty seven fathoms west of Engine Shaft.

A "drying and changing house" had been provided at Wheal Anna Maria. It was forty two feet long and twenty four feet wide. It was lit and ventilated from the roof. No provision was made for washing. It was also noted that "a large house is specially devoted for those working on the Anna Maria dressing floors to dine in". This building was "furnished with table and seats, a large oven for warming dinners, and a boiler for water for making tea or coffee".

At the time of Twitte's report to the Commission, Wheal Josiah was the main focus of operations, the Main Lode having been opened down the 212 level. Whilst the Main Lode was of prime importance a great deal of effort was also being put into developing South Lode by means of crosscuts driven from the Main Lode. Presumably the proximity of South Lode to the Main Lode meant that it was not thought worthwhile to sink shafts on South Lode or to duplicate hoisting and pumping facilities.

Although Field Shaft was considered part of Anna Maria it also constituted the western boundary of Wheal Josiah. Twitte noted that numerous points were being developed eastwards from Field's Shaft: The 130 was being driven on South Lode by four men, the 115 driven by four men, the 103 fathom level, the 90 fathom level, a winze was being sunk from the 90 by four men to communicate with a raise in the back of the 103 also being worked by four men, the 80 fathom level, the 70 fathom level, the sixty fathom level, and a crosscut south from the 70 which cut a minor lode referred to by Twitte as "Frank's Lode".

Richards' Shaft was, at the time of Twitte's report being sunk below the 212 by nine men. It was used both for pumping and hoisting. Surprisingly, given the importance of Richards' Shaft, kibbles drawn by a six inch flat rope were in use. West of the shaft James' crosscut was being driving from the 90 west to communicate with South Lode. The 144 west was within one or two fathoms of communicating with the 144 east of Field Shaft. A rise was being put up from the back of the 158 to communicate with the 144. Work on the 212, the deepest level in the mine had been suspended due to the high temperatures which Twitte quoted as ninety degrees Fahrenheit whilst James Richards quoted seventy degrees. Whatever the exact temperature, ventilation in these deep levels was evidently a problem and Twitte recorded that:

> A very powerful double piston air-machine is now being made, which will be placed in Richards' Shaft under the 144, and will be worked by the pumping rods, to ventilate these deep workings (see Note 4).[33]

Captain James Richards, in his evidence to the Commission given on Tuesday. 26th May 1863 noted that a winze was being sunk to improve ventilation at depth:

> We have an air-machine now working in the shaft to supply the winze with air..... the temperature is nothing like 70° now.[34]

In addition to ventilation issues, working at depth also caused other problems. Thomas Morris in his evidence to the Commission, also given on the 26th May, commented that it was taking half an hour for the men to reach the 212 on ladders, and the same to exit. Given that the time spent climbing ladders was included in the miners eight hour shift, the company was losing an hours work a day. Likewise Morris agreed with Sir Philip Egerton, who was questioning him, that the physical exertion involved in long ladder climbs reduced the men's capacity for physical work. To ameliorate the problem the Commission was informed that, should the 212 prove productive, the company were considering installing a man engine, Morris commenting that:

> At the 212 fathoms it has been a query with us whether we should abandon it, till lately; but prospects are now so good that the probability is that we shall put a man-engine down there, and a perpendicular shaft and a steam engine.

Hitchins' Shaft, recorded as being twelve feet by six feet, was 170 fathoms deep and was used for both pumping and hoisting. Like Richards' Shaft a kibble and six inch flat rope was in use. South Lode was largely, although not exclusively, being developed from Hitchins' Shaft. On the 70, Wilken's crosscut had been driven to cut South lode thirty five fathoms distant, three of which had been driven at the time of Twitte's report. On the 60 a crosscut had been driven on a crosscourse, cutting South Lode after thirty fathoms drive both to the east and west were being made on the lode. Trevena's crosscut on the 50 had cut South Lode after twenty five fathoms from which a level was being driven both east and west. Trevena's crosscut continued north of the Main Lode as Hawk's crosscut where after twenty five fathoms it had cut a "north part of the Main Lode". South Lode was cut on the 40 fathom level by Bennett's crosscut from which lode drives extended both east and west. South Lode 30 fathom level was being worked via Burrow's rise from the 40 and had been driven five fathoms to the west. Finally Steven's crosscut on the 80 cut South Lode after thirty fathoms and was being driven west on lode.

Old Agnes Shaft was in use for ventilation and as a ladderway down to the 103. New Agnes Shaft had reached the 184, it was twelve feet by seven feet in size. The shaft was perpendicular to the 115 and then continued on the underlie. New Agnes was used for both pumping and hoisting. Two kibbles were used in the shaft, drawn by a four inch flat rope. The Main Lode was being developed to the east of New Agnes shaft at three points including: the 137 east, the 144 east and the 130 east. The ground in the 130 east was so soft that it did not require any powder. A rise was being put up from the end of the 130 to communicate with the 124 in Wheal Emma.

At surface the Wheal Josiah drying and changing house was eighty one feet long and fifteen feet wide. As with the Anna Maria dry it was lit and ventilated from the roof. Whilst the floor was stone it was provided with wooden platforms for the men to stand on when changing. Wooden pegs were placed around the walls for the men to hang their clothes on. The dry was heated by a fire and a heated tube running lengthways along the length of the house on which the miners could dry their wet working clothes. Twitte reports that two men were employed in the Wheal Josiah dry "to attend to this house, keeping it clean and drying the clothes". The men were not allowed to keep their "hutches" (presumably powder cans) in the dry, Twitte noting that another house was "provided for the purpose, into which no candle is admitted, it being lighted by a lamp outside the window to avoid the risk of explosion". Further "houses" were provided on Wheal Josiah for lamp men and watch men.

Eastwards at Wheal Emma "Incline shaft" was fourteen feet by seven feet and had reached eight fathoms below the 162. Inclined Shaft was used for both pumping and hoisting. A double skip road had been installed, "the skips being drawn by wire ropes over timber guides". In his evidence to the Kinnaird Commission, Thomas Morris commented that "steel" rope had only recently been introduced on the mine and it was found that it "answers very well". A number of levels were being driven from Inclined Shaft, these included: the 162 east, the 150 east and west, the 124 west, the 50 west, Parson's crosscut driven south from the 47 and Barkell's crosscut driven south on the 55. The 124 west had, at the time of Twitte's report, almost communicated with the 130 east of Agnes Shaft.

At Thomas' Shaft six men were sinking below the 145. The twelve feet by seven feet shaft, was vertical to the 50 and then sunk on lode to the 145. The shaft was used for pumping and hoisting, drawing being done by kibbles and chains. Several drives were being worked at the time of the report including: the 130 east, the 60 east, a 30 fathom crosscut driven by two men south from the shaft, and the 20 fathom level.

Twitte was also able to report progress in the Eastern Ground:

> At a distance of 660 fathoms east of the old boundary a cross cut was commenced from the adjoining valley, which cut the lode 25 fathoms south. Driving was continued of the course of the lode for 110 fathoms, when a cross course was met with; through which a cross cut was driven 25 fathoms long, and the lode again met with, and followed upon. This level is 100 fathoms in serving as an adit, and is still driving. Several air shafts have been opened from the surface to ventilate these workings, and there is one now open within 20 fathoms of the end.

At the 1863 Annual General Meeting held on 13th May the Directors were in self congratulatory mood, commenting that this was one of the most successful years of the mine's operation; "the increase in dividends, the splendid discoveries in the mines, and the relative diminution of expenditure, are substantial matters for congratulation". The Company also announced that they had purchased a second locomotive from Messrs. Gilkes, Wilson, and Co. at a cost of £1,000, noting that "the first one having proved itself to be an admirable and effective engine in all respects".[35]

The Directors were also able to tell the shareholders about the newly established copper precipitation works.[36] The aim of the works was to recover the considerable quantities of copper held in solution in the mine water; this also had the added bonus of reducing the amount of toxic materials entering the Tamar. This works was probably located below the main tailings dump at Wheal Anna Maria.

1864

The May 1864 meeting was informed of the recent discovery of a new lode "unworked and whole to surface".[37] The new discovery would be named New South Lode and proved to be of major importance during the latter years of the mine's life. New South Lode lies parallel to and one hundred fathoms south of the Main Lode; the main body of ore being located in the western part of Wheal Emma.

The Company's lease from the Duchy of Cornwall to abstract water from the Tamar was due to expire in 1865. The supply of water from the River was so absolutely fundamental to the economical working of the mine that the Directors took the matter in hand over a year prior to the expiry. They were able to announce a satisfactory conclusion to their negotiations at the 1864 Annual General Meeting held on May 11th, the new lease running for twenty two years from Lady Day 1864.[38]

Whilst the Company had managed to secure its water lease it could still be held hostage by the vagaries of the weather. The early summer of 1864 was a particularly dry one, so much so that there was not enough water to supply the mine's pumping wheels. The *West Briton* of the 5th of August recorded the extraordinary measures taken by the Company to keep the mine pumped and maintain production:

> Owing to the extraordinary dryness of the season the supply of water from the River Tamar has become insufficient for the purpose of moving the powerful pumping machinery of these mines, and the water underground has risen to such an extent as to render it certain that in the event of the continuance of the drought a large number of the many hundreds of persons engaged must be thrown out of employment. In this emergency it was decided to obtain auxiliary power by the removal of a steam engine of 40-inch cylinder, with bobs, rods, and other necessary attachments and connections, from the lower part of the mines to the main and most central shaft at Wheal Anna Maria, and this work has been completed, and the engine set in motion within the short space of a fortnight – a feat in mining

engineering which if equalled, has probably never been surpassed.[39]

In December 1864 the *Tavistock Gazette* published a comprehensive description of the mine running over four issues. Of particular interest was a description of the newly established precipitation works which was undergoing considerable extension:

> One of the surface operations at Devon Great Consols is peculiarly interesting, as an inexpensive and effective application of a scientific process. We allude to the precipitate works, which are carried on more efficiently, and it is said to a larger extent than at any other place. The water from copper mines holds that metal in chemical solution, which is the cause of their poisonous character. By the process to precipitation the copper is taken from the water in a marketable state, wit the additional advantage of rendering the fluid less hurtful if not entirely innocuous. Where the water is suited for precipitation all that is necessary to be done is to let it flow over pieces of iron. The copper is then immediately deposited, and the iron taken up instead, in its turn to be thrown down as an ochreous oxide. Usually this operation is conducted in shallow pits called strips, divided breadth ways into compartments. The great objection to this plan is that by it the ochre, which forms most rapidly inn summer, and the copper become mixed, and, in much as their specific gravity is much the same, cannot be separated. Captain Isaac Richards has cleverly contrived a mode of proceeding which entirely meets and overcomes this difficulty. He has erected decagonal wooden cisterns, several feet across, on the tiled bottom of which the iron is placed. In each cistern is what he calls a sprinkler, a hollow wooden framework poised on a pivot, with several radiating arms, nearly as long as the cisterns radius. The water having been filtered is conveyed by a launder to the funnel-shaped centre of this framework, into the arms of which it flows, and issuing thence from holes pierced in their sides gives to the machine a rotary motion by its impingement against the air. In this way the water is sprinkled equally over the whole of the iron in the cistern, without being permitted to lie upon the metal, flowing out at the same rate at which it enters, and washing with it the deposited copper, which falls free from any ochreous admixture into a pit immediately outside the place of exit. The water is run in succession over two sprinklers, and is then filtered from the iron in solution in an ochre bed.[40]

It required three tons of iron to recover one ton of precipitate. The best grade precipitate contained about fifty five per cent copper with a market value of about £48 per ton, compared with £5 5s per ton for "common ore".[41]

1865

1865 was a significant year for Devon Great Consols, being the mine's 21st birthday. At the 21st Annual General Meeting in May of that year the Directors were, as usual, in self congratulatory mood. Captain James Richards, in his report to the A.G.M informed the meeting that the Main lode continued rich and he had high hopes for the deeper ground in Wheal Emma. Elsewhere in the mine Richards spoke well of the prospects of New South Lode.[42]

The Company was also pleased to announce the erection of a man engine at Hitchins' Shaft, a second man engine was under construction at Wheal Emma (see Note 5).[43] The introduction of man engines on the mine was an important technological step forward, making the deeper sections of the mine much more accessible, the Wheal Josiah man engine working to a depth of 170 fathoms. The *Tavistock Gazette* described the Wheal Josiah man engine as "a humane and scientific contrivance".[44] Whilst the introduction of man engines was certainly a huge boon to the miner who was spared the misery of long ladder climbs to and from the deep sections of the mine there is no doubt that they added an extra dimension of danger to the miner's already dangerous life:

> On Thursday last as a miner named Sleman of Gunnislake was ascending the man-engine, at Devon Great Consols, he received owing to insufficient attention to the working of the machine a rather severe injury, from which, however, we are glad to report he is slowly recovering.[45]

> A man named Richard Horn was seriously injured at Devon Great Consols, on Tuesday last, whilst descending the shaft by the man engine, at Wheal Josiah. It appears to have been his first attempt at descent by the engine, and consequently he felt somewhat nervous. Some way down he got confused and allowed the step to leave before taking his position, and his attempt to reach it was unsuccessful, which left him sitting on the landing stage. The next step of the engine immediately came in contact with his body, crushing his stomach and legs in a shocking manner. Happily the step gave way in the collision, or the poor man must have been killed.[46]

In June 1865, 33 men were employed underground at Wheal Maria and Wheal Fanny, 76 underground at Anna Maria, 201 underground at Wheal Josiah and

76 underground and at Wheal Emma and Wheal Thomas 122. At surface there were 225 carpenters and sawyers, 41 smiths and foundry men, Crossman's and Glanville's dressing "pair" (pare) comprised 216 individuals, 231 were employed at surface at Wheal Anna Maria and Wheal Josiah whilst 76 were engaged at Wheal Emma and Wheal Thomas.[47]

The Company decided to celebrate its "coming of age" in some style. On Saturday 17th June 1865 the mine's 1,200 employees were entertained to "a good plain old English dinner of roast and boiled beef and plum pudding", serenaded by the Devon Great Consols Band under the leadership of Captain Cock. Catering was on a gargantuan scale, the *Tavistock Gazette* noting that two thousand pounds of beef, one thousand three hundred pounds of bread, one thousand and eight pounds of plum pudding, eight hogsheads of beer, four of cider and one of lemonade were provided. The men were also provided with a "long pipe" and half an ounce of tobacco. After dinner sports were held with "donkey races, foot races.... bobbing for money, jumping in sacks, driving wheelbarrows blindfold". The *Gazette's* correspondent was astounded that anyone was able to run after the dinner they had just consumed:

> Think of a man running a steeple chase with a pound of cold plum pudding on his stomach, to say nothing of perhaps more than an equal quantity of beef! With such facts before us we cannot believe in the degeneracy of the race.[48]

The sports themselves were watched by between three and four thousand people. The games over, around 200 of the mine's female employees sat down for tea in the carpenters shops which had been "tastefully fitted up for this purpose". So ended a red letter day in the history of the mine, made all the sweeter for the workforce who also received a full days pay.[49]

1866

The celebration of the Devon Great Consols twenty first anniversary was hardly over when a darker episode in the mine's history began to unfold. In early 1866 local newspapers began to report the stirrings of industrial unrest in both the Liskeard and Tavistock mining districts. At the heart of the unrest was the Miners' Mutual Benefit Association.

The Association was formed by miners in the Liskeard district, their inaugural meeting being held on Monday 5th February 1866, a second meeting being held on Saturday 10th.[50] As the name suggests the organisation was a mutual benefit association, the need for this arising from a general dissatisfaction amongst miners with the practice on behalf of the mines of deducting money for "Doctor and sick club". This was felt by many to be a worthy endeavour, the *West Briton's* correspondent noting the it indicated that the miners were "becoming more thoughtful, prudent and thrifty – qualities which are truly commendable",[51] whilst a *Mining Journal* correspondent commented that "a Miners' Mutual Benefit Association properly constituted.....would be one of the greatest boons which the miners could be offered".[52]

However the aspirations of the Association went beyond medical and welfare benefits, they wanted to become involved in setting of working pitches; the following clause being included in their regulations:

> In every mine where there are fifty men working there shall be a committee formed of nine men, to be chosen by a majority, to whom miners shall go when they deem the price offered by the mine agents for doing certain work insufficient, and the committee shall see the place in dispute and decide whether the price offered is sufficient or not, and if it is the men must bear their own responsibility, but if not, the committee shall consult the agents and ask them to advance, and in the case of the agents refusing to do so at the expiration of a fortnight, such committee shall correspond with the general secretary who shall call a delegate meeting to decide what steps shall be taken.[53,54]

Under the Association's regulations the delegate meeting would have the authority to suspend labour, in other words to call a strike. There was also provision in the regulations to provide financial support to miners who withdrew their labour.

The use of strikes was no idle threat; in the Liskeard district, by the end of February 1866, all the men at both East Caradon and Marke Valley mines were out on strike as a direct result of three pitches being refused and the price offered not being increased.[55,56]

Having established itself in the Liskeard district, the Association set up a branch in the Tavistock district. An initial meeting took place in Gunnislake on Friday, 16th February. A second, better reported, meeting was held on Saturday 17th February in the Tavistock Temperance Hotel. The Tavistock men were addressed by Henry Cliff of South Caradon Mine. Evidently Cliff was a persuasive man; at the conclusion of the meeting 172 men enrolled as members of the Association.[57]

Whilst the Liskeard district was the initial focus of industrial unrest tensions were starting to run high in the Tavistock district: Anonymous threats had been sent to Captain James Richards at Devon Great Consols. At Drakewalls, on the Cornish bank of the Tamar, a miner who had not joined the Association, named George Lane was placed on a rough fir pole and paraded from Drakewalls Mine down Sand Hill to Old Gunnislake Mine by which time a crowd numbering three hundred had gathered. Fortunately for Lane the police intervened.[58]

In response to the growing unrest the management of all the mines in the district, foremost amongst whom was Captain James Richards, collectively placed the following notice in the *Tavistock Gazette* of Friday, March 2nd 1866:

> Proposed Miners Association
> The Managers of the whole of the Mines in the Tavistock District, with the concurrence of the proprietors have come to the following resolution in reference to this movement.
>
> Notice
> Observing that Public Meetings are being held in this and other districts for the purpose of forming a "Miners Association", and believing that if carried into effect, the code of rules submitted at these Meetings would subvert the authority vested in the Mine agents, and at the same time cause an irreparable injury to Mining enterprise, we, the undersigned, deem it necessary in the interests of our employers, as well as the working Miner, and for the public good, to determine that as any person joining that Association, and subscribing to the annexed code of rules, cannot consistently act in accordance with the rules already established for the regulation of Mining, it will be desirable to withhold employment from all persons who shall become members of that Society, and we accordingly hereby agree to carry this determination into effect. Signed.....

Without a doubt the threat to "lock out" Association men was intended to bring matters to a head. The timing of the notice, coming a day before the Devon Great Consols bimonthly setting day, due to be held on Saturday 3rd March, can hardly have been a coincidence.

That Friday evening a further miners meeting was held at the Tavistock

Temperance Hall. Messrs. Rowe, Criper, Vogwill, Rowse and Spencer addressed the assembled miners. Whilst the meeting reiterated that their cause was just it was felt that approaches should be made to the management at Devon Great Consols to ascertain whether an amicable agreement could be reached. To this end the meeting authorised Dr Harness, Mr Criper and Mr. Spencer to approach the agents at the mine with a view to resolving the situation.[59]

Whilst the members of the Association were attempting to diffuse potential trouble the agents on the mine had worked themselves up into an almost hysterical state. They deemed the situation to be so serious that on Friday March 2nd 1866 the following letter was presented to local magistrates:

> We the undersigned agents of the Devon Great Consols Mines, hereby declare that we feel ourselves to be in the position of most imminent danger; and we fear from the alarming reports that are hourly reaching us the threatening letters received, and the general tone of conversation of the men, that unless a strong military power is present tomorrow for our protection, serious consequences will ensue; and we are the more fearful of the result in consequence of the report in circulation that large bodies of men from a distance have resolved to attend here tomorrow to support the men who have joined the "Miners Association" in this neighbourhood.
> Dated Devon Great Consols, March 2nd, 1866
>
> William Clemo, Thos. Williams, Joseph Richards, Henry Rodda, James Bunt, Y. B. Barnett, Thos. Youren, W. H. Barnett, B. B. S. Richards.[60]

The Chief Constable of Devon, Gerald de Courcey Hamilton, in consultation with the magistrates, decided to deploy both a large number of police and the military. On the morning of Saturday 3rd March 150 soldiers of the 66th Regiment and 131 policemen were mobilised and 200 special constables were sworn in. Initially gathered at the Bedford Hotel in Tavistock the soldiers were moved up to the Harvest Home at Lumburn, twenty minutes march from the mine. The policemen and specials were distributed across the mine, the highest concentration being in the vicinity of the Count House. As if this force was not enough a body of marines was in place on the River Tamar ready to respond to incidents either at Devon Great Consols or at Hingston Down Mine on the Cornish bank of the river (see Note 6).[61,62]

On Saturday morning the Association's delegation comprising Dr Harness,

Messrs. Criper and Spencer and two members of the Association's committee approached the Devon Great Consols management to see if the lock out could be averted. Captain Richards refused to see the Association's committee members, however a meeting did take place between Harness, Criper and Spencer and W. A. Thomas, Thomas Morris and James Richards. The delegation made the point that the rule about committees had been misunderstood by the newly signed up members of the Association and they believed that they would be willing to remove it. They suggested that the lock out be postponed to give the Association members time to amended their rules. This proposal was rejected by the Devon Great Consols management who argued that they must stand by their decision at least until they had consulted with other mines managements; a meeting of "proprietors, managers and agents of mines" being scheduled for Wednesday, March 7th.[63]

Having met with the Association's delegation W. A. Thomas, T. Morris and Captain James Richards held a council of war. After "long and anxious deliberation" they decided that they would go ahead with the setting as usual at 1 o'clock.[64] As one o'clock approached a crowd of 1,200 to 1,300 gathered in front of the Count House. The police drew themselves up into two "divisions". One o'clock came and went. At half past one Captain James Richards appeared at the window of the Count House accompanied by W. A. Thomas, T. Morris and the County Magistrates: J. Gill, W. P Michell and J. Carpenter-Garnier. Before Captain Richards commenced the setting, W. A. Thomas addressed the gathered throng. Speaking at some length Thomas reiterated that no man who was a member of the Association would be permitted to take a bargain on the mine. Thomas went on to say that he and his fellow directors had no objection to mutual benefit societies but that any attempt to dictate terms to the management of the mine was wholly unacceptable. Thomas was willing to employ members of the Association provided the objectionable aspects of the Associations regulations were expunged. Having had his say W. A Thomas handed over to Captain Richards who continued with the setting. The setting proceeded with "the upmost order" however out of the twenty four bargains that were offered only four were let.[65,66]

At this point W. A. Thomas lost his usual composure and demanded to know what was "the meaning of all this nonsense? Why did they not take the bargains and go to work like men"?[67] After a degree of to and fro between Thomas and various members of the crowd, notably William Cater and Thomas Ennor, W. A. Thomas made a small concession; he would keep the bargains open until the following Monday. However beyond that Thomas stood his ground, stating that until he was satisfied that the rules of the Association had been altered to his satisfaction no bargains would be set. With that Thomas and his cohorts withdrew. Although the business of the day was officially over the miners did not leave, choosing to linger around the Count House. During this time Thomas went freely amongst the miners chatting with them. By four o'clock the miners had dispersed; what must have been an extremely tense day had ended in something of a stalemate. As the setting had passed off peaceably the soldiers at the Harvest Home were stood down as were the majority of the police and specials, however eighty policemen were quartered on the mine.[68,69]

The following week both sides regrouped and considered their positions. The Association men held a meeting on Monday the 5th March in the New Hall in Tavistock attended by "not less than eight hundred men". The meeting reviewed the rules of the Association and agreed to remove the appointment of committees on mines, and payment to members who suspended their labour, however they voted to keep the rule binding the members not to take a pitch refused by an Association member. On the Tuesday morning Messrs. Criper and Spencer (Dr Harness being out of town), went up to the mine and informed the management that the Association had amended its rules.[70] The Association's delegates were informed that a meeting of proprietors and agents was to be held the following day.[71]

The meeting of "proprietors, managers and agents of mines" was held on 7th March 1866 at Chubb's Hotel in Plymouth. Every mine in both the Liskeard and Tavistock districts was represented The general mood of the meeting was very much against the Association, any attempt to interfere in the setting of the pitches being seen as prejudicial to the proper management of mines. The meeting reviewed the amended rules of the Association and took objection to rule 9: no member of the Association "should be allowed to take any person to work with them who was not a miner", rule 10: "That any member taking a pitch or bargain belonging to another member shall be excluded from, and forfeit all rights and claims to the Society" and rule 11: "If at any time their be a pitch or bargain lying idle not having been refused because the price was not sufficient, such pitches or bargains shall be drawn for when more than one party want it". The meeting unanimously resolved that no man who subscribed to rules 9, 10 and 11, would be allowed to "settle in any pitch or bargain." The meeting did recognise that there was a need to address the health and welfare issues facing the miners and it was proposed that a committee should be set up to consider setting up a miners benefit society, albeit one that they controlled.[72,73] A resolution to this end being passed at a further meeting held on the 20th March 1866.[74]

On Tuesday 12th March 1866 two thousand miners gathered for a meeting on Kit Hill at which they determined to stand by the Association.[75] Whilst the Liskeard men held firm in maintaining their action, the Tavistock men proved not to be so staunch. The *West Briton* of 16th March 1866 reporting that at Devon Great Consols the men had returned to work, the bargains being taken partly by non Association men and partly by men who had been members but had given up their membership.

With the dispute safely over the management were able to be bullish about the whole affair:

> The most important occurrence since the last annual general meting is the late dispute with the miners about certain objectionable rules of a proposed association, to which many of them had given their adherence, - Rules which were calculated to interfere with the due authority of the agents, and the independence of the miners themselves, and which they foolishly imagined would enable them to exact higher wages, forgetting that the ancient custom of submitting all bargains to public competition left them free to accept or refuse them, and that any interference with those who are willing to work is not only unjust but illegal. The unflinching determination of the directors, supported by the civil authorities of the county, to resist any such innovation, sufficed to crush a senseless agitation, instigated by a few idle and unprincipled individuals.[76]

Perhaps a more balanced epitaph was supplied by an anonymous *Mining Journal* correspondent signing himself simply as "A. Miner":

> Happily the dispute between the Miners and Mine Adventurers is at an end, and I think all who have watched the progress of the struggle must admit that the men have shown quite as much wisdom and moderation as their employers; and have certainly suffered no greater defeat.[77]

The 1866 Annual General Meeting was informed that an "influential committee" had been formed with a view to forming a miners' benefit society acceptable to management. It was the intention, the meeting was informed, to name the society "The Royal United Miners Association". The Company resolved to subscribe £500 towards the establishment of the society.[78] It is arguable that the establishment of the Royal United Miners Association was never a serious proposition. Certainly by the time of the 1867 Annual General Meeting the idea had been abandoned, the Directors citing a failure to get the "western mines" to cooperate and the depressed state of mining in general.[79]

Whilst the dispute between the workers and the company was played out very publicly another, ultimately much more important, development was quietly taking place: the introduction of arsenic production on the mine.

In addition to copper ore the lodes at Devon Great Consols were also phenomenally rich in arsenopyrite or mundic as it was termed. Whilst the mine had made small sales of mundic as early as 1852 during the early years of the mine's history arsenic was considered, at best, an insignificant by product and, at worst a contaminant, and it was usually left *in situ* underground. By the mid 1860s, faced with falling copper prices and the lodes becoming less productive at depth, the mine began to look at arsenic in a different light. As the chemical industry developed during the nineteenth century the properties of arsenic began to be better understood as did its myriad uses including use in glass making, the manufacture of shot, tanning, dye making for fabrics and wall papers and most importantly as a pesticide. Fortunately for mines such as Devon Great Consols the late 1860s and early 1870s saw a huge growth in the demand for arsenic both for dye making, Germany being a particularly prolific user, and for pesticides, America using huge amounts of arsenic based pesticides to combat the Colorado Beetle.[80]

As early as 1864 it was proposed to erect a "calcining house" on the mine to treat ore with a high arsenopyrite content. The ore would be roasted to drive off the arsenic and sulphur; the residue would then be washed so that the copper could be recovered by precipitation.[81] At this stage it would appear that the Company had no intention to recover the arsenic but rather to recover copper from lower grade halvans.

Writing in 1896 Isaac Richards stated that arsenic production commenced at the mine in 1864.[82] However it would seem that time was playing tricks with Captain Richards' memory, it was not until September 24th 1866 that an agreement to construct an arsenic works at Anna Maria was signed between John and William Alexander Thomas on behalf of Devon Great Consols and William, Duke of Bedford. In return for permission to build the arsenic works the Duke of Bedford was to receive a rent of 1/12th of marketable products from the works.[83]

The September 1866 agreement contained a number of clauses, laying out in some detail the standards to which the works should be constructed:

2. In the construction of the intended works the chambers and flues connected with the roasting kilns shall be carried to distance (measured horizontally) of at least six hundred feet before uniting with the stack. The section of main chamber and first length of flue being of the length of ninety feet shall be twelve feet in height and six feet in width as shown in the said plans. A reduction of area shall be allowed after the first length of flue but no part of the flue shall be of less dimensions than four and a half feet high by three feet wide. The walls of the chambers and flues shall be solidly built of the thickness of at least two and a half feet of masonry where the flue is of the greatest dimensions and nowhere less than two feet of masonry covered with cement or puddle.

3. The precipitation of the arsenic sulphur gases and volatile substances which shall pass beyond the main flue and chambers shall be effected by means of waterfalls or water showers as shown in the said plans and in the building of the smaller flue provision shall be made for not less than three such water falls between the point where it leaves the main flue and its junction with the stack and cisterns shall be provided in the bottom of the smaller flue for the reception of all substances precipitated as above mentioned with means at the side for their cleansing and refitting where necessary.

4. The refinery shall be connected with the stack by flues of dimensions not less than those mentioned with regard to the flues from the roasting kilns and the volatile substance in the smaller flue communicating with the refinery shall be precipitated by means of water falls or water showers and such flue shall be provided with cisterns in like manner as above stipulated with regard to the smaller flue communicating with the roasting kilns.

5. The flues from the roasting kilns and from the refinery shall communicate separately with the stack and at or near their respective confluence therewith each flue shall be furnished with a damper for the purpose of regulating the draft and which will also afford the means of preventing any matter however volatile escaping beyond the flues.

6. That regard being had to the frequent and dense fogs which rise to a

> great height from the Tamar Valley and to the risk of damage from rapid condensation so occasioned and with a view to obviate such risk by the abundant dilution of the smoke with the atmosphere before it comes into contact with said fogs the stack shall be of the height of at least one hundred and twenty feet.[84]

1867.

Josiah Paull inspected the mine on 18th and 19th February 1867. In his report dated 28th February he notes that "a few men and boys" were employed extracting ore at Wheal Maria. During April and May a 20 fathom level was driven between Gard's and Morris' Shafts. This level was driven on "one of the South Lodes" opening up some productive ground.[85]

At Wheal Fanny "a few hands" were employed extracting ore throughout the year. During June and August a rise was driven on the 35 fathom level east of Ventilating Shaft, this communicated with the 30 in Anna Maria in August 1867. In December it was noted that reserves in Wheal Fanny were becoming very low.[86]

At Anna Maria during February a few men were driving Jeffrey's crosscut on the 80 with the intention of cutting New South Lode. That said by April 1867 Paull noted that work on the crosscut had been suspended and work did not recommence that year. Between April and October a series of exploratory rises were put up above the 70 with little success. In spite of limited exploration Anna Maria continued to return "a great deal of ore".[87]

During 1867 no exploration was carried out on the Main Lode between Field Shaft and Hitchins' Shaft at Wheal Josiah. East of Hitchins' Shaft, Josiah Paull, in his annual report of December 1867, notes that "some spirited trials have been made and are still continued".[88] Productive ground being encountered on the 103, 115, 130 and 114 fathom levels. East of Agnes Shaft the 80 and 90 fathom levels had been driven on the north part of the Main Lode with little success. At greater depths the ground between the 157 and 170 fathom levels had proved interesting. It was noted that "the lode is large and produces a very fair sample of tin ores".[89] Evidently the lure of tin had not reached the proportions it would in later years and operations were suspended in October although Paull did comment that "the necessity for further trials on this piece of ground will not be lost sight of" (see Note 7).[90] Finally, on the Main Lode, Richards' Shaft was below the 224 but, according to Paull, "little has been done this year".[91]

Figure 14. Captain Isaac Richards standing beside an Oxland and Hocking calciner.

Figure 15. The Wheal Emma section of the mine; Inclined Shaft is in the centre.

Figure 16. Morwellham Quay, around 1902-1905.

Figure 17. The Wheal Anna Maria counthouse.

Figure 18. The Bedford Estate's Garrett steam lorry, 1925-1930.

Figure 19. The waterwheel at the New Dressing Plant, 1938.

Limited work had been carried out on the Middle Lode, an offshoot of South Lode, at Wheal Josiah, a few fathoms having been driven on the 103.[92]

In his report to the May 1867 Annual General Meeting Captain James Richards highlighted the "extraordinary productiveness" of South Lode at Wheal Josiah. Richards was keen to point out that South Lode was looking promising, presumably in contrast to the Main Lode which was showing some signs of failure at depth.[93] The 90 on South Lode in the vicinity of both Richards and Hitchins' Shafts cut some very productive ground as did the 115. In August the 90 cut the intersection between South and Main Lodes in the vicinity of Agnes Shaft. Winzes sunk below the 115 cut, in the words of Richards in May 1867 report, "a fine course of ore".[94] In spite of Captain Richards enthusiasm, the lode on the 130 was giving some cause for concern, being described by Josiah Paull as "unproductive... although it is by no means .... unpromising".[95]

The Main Lode at Wheal Emma received considerable attention at depth; the 162, 175 and 190 fathom levels having been driven "steadily" east of Inclined Shaft during 1867. Both the 162 and 175 had been temporarily suspended by the end of December 1867 due to poor ventilation. Evidently the company held out hopes for this section of the mine as they were engaged in sinking Inclined Shaft below the 190 with nine men. In December 1867 it was hoped that they would be deep enough for a 205 fathom level "in about four months". Thomas' Shaft was also being sunk below the 145, the intention being to communicate with the 175. It was anticipated that this work would take about a year.[96]

Progress was also being made on the exploration and development of New South Lode; the lode was being developed in two, more or less independent, sections from Wheal Josiah and Wheal Emma respectively. In February 1867 two men were driving the adit level (60 fathoms) on the Wheal Josiah section of New South Lode. Paull observed that "the lode is a little larger than it has been with capel mundic and a little copper ore". This work continued throughout most of the year, however work was suspended by December due of "defective" ventilation. To remedy this a rise was be put up from adit to surface. In February four men were employed in "making good the new shaft from surface to this level", this shaft would subsequently be known as Counthouse Shaft. By April a "drawing machine had been set up on the shaft and Paull expressed the hope that the shaft would be sunk below adit "in a little time". By the end of the year the shaft was being sunk below the 70 fathom level by six men.[97]

During February 1867 in Wheal Emma a winze on New South Lode was being sunk on the 50 fathom level west of Inclined Shaft. The lode, as cut by the, winze, was about two and a half feet wide and comprised of capel and mundic with a little copper ore. Paull notes that the winze was suspended by April 1867. To develop the lode below the 50 crosscuts were being driven south from the Main Lode on both the 75 and 100 fathom levels during February.[98]

In his April 1867 report Josiah Paull noted with pleasure that the number of men employed on exploratory work had increased. Paull ascribes this increase to the fact that Captain James Richards was taking a more active role in the direction of the underground works, an aspect of the mine that had previously been almost entirely under the control of William Clemo. Paull is less than complimentary regarding Clemo's ability:

> Capn. Clemo to whom the underground management has been left of late years although an excellent man to take away ore when discovered possesses no spirit and has no faith in an outlay for the purpose of making with it.[99]

The 1867 Annual General Meeting, held on May 14th was informed that the arsenic works would be completed and in production towards the end of the year.[100] In December 1867 Paull was able to report that works were "all but completed" and arsenic production had commenced.[101]

Copper precipitation was still an important, albeit small, part of the mine's activity. During 1867 precipitation works were being operated at Wheal Maria, Wheal Josiah and Wheal Emma. Although due to the low price of copper the company were stock piling the precipitate until such time as the market conditions might improve. In December 1867 Paull estimated that between seventy and eighty tons, worth between £25 and £30 per ton, were in stock.[102]

1867 also saw the mine producing ochre for the first time. In effect this was a by-product of the precipitation process. Paull noted that "although its value is not great nor the quantity large at present it may become so in the course of time". During 1867 the mine produced 33 tons, 18 cwt. of ochre worth £56 8s 9d.[103]

1868.

During 1868 activity at both Wheal Maria and Wheal Fanny was limited to extracting remaining ore. Josiah Paull was of the opinion that this could continue "some time longer" at Wheal Maria, whilst Wheal Fanny was almost exhausted.[104]

Anna Maria continued productive throughout 1868 although no exploratory work was undertaken.[105] At Wheal Josiah some work had been done on the Main Lode at depth. The 103 west of Agnes Shaft and the 130 east of Hitchins' receiving attention. Additionally deep exploration on the north part of the Main Lode had proved unproductive during the year.

Both the 103 and 115 fathom levels on Middle Lode had been driven a short distance with little result. With regard to the 103 Paull commented:

> on the whole the driveage has laid open but little ground that will pay for working away.[106]

On South Lode the 115 east of Hitchins' Shaft had been driven throughout the year by four men but cut little good ore ground. The 130 was driven both east and west of Richards' Shaft with little result although the end in the 130 east of Hitchins' was starting to look promising at the end of December 1868.[107] Gilson Martin, in his report to the Duke of Bedford dated 31st December 1868, expressed concerns regarding the future of South Lode which:

> Has probably yielded ore to the value of £600,000 is rapidly declining in value and its working have now reached the depth at which the Main Lode became unproductive in the same part of the mine. This lode is but a part of the Main Lode and has its separate position in Wheal Josiah only, I can therefore hold out but little hope of future discoveries from it.[108]

Exploration and development on the Main Lode continued at Wheal Emma. Crosscuts were driven northwards on the 162 and 175 fathom levels to explore the northern part of the lode which was found to be poor. The 175 was also driven east of Inclined Shaft, the ground proving reasonably good, at one point being worth four tons per fathom. The 195 east of Inclined Shaft also received considerable attention during 1868 having been driven "a great many fathoms this year" albeit with little result. A winze linking the 175 and 190 on the other hand opened up "many fathoms of profitable ground". Inclined Shaft had reached the 205 in May 1868 and sinking continued deeper during the rest of the year, six men being engaged on the task. The 205 had been driven a few fathoms east of Inclined Shaft however the lode proved "barren". The sinking of Thomas' Shaft had proceeded rapidly, nearly thirty fathoms had been sunk during the year and, in late December, it was anticipated that the shaft would communicate with the 175 in little over a month. East of Thomas' a rise had been put up between the 100 and 87 fathom levels the intention being to continue it to surface. The purpose of this rise was to improve haulage from the 145 to surface and facilitate exploration of the Eastern Ground at depth. Between January and August 1868 the 190 east of Thomas' Shaft was driven through an untried piece of ground. The lode was described as being of "great width" yielding ore in "paying quantities". In addition to copper this section of the Main Lode was also worth twenty tons of mundic per fathom. Although idle in December 1868 Paull opined that work would be resumed "at no distant date".[109]

Deep adit on New South Lode in Wheal Josiah was driven about a "dozen fathoms" in the course of 1868. The 70 had been driven east and west of the "Engine Shaft" (*i.e.* Counthouse Shaft). The 70 west had communicated with "one at the same depth driven from a point further west". Engine Shaft reached the 82 fathom level in April and the 82 was driven east and west of the shaft during the remainder of the year.[110]

At Wheal Emma a rise on New South Lode between the 75 and 50 fathom levels was "made good" to surface; the new shaft was known as Railway Shaft. In December 1868 six men were engaged in sinking the shaft below the 75. New South Lode in Wheal Emma was proving to be highly promising: Railway Shaft was cutting good ground as was the 75 which had been driven "many fathoms" west of the shaft. The 100 fathom level crosscut, known as Alford's, from Inclined Shaft had made good progress throughout the year cutting a number of "small lodes or branches". It was not thought that any of these were New South Lode so work continued driving south.[111]

In terms of mining 1868 must have been a disturbing year for the company. New discoveries were thin on the ground, Josiah Paull observing that "they are less by far than in any other year of the company's existence".[112] The Main Lode was showing signs of exhaustion as was South Lode. Writing in late December 1868 Gilson Martin commented "that an end must come to the returns from those parts of the mine at no distant date".[113] Whilst New South Lode was showing promise it was a long way from full production. New South Lode aside, the only aspect of the mine's operation from which the directors, shareholders and the estate could draw comfort from was the newly established arsenic works.

On January 31st 1868 Josiah Paull noted that the arsenic works was in full operation and "many tons of refined arsenic are already made".[114] As with any newly introduced process there were some initial issues to be overcome. By February it had become apparent that the throughput of the works was only half of what had been calculated. The reason was that the ore had to be roasted at a lower temperature than anticipated to allow "a perfect sublimation of the arsenic".[115]

At the end of March 1868 Paull recorded that plans were being drawn up for additions to the works, this presumably refers to additional calciners.[116] By the time of the 1868 Annual General Meeting, held on May 14th the "reduction and arsenic works" were noted as being in production and capable of producing fifty tons of arsenic a month, the works having cost £1,382 to date. The Directors were of the opinion that the arsenic works needed to be considerably enlarged to keep pace with the increasing amount of lower grade ores being raised.[117] Work to extend the works was soon in progress "a large force of masons" being at work by the end of May. During June 1868 advantage was taken of the particularly fine weather to "push forward" the extension of the arsenic works. Work continued apace through the summer of 1868, although poor weather during September slowed progress. In September 1868 Isaac Richards, who was directing operations, estimated that the work would be completed by the end of the year. In his final report of the year, dated Paull commented:

> These are now all but complete and as arsenic works they are the largest and best erected in England.[118]

As part of the new additions to the arsenic works "a new contrivance for assisting the fume in the flue" known as "water splashes" were installed. The water splash consisted of a self acting "water box" which emptied thirty gallons of water into the flue every two or three minutes. The water cooled stones placed in the flue, the arsenic being deposited on the "damp and cold surface of the stones".[119]

Figure 20. Workers in the DGC arsenic reduction works, 1893.

Gilson Martin's report dated 31st December 1868 contains extensive details of the works as built:

Two years ago this company commenced very extensive works for manufacturing arsenic from Mundic Ores a large portion of which was comparatively worthless. It has spent several thousand pounds and the works are now in full operation. 473 tons have now been sold and a contract is now entered into to supply 150 tons per month for 12 months. The price is about £6 per ton delivered on the quays......

The Mundic Ores containing about 30 per cent Arsenic and generally a small quantity of Copper after being ground at the crushing mill sufficiently fine to pass through a wire sieve the meshes of which are about half an inch square are taken to the calcining furnace which may be described as an oven, 15 feet long, 9 feet wide and 3 feet high, they are spread over the bottom of the furnace and exposed to the heat of a fire which is placed at one end and when heated through the arsenic immediately passes of in fumes at the other in this state it passes through a series of flues the length of which is 2591 feet & and the cubical area 44,367 feet it will readily be understood that directly the fume reaches an atmosphere sufficiently cool the precipitation of arsenic to the bottom of the flue commences; Openings at the sides of the flues are made at convenient distances and the Arsenic extracted through them, it is subsequently refined of the coal dust & other impurities which it might have gathered in the first process in a second furnace, the purified arsenic is driven off in fume and again precipitated in a second flue the total length of which is 2,054 feet and the cubical area 40,184 feet. The flues from the calcining furnaces have an average rise from a horizontal line of 1 in 36 & those from the refining furnaces have various elevations two being 1 in 9 and another 1 in 14. There are 5 calcining & 3 refining furnaces, the total length of flues is 4,645 feet & the cubical area 84,551 feet, these end at the bottom of a chimney or stack 120 feet in height.

To prevent the escape of fume and gases injurious to woods & herbage the Co. are bound to their working agreement with the Duke to use an injection of water from the top of the main flues in a copious shower is turned on in this way through which the smoke must necessarily pass on its way to the stack, It is hoped and believed that the damage to the surrounding woods and crops will thus be reduced so as not to be observable. The residuum or calcined ores having been reduced in weight & brek by the separation of arsenic before contained in them are afterwards sent to the quays at Morwellham & sampled and sold to the copper smelter.[120]

Figure 21. Man shovelling refined arsenic into the DGC grinder.

The calcining furnaces referred to by Martin in his report would have been the ubiquitous Brunton. The Brunton calciner had been around since the 1820s and examples would continue to be used in Cornwall until the 1950s (see Note 8).

It was estimated that the extended works would be able to produce between 1,500 and 1,600 tons annually. In addition to arsenic an estimated 3,000 tons of calcined low grade copper ores would also be produced, effectively as a by product. After further treatment the low grade copper ore would be sold to the smelters.[121]

Although the production of arsenic would become, in effect, the saviour of the mine in latter years there was a down side. The company were certainly aware of the health issues involved in arsenic production. Josiah Paull's report of April 1868 contains the following:

> The occupations of roasting and refining are to the men engaged about the most unhealthy that can be conceived.[122]

"The precipitation of copper on iron from the mine water" continued apace across the mine. However due to the low price of copper the mine had made no sales of precipitate during the year, preferring to continue their policy of stock piling pending an improvement in the market. It was estimated in December 1868 that the mine held stocks of more than one hundred tons of precipitate.[123]

1869

The Final Annual General Meeting of the 1860s was held on May 11th 1869. The report of the 1869 Annual General Meeting contained a melancholy item, namely the death of John Thomas, W. A Thomas' brother and one of the original shareholders. The Directors who faced their shareholders at this meeting were not able to muster their customary optimism:

> The continued and increased depression of the copper market, together with the absence of any valuable discovery in the mines during the last twelve months, deprives the directors of the pleasure of the usual felicitations on the remarkable success which has attended the operations of this company for 25 years. The directors have strained every point compatible with prudence, in the hope of maintaining the previous rate of dividend, and although they have drawn considerably from the reserves of ore , and sold a greater number of tons containing more fine copper, yet the decline in the market value shows a deficit of about £5,000..... The directors have given their earnest attention to a possible reduction of expenses, but are convinced that little can be accomplished on this head without impairing the efficiency of working of the mines.[124]

With regard to the future prosperity of the mine all the Directors were able to offer was a vague possibility that, if "explorations are prosecuted with vigour" new discoveries would be forthcoming.

During the year no exploration was carried out in Wheal Maria, Wheal Fanny and Wheal Anna Maria. Wheal Maria was being worked by a few tributers. In Wheal Fanny little ore had been raised either "from tutwork or tribute working". Wheal Anna Maria continued to be productive however the lack of exploration meant that "its returns have therefore been seriously lessened".[125]

From August to December 1869 some exploratory work had been carried out on the northern part of the Main Lode at Wheal Josiah, paying ground having been found on the 50 east of Agnes Shaft. On the Main Lode proper, no levels had been driven. All the levels down to the 144 communicated to the west with Anna Maria. To the west all the levels down to the 115 communicated with Wheal Emma. Josiah Paull was of the opinion that the appearance of the Main Lode at depth in Wheal Josiah presented little inducement for further development.[126]

Between April and August four men were employed on the 130 on Middle Lode with little result.[127] The news from South Lode at Wheal Josiah was disappointing. The 115 east of Hitchins' Shaft had been driven throughout the year without opening any profitable ground. The 130 east and west of Hitchins' Shaft had also received attention during the year with limited results. Paull in his annual report of December 1869 commented:

> I think it must be admitted that the South Lode is now proved to be of no value from a point about 70 fathoms east of Hitchins' Shaft to its junction with the Main Lode a little to the east of Agnes Shaft.[128]

At Wheal Emma the 175 fathom level east of Inclined Shaft communicated with Thomas' shaft in February 1869. Similarly the 190 linked the two shafts in May. The 205 east of inclined shaft was driven "a great many fathoms" until it was suspended in September. Unfortunately these deep drives opened up little good ground. In October work started driving the 112 fathom level east of Inclined Shaft. The intention was to open up the unproved ground between Inclined Shaft and Thomas' from the 47 to the 112. Inclined Shaft itself continued to sink below the 205 until it was suspended in June, the lode not being of "much promise".[129] Summing up the year's progress on the Main Lode at Wheal Emma Josiah Paull commented:

> In Wheal Emma the works on this Lode have been carried on with a spirit deserving of great success but they have met with none.[130]

New South Lode received considerable attention during 1869 At Wheal Josiah, Engine Shaft had reached the 100 fathom level in October. The lode below the 82 was described as presenting "but few promising features, but in the bottom of the shaft it looks stronger and better". The 82 was driven several fathoms east of the shaft, however initially promising ground did not live up to expectation and work was suspended in July. The lode on the 100 east was noted as being small and unpromising. No work had taken place on Deep Adit during the year. In December Paull noted that New South Lode in Wheal Josiah did not seem to improving with depth.[131]

At Wheal Emma Railway Shaft had been sunk "with spirit" below the 75, reaching the 100 in October. New South Lode, in this section of the mine, was recorded as being "large and very promising". The 100 east of Railway Shaft had been driven through rich ground, yielding eight tons of copper ore per fathom. At year's end the 100 east was close to communicating with the 100 crosscut west of Inclined Shaft. One slight cloud on the horizon was the growing realisation that there was little good ore ground at shallow depth. To unlock the potential riches of New South Lode the company would have to go deep. To speed up the development of New South Lode a new shaft which would be given the imaginative name "New

Shaft" was begun in September. At the end of December twenty six men were engaged on the shaft working on the 50, 75 and 100 fathom levels. Such was the urgency to explore and bring New South Lode into full production the agents had promised to make New Shaft "good" to the 100 fathom level in time for the May 1870 Annual General Meeting.[132]

To expand the potential ore ground the Company announced at the May 1869 Annual General Meeting that it had acquired the lease, including "valuable plant and machinery", of an "adjoining mine" "contiguous to the lodes going eastward from the old set".[133] The mine acquired by the Company was Colcharton, something of a dubious prospect. In 1868 Gilson Martin, in his report to the Duke of Bedford, dismissed the mine as of little or no value.[134] In spite of having reached a depth of 65 fathoms little if any ore had been raised. The "valuable plant and machinery" would seem to have been a 30-inch rotary engine which, in 1863, served for pumping, hoisting and crushing.[135] By the end of the year twelve men were undertaking exploratory work on the mine.[136]

In spite of the potential of New South Lode the share markets were loosing confidence in the mine. On 29th December 1868 the lowest price being quoted for a Devon Great Consols share was £320, on 28th December 1869 shares were being quoted at £120, a drop of £200 per share; £204,800 had been wiped off the share value of the mine in a year![137]

Whilst the prospects for copper were disappointing the Directors were at least able to be more positive regarding arsenic. The expansion of the arsenic works discussed at the 1868 Annual General Meeting had been completed at a cost of £2,538 3s 9d. The works were now capable of producing one hundred and fifty tons of arsenic a month. The Company had also managed to enter into a twelve month contract to supply their arsenic. One suspects that the contract was not as good as the Directors would have liked as they commented that once the quality of their arsenic was better known it would command a higher price.[138]

The start of 1869 saw the arsenic works in full production. Isaac Richards was still experimenting with and fine tuning the process. For example in February 1869 it was noted that:

> the kilns have lately been thrown down and the mundic and low quality copper ore will in future be passed through the crushing mills preparatory to being roasted in flat bottomed furnaces.[139]

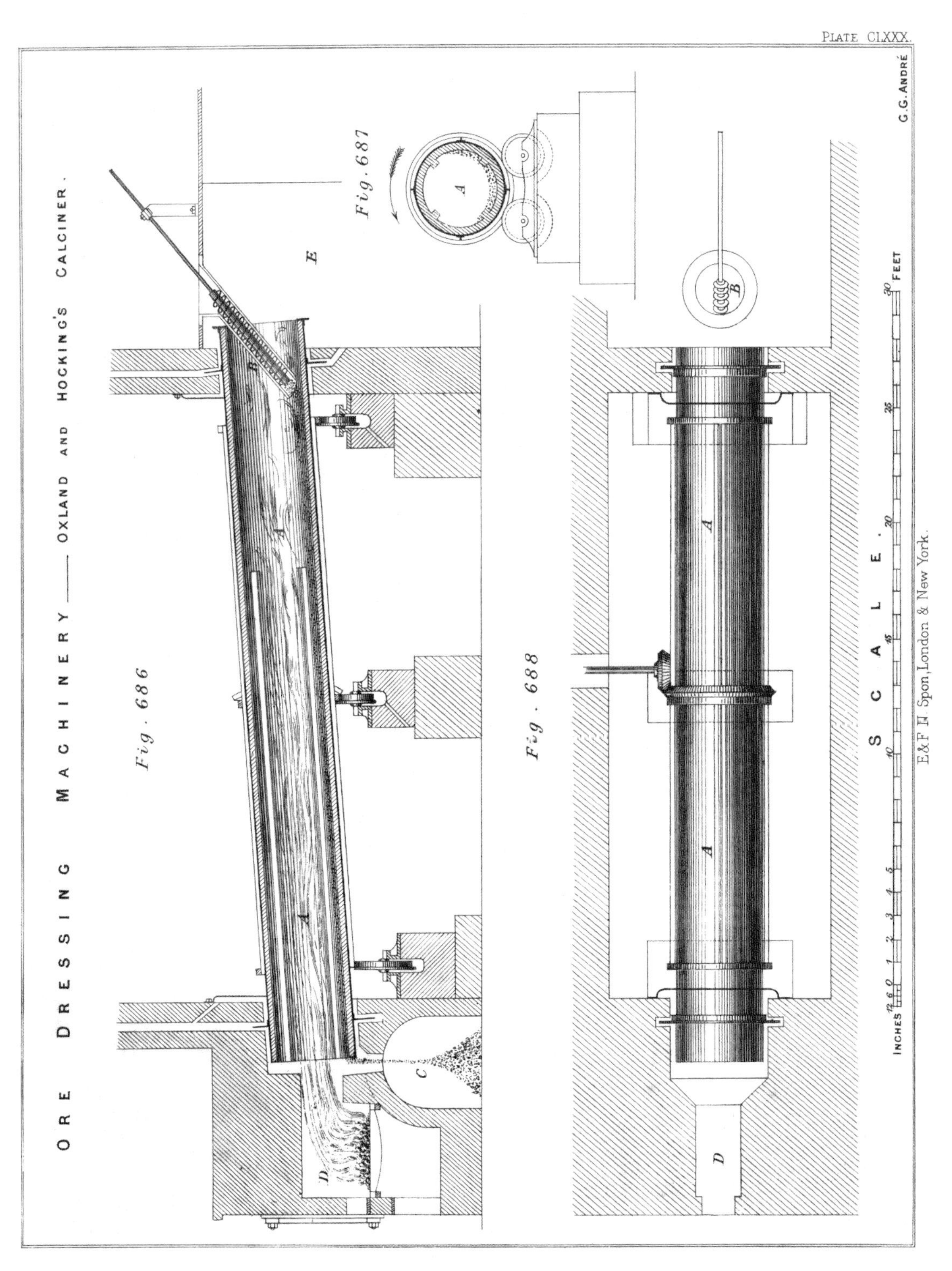
PLATE CLXXX.
ORE DRESSING MACHINERY —— OXLAND AND HOCKING'S CALCINER.
G.G. ANDRÉ
Fig. 686
Fig. 687
Fig. 688
SCALE.
INCHES
FEET
E&F N Spon, London & New York.

Figure 22.

In June 1869 Paull reported that "a new circular furnace which is to revolve on an axis" was being erected at the arsenic works. The "circular furnace" was Oxland and Hocking's calciner. The "Oxland tube" was at the time of its introduction at Devon Great Consols, a new piece of technology having only been developed a year or two previously (see Note 9). At the time it was felt that the "Oxland tube" would allow the mine to economically process "arsenical pyrites of low produce". The Devon Great Consols tube was thirty feet long, three feet in diameter and driven by water power. By the end of August 1869 the Oxland tube had been started and although there were some initial problems Isaac Richards expressed the fullest confidence in it. In a report dated 23rd September Josiah Paull was able to report that the "revolving furnace" was at work commenting:

> it fulfils the calculations of its inventors both as to dispatch and the perfect sublimation of arsenic and sulphur.[140]

Unfortunately the sulphur and the arsenic were condensing at the same point, rather than the sulphur passing out of the stack. Paull noted that "this defect was receiving the attention of Mr Oxland and Capt. Isaac Richards".[141] By late October the problem had not wholly been resolved, the arsenic not being quite as pure as desired. Interestingly it was felt that the "Oxland tube" was much less injurious to the workmen's health than "kilns and reverberatory furnaces".[142]

In his final report of 1869 Paull commented that "the arsenic works have been carried on with great activity the make of refined having amounted to 1700 tons".[143] Copper precipitation continued throughout 1869, however no sales were made. The company was continuing its policy of stockpiling its precipitate against better times. At the end of the year the company had 180 tons of 40% precipitate in stock.[144]

The situation facing the mine at the end of the decade can hardly have been more different from that at the beginning of the decade. At the start of the decade the mine, as we have seen, saw itself as entering a comfortable maturity, the richness of the Main Lode was a given, subsidiary lodes, in particular South Lode held out prospects of future prosperity. The mine could look forward to the future with equanimity. However by the end of the decade copper prices were falling, the Main Lode was showing signs of exhaustion, South Lode looked as if it was going the same way. The Company's relationship with its workforce, hitherto characterised by a degree of philanthropy, had become muddied and complicated by the 1866 lock out. To give the Company credit its Directors responded well to the changing situation, particularly the decline in copper; the transition to arsenic production ensuring the survival of the mine for the next three decades.

Chapter 6 Notes

Note 1

In 1860 Salmon noted the following waterwheels on the mine: The two Great Wheels 40′ x 12′ pumping Anna Maria Engine Shaft, Fields Shaft, Richards' Shaft and Hitchins' Shaft. Agnes Wheel 32′ x 10′ pumping Agnes Shaft. Plunger 32′ x 16′ pumping water up to the "Great Reservoir" at the top of the mine. 30' x 8′ wheel pumping water from the river up to the halvans floors. Wheal Maria Wheal 50′ x 4′ pumping Gard's Shaft at Wheal Maria and Western Shaft at Wheal Fanny. 35′ x 4′ at Foundry. Wheal Fanny stamps wheel 35′ x 4′. Wheal Fanny drawing machine 32′ x 4′. Wheal Josiah saw mill wheel 32′ x 4′. Wheal Maria grinder wheel 50′ x 4′. Wheal Thomas pumping wheel 36′ x 4′.[145]

Whilst water was the dominant motive power there were also a number of steam engines on the mine as recorded by Salmon in 1860: Wheal Emma 40″ pumping Thomas' and Incline Shafts. Anna Maria 36″ old dressing floors crusher. Anna Maria 30″ halvans crusher. Wheal Josiah whim 30″ & 16″ combined Sims engine. Anna Maria 24″ whim. Wheal Emma 22″ Whim. 22″ at Morwellham incline head.[146] By this time the Morris' Shaft engine was out of use.

Note 2

Josiah Paull's reference to the South Fanny workings in his December 1862 report is the earliest use of the name "South Fanny" that the current author has found. Interestingly Paull refers to these workings as "the above mine formerly known as South Fanny".[147]

Note 3

Twitte records "Jeffries" crosscut as being driven on the 70; this is an error. The various Bedford Estate mining reports produced by Josiah Paull all note that this crosscut was driven on the 80 as does the working plan.

Note 4

In his evidence to the Kinnaird Commission James Richards commented that in addition to mechanical ventilation "waterfalls" had also been used. Richards informed the Commissioners that water was taken from "the lift in the shaft if we have not it otherwise".[148]

Note 5
The Wheal Emma man engine was located on Inclined Shaft

Note 6
The *Tavistock Gazette's* correspondent seemed rather amused by the presence of the military and was inclined to be satirical:

> The arrival of the soldiers created a great sensation at Tavistock. They were kept at the railway station for some time, and then marched to Lumburn, and then, to the great consolation of the men, marched back to the Tavistock Hotel to dinner.[149]

Indeed the whole operation came in for a degree of scorn:

> the enormous force of police and military had no more serious duty to perform than to keep themselves warm, look at each other and eat their dinner. On Friday a large body of police arrived on the mine, and on Saturday morning a large addition was made to their number. More than a hundred special constables were on the ground. These were ill prepared for a conflict, as except an occasional walking stick, we observed no defensive weapon among them. Doubtless they relied either of the moral effect of their presence to control disorder, or as someone wickedly suggested, on their heels to ensure their own safety. However, they proved their true British blood, by grumbling enough at the duty assigned to them, and their consequent absence from their business".[150]

Note 7
In spite of Paull's comment that further trials were necessary this area does not appear to have been reinvestigated for tin during the 1870s and 1880s when the mine was attempting to reinvent itself as a tin producer.

Note 8
When William Brunton patented his calciner on 21st February 1828, its suitability was soon appreciated; John Taylor in his *Records of Mining*, published in 1829 comments on "the peculiar excellence of this furnace".[151] At its most basic the Brunton comprised a furnace with a revolving, concave bottom about sixteen feet in diameter. The ore is fed into the furnace from above forming a conical heap in the centre of the rotating plate (temperature here being 1,100° F) and from there is gradually worked out to the edge (temperature here being 1,000° F); this

is facilitated by cast iron "flukes" or rakes built into the roof of the furnace. The flukes also served to stir or "rabble" the roasting ore allowing an even roasting. On reaching the edge of the rotating plate the now calcined ore fell off the edge and into a cooling chamber or pit. The plate itself rotates at a speed of between three and five revolutions an hour, the exact rate being dependent on the grade of the ore. A typical throughput for a Brunton would be about six tons in twenty four hours. The fumes given off by the calcining process were drawn off into the flue system.

Note 9

Oxland and Hocking's calciner consisted of an inclined wrought iron tube about thirty feet long and four feet in diameter lined with fire bricks. A furnace or fire place located at the bottom end of the tube provided heat for the process. The tube was supported on rollers and geared to rotate slowly. Ore was first dried on iron plates and then fed into the upper end of the tube, the rotation of the tube causing the ore to slowly work its way down to the bottom end of the tube. The calcined ore being discharged from the bottom end of the tube into a small chamber located between the tube and the fire place. Having a larger throughput and a greater fuel economy than the Brunton it is not surprising that a number of mines, Devon Great Consols included, adopted the "Oxland tube".[152,153] However the Oxland tube did not render the Brunton obsolete; its primary flaw was that the constant rolling of the ore created large amounts of dust which contaminated the arsenic soot and led to a reduction in recovery rates.[154] It is perhaps not surprising that the mine adopted Oxland and Hocking's calciner given that the Oxlands were significant shareholders.

Chapter 6 references

1. *Mining Journal* 27 July 1860
2. *Mining Journal* 27 July 1860
3. *Mining Journal* 27 July 1860
4. Devon Record Office document 8187, Bedford Estate Mine Reports 1860-1862
5. *Mining Journal* 27 July 1860
6. *Mining Journal* 27 July 1860
7. Devon Record Office document 8187, Bedford Estate Mine Reports 1860-1862
8. *Mining Journal* 27 July 1860
9. *Mining Journal* 23 February 1861
10. *Mining Journal* 5 January 1861
11. Devon Record Office document 8187, Bedford Estate Mine Reports 1860-1862
12. *Mining Journal* 27 July 1860

13. Devon Record Office document 8187, Bedford Estate Mine Reports 1860-1862
14. Devon Record Office document 8187, Bedford Estate Mine Reports 1860-1862
15. *Mining Journal* 23 February 1861
16. Devon Record Office document 8187, Bedford Estate Mine Reports 1860-1862
17. *Mining Journal* 23 February 1861
18. *Mining Journal* 23 February 1861
19. Devon Record Office document 8187, Bedford Estate Mine Reports 1860-1862
20. *Mining Journal* 18 May 1861
21. Devon Record Office document 8187, Bedford Estate Mine Reports 1860-1862
22. Devon Record Office document 8187, Bedford Estate Mine Reports 1860-1862
23. Devon Record Office document 8187, Bedford Estate Mine Reports 1860-1862
24. Devon Record Office document 8187, Bedford Estate Mine Reports 1860-1862
25. Devon Record Office document 8187, Bedford Estate Mine Reports 1860-1862
26. Devon Record Office document 8187, Bedford Estate Mine Reports 1860-1862
27. Devon Record Office document 8187, Bedford Estate Mine Reports 1860-1862
28. Devon Record Office document 8187, Bedford Estate Mine Reports 1860-1862
29. Devon Record Office document 8187, Bedford Estate Mine Reports 1860-1862
30. *Mining Journal* 17 May 1862
31. Kinnaird, 1864, *Report of the Commissioners appointed to inquire into the condition of all mines in Great Britain*, H.M.S.O.
32. Devon Record Office T1258M/E44 a-b, Devon & Cornwall Estates, reports on the mines & quarries 1868 &1869.
33. Kinnaird Report, 1864, *op. cit.*
34. Kinnaird Report, 1864, *op. cit.*
35. *Mining Journal* 16 May 1863
36. *Mining Journal* 16 May 1863
37. *Mining Journal* 14 May 1864
38. *Mining Journal* 14 May 1864
39. *West Briton* 5 August 1864
40. *Tavistock Gazette* 2-23 December 1864
41. *Tavistock Gazette* 2-23 December 1864
42. *Mining Journal* 20 May 1865
43. *Mining Journal* 20 May 1865
44. *Tavistock Gazette* 23 June 1865
45. *Tavistock Gazette* 16 February 1866
46. *Tavistock Gazette* 22 July 1870
47. *Tavistock Gazette* 23 June 1865
48. *Tavistock Gazette* 23 June 1865
49. *Tavistock Gazette* 23 June 1865

50. *Tavistock Gazette* 9 February 1866
51. *West Briton* 2 March 1866
52. *Mining Journal* 3 March 1866
53. *Tavistock Gazette* 23 February 1866
54. *West Briton* March 9 1866
55. *West Briton* March 2 1866
56. *West Briton* March 9 1866
57. *Tavistock Gazette* 23 February 1866
58. *West Briton* March 9 1866
59. *Tavistock Gazette* 9 March 1866
60. *West Briton* 9 March 1866
61. *Tavistock Gazette* 9 March 1866
62. *West Briton* 9 March 1866
63. *Tavistock Gazette* 9 March 1866
64. *Tavistock Gazette* 9 March 1866
65. *Tavistock Gazette* 9 March 1866
66. *West Briton* 9 March 1866
67. *West Briton* 9 March 1866
68. *Tavistock Gazette* 9 March 1866
69. *West Briton* March 9 1866
70. *Tavistock Gazette* 9 March 1866
71. *Tavistock Gazette* 9 March 1866
72. *Tavistock Gazette* 9 March 1866
73. *Mining Journal* 10 March 1866
74. *Mining Journal* 24 March 1866
75. *Tavistock Gazette* 16 March 1866
76. *Mining Journal* 19 May 1866
77. *Mining Journal* 7 April 1866
78. *Mining Journal* 19 May 1866
79. *Mining Journal* 18 May 1867
80. Barton D. B., 1970, *Essays in Cornish Mining History*, Vol. 2, D. Bradford Barton Ltd.
81. Tavistock Gazette 2-23 December 1864
82. *Tavistock Gazette* 11 December 1896
83. Devon Record Office document L1258/MC1/, 1866 Arsenic agreement
84. Devon Record Office document L1258/MC1/, 1866 Arsenic agreement
85. Devon Record Office document 8187, Bedford Estate Mine Reports 1867-1872
86. Devon Record Office document 8187, Bedford Estate Mine Reports 1867-1872
87. Devon Record Office document 8187, Bedford Estate Mine Reports 1867-1872

88. Devon Record Office document 8187, Bedford Estate Mine Reports 1867-1872
89. Devon Record Office document 8187, Bedford Estate Mine Reports 1867-1872
90. Devon Record Office document 8187, Bedford Estate Mine Reports 1867-1872
91. Devon Record Office document 8187, Bedford Estate Mine Reports 1867-1872
92. Devon Record Office document 8187, Bedford Estate Mine Reports 1867-1872
93. *Mining Journal* 18 May 1867
94. *Mining Journal* 18 May 1867
95. Devon Record Office document 8187, Bedford Estate Mine Reports 1867-1872
96. Devon Record Office document 8187, Bedford Estate Mine Reports 1867-1872
97. Devon Record Office document 8187, Bedford Estate Mine Reports 1867-1872
98. Devon Record Office document 8187, Bedford Estate Mine Reports 1867-1872
99. Devon Record Office document 8187, Bedford Estate Mine Reports 1867-1872
100. *Mining Journal* 18 May 1867
101. Devon Record Office document 8187, Bedford Estate Mine Reports 1867-1872
102. Devon Record Office document 8187, Bedford Estate Mine Reports 1867-1872
103. Devon Record Office document 8187, Bedford Estate Mine Reports 1867-1872
104. Devon Record Office document 8187, Bedford Estate Mine Reports 1867-1872
105. Devon Record Office document 8187, Bedford Estate Mine Reports 1867-1872
106. Devon Record Office document 8187, Bedford Estate Mine Reports 1867-1872
107. Devon Record Office document 8187, Bedford Estate Mine Reports 1867-1872
108. Devon Record Office document T1258M/E44 a-b, Report on the Mines 1868
109. Devon Record Office document 8187, Bedford Estate Mine Reports 1867-1872
110. Devon Record Office document 8187, Bedford Estate Mine Reports 1867-1872
111. Devon Record Office document 8187, Bedford Estate Mine Reports 1867-1872
112. Devon Record Office document 8187, Bedford Estate Mine Reports 1867-1872
113. Devon Record Office document T1258M/E44 a-b, Report on the Mines 1868
114. Devon Record Office document 8187, Bedford Estate Mine Reports 1867-1872
115. Devon Record Office document 8187, Bedford Estate Mine Reports 1867-1872
116. Devon Record Office document 8187, Bedford Estate Mine Reports 1867-1872
117. *Mining Journal* 16 May 1868
118. Devon Record Office document 8187, Bedford Estate Mine Reports 1867-1872
119. Devon Record Office document 8187, Bedford Estate Mine Reports 1867-1872
120. Devon Record Office document T1258M/E44 a-b, Report on the Mines 1868
121. Devon Record Office document 8187, Bedford Estate Mine Reports 1867-1872
122. Devon Record Office document 8187, Bedford Estate Mine Reports 1867-1872
123. Devon Record Office document 8187, Bedford Estate Mine Reports 1867-1872
124. *Mining Journal* 15 May 1869
125. Devon Record Office document 8187, Bedford Estate Mine Reports 1867-1872

126.Devon Record Office document 8187, Bedford Estate Mine Reports 1867-1872
127.Devon Record Office document 8187, Bedford Estate Mine Reports 1867-1872
128.Devon Record Office document 8187, Bedford Estate Mine Reports 1867-1872
129.Devon Record Office document 8187, Bedford Estate Mine Reports 1867-1872
130.Devon Record Office document 8187, Bedford Estate Mine Reports 1867-1872
131.Devon Record Office document 8187, Bedford Estate Mine Reports 1867-1872
132.Devon Record Office document 8187, Bedford Estate Mine Reports 1867-1872
133.*Mining Journal* 15 May 1869
134.Devon Record Office document T1258M/E44 a-b, Report on the Mines 1868
135.Jenkin A. K. H,. 1974, Mines of Devon, Vol. 1 The Southern Area, David & Charles.
136.Devon Record Office document 8187, Bedford Estate Mine Reports 1867-1872
137.Devon Record Office document 8187, Bedford Estate Mine Reports 1867-1872
138.*Mining Journal* 15 May 1869
139.Devon Record Office document 8187, Bedford Estate Mine Reports 1867-1872
140.Devon Record Office document 8187, Bedford Estate Mine Reports 1867-1872
141.Devon Record Office document 8187, Bedford Estate Mine Reports 1867-1872
142.Devon Record Office document 8187, Bedford Estate Mine Reports 1867-1872
143.Devon Record Office document 8187, Bedford Estate Mine Reports 1867-1872
144.Devon Record Office document 8187, Bedford Estate Mine Reports 1867-1872
145.*Mining Journal* 27 July 1860
146.*Mining Journal* 27 July 1860
147.Devon Record Office document 8187, Bedford Estate Mine Reports 1860-1862
148.Kinnaird Report, 1864, *op. cit.*
149.*Tavistock Gazette* 9 March 1866
150.*Tavistock Gazette* 9 March 1866
151.Taylor J., 1829, *Records of Mining*, Part 1. John Murray London.
152.Miners Association of Devon & Cornwall, Annual Report 1868
153.Miners Association of Devon & Cornwall, Annual Report 1877
154.Toll R. W. 1938, The arsenic industry in the Tavistock district of Devon, *Sands Clays & Minerals*, April 1938, pp 224-227.

Chapter 7

The 1870s: Reconstruction and reinvention

"The search for tin may therefore be said to be now in hand"
– Josiah Paull, November 1872.

1870

As the new decade started the Directors were well aware that falling copper prices and the exhaustion of the principal lodes meant that they could no longer take the mine's traditional prosperity for granted. Whilst they had no influence over the price they received for their copper they did feel that they could influence the costs of production. To that end Captain James Richards was invited to a conference of the principal shareholders held in London in January 1870 to discuss the issue. One area in which the Directors felt they could make savings was exploration. Richards argued that the short term expediency of reducing costs would have a long term impact on the productivity of the mine. He made a strong case in favour of continued exploration and development, arguing that if such work stopped the mine would have, at best, a life of three years based on known reserves. Expenditure on development and exploration might, Richards suggested, be reduced once the development work on New South Lode had become sufficiently advanced to prove its value.[1]

The first Annual General Meeting of the 1870s was held, as usual, in the Company offices at Gresham House, Old Broad Street, London. The meeting took place on Monday, May 30th, with a somewhat sombre W. A. Thomas taking the chair. The Directors informed the gathered shareholders that the situation was no better than it had been the previous year: no new discoveries had been made, resulting in a reduction of the mine's reserves. These problems had been compounded by the continuing low price of copper.[2]

The Directors outlined the details of their January meeting with Richards, informing the meeting that they had found it impossible to reduce costs in any significant way. Apart from production costs the Directors had also attempted to reduce the royalty paid to the Duke of Bedford for arsenic and, in the light of falling copper prices, on copper. The response from the Duke was described at "curt and unsatisfactory". The Directors held out hopes that the Duke would consider reviewing his royalties thus demonstrating the supremacy of optimism over experience.[3]

In his report Captain Richards drew the meeting's attention to the western end of the sett. In the neighbouring sett of West Maria and Fortescue "a fine course of ore" had been encountered on the 60 fathom level. The proximity of this find to the Company's ground had not escaped Richards who expressed the intention to start exploratory work to prove the eastward extension of the ore body into Wheal Maria.[4]

In spite of the Directors' assertions that they found it impossible to reduce costs the May 1870 meeting saw a significant change in policy regarding exploration and development. Throughout the mine almost all exploration was suspended or, at best, scaled back. This was much to the chagrin of Josiah Paull who, in his role as the Bedford Estate's mineral agent, felt that this was purely an attempt to cut costs and certainly did not serve the best interests of the Estate. The only exception was New South Lode at Wheal Emma into which the mine poured considerable resources in order to prove its reserves and bring it into production as quickly as possible.[5]

During 1870 no exploratory work was undertaken on the Main Lode west of Wheal Emma. Wheal Maria was felt to be exhausted; at year end reserves were estimated at about 70 tons. The situation at Wheal Fanny was slightly better with reserves of 700 tons. Anna Maria was still productive with eighteen men employed and estimated reserves totalling some 3,000 tons. Worryingly no tutwork operations had been carried out on the Main Lode at Wheal Josiah during the year.[6]

Limited exploratory work had been carried out on South Lode, both the 115 and the 130 having been extended during the year.[7] At the May Annual General Meeting Captain Richards highlighted Painter's winze in the bottom of the 130 which was proving rich. This was significant as, to date, this was the deepest productive ore ground encountered on South Lode.[8] By December 1870 the

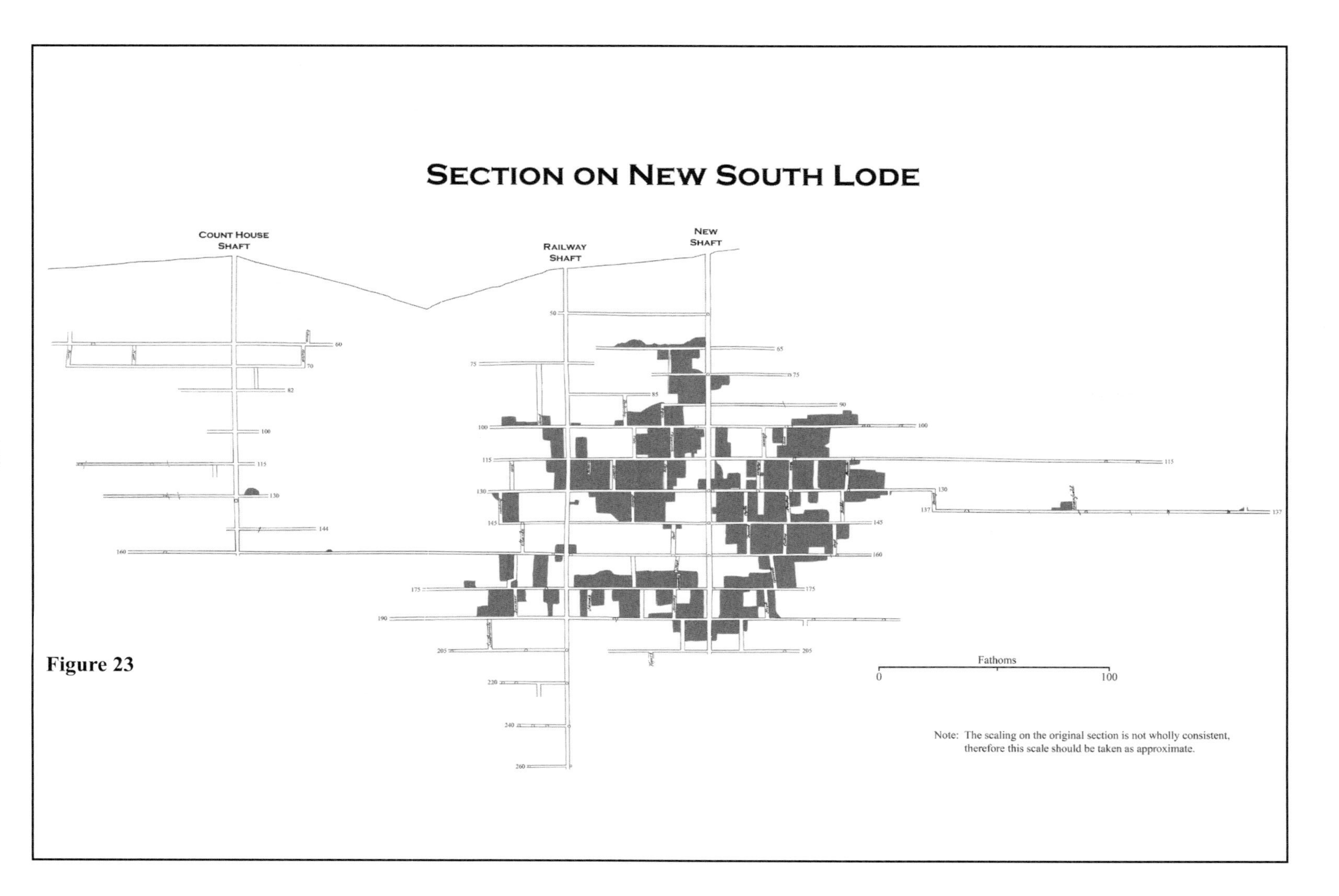

Section on New South Lode
Count House Shaft
Railway Shaft
New Shaft
Fathoms
0
100
Note: The scaling on the original section is not wholly consistent, therefore this scale should be taken as approximate.

Figure 23

winze was seven fathoms below the 130 and was still in very good ground; being worth fourteen tons of good ore per fathom. The deeper levels on South Lode, the 141 east of Field Shaft and the 144 east of Hitchins' Shaft, were not cutting such good ground.[9]

At Wheal Emma the 112 drive intended to open up the untried ground between Inclined Shaft and Thomas' Shaft had been suspended. Work on the 201 east of Thomas' was suspended at the end of May, the lode being "irregular and of no value" having been "disordered" by a "powerful crosscourse". However Paull was of the opinion that if the crosscourse was passed the lode would improve. During the year Thomas' Shaft had been sunk from the 201 to the 216. East of Thomas' the "long rise" had been "made good" to surface in April. This put the company in a strong position to develop the Eastern Ground at depth, Paull noting that "it can at once be made available for pumping and haulage to the depth of the 145 fathom level".[10]

At the time of the May meeting, Alford's crosscut on the 100 fathom level west of Inclined Shaft was being extended south to explore the ground between New South Lode and Wheal Thomas.[11] However soon after the meeting work was suspended, much to the annoyance of Josiah Paull.[12]

On New South Lode at Wheal Josiah work on the 100 fathom level had been suspended in July. The lode was noted as being generally large and Paull was of the opinion that there was no good reason to suspend work. Engine Shaft was sunk between the 100 and 115 fathom levels, the intention being to start sinking below the 115 in January of the following year. In October work started driving the 115 east and west of the shaft.[13]

Most of the mine's exploratory efforts during 1870 focused on opening up New South lode at Wheal Emma as quickly as possible. Railway Shaft had reached the 115 fathom level in October.[14] Much to the credit of the agents and men New Shaft was down to the 100 fathom level (see Note 1) by the time of the Annual General Meeting in May as promised the previous year.[15] Josiah Paull was extremely impressed:

> to sink a shaft through the whole depth in eight months from the time of starting it is a very creditable performance and as such has not been excelled in the experience of most miners.[16]

Figure 24.

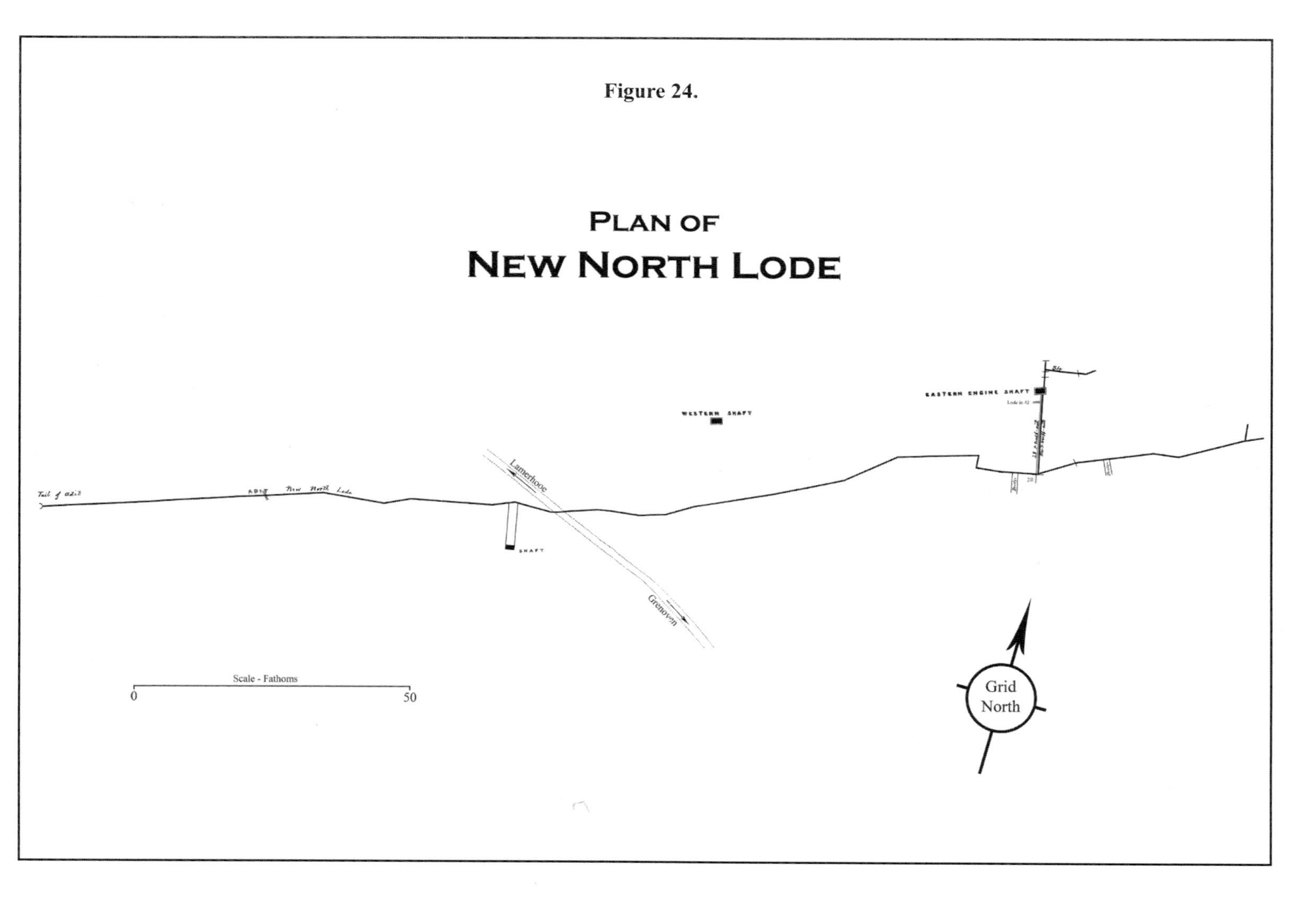
PLAN OF
NEW NORTH LODE
EASTERN ENGINE SHAFT
WESTERN SHAFT
SHAFT
Lamerhooe
Grenoven
Grid
North
Scale - Fathoms
0
50

During the year driveages were pushed forward on the 75, 85, 100, and 115 fathom levels.[17] In June 1870 work was suspended at Colcharton and the mine was allowed to flood.[18]

In summing up the year's activities at the mine Josiah Paull was fairly scathing, particularly with regard to the suspension of exploratory work:

> I have been deeply concerned to see that all works of this kind are cut short by a company who have obtained unprecedented profits from mining in His Grace's lands and the more so as the agents of other mines on the Estate taunt me with it when I call on them to do more in their concerns, indeed it cannot be expected with reason in non – paying mines much new ground should be explored from the pockets of the Shareholders if those who have taken nearly all the profits made in the district are the first to lose heart and direct in the direction of dividends the money which should be spent in the vast area of untried ground within this sett.[19]

Whilst Paull accepted that the company had made considerable efforts to open up New South Lode, he was less than sanguine regarding the results:

> I must except in my remarks on the general slackness of exploration work the trials made and making on the New South Lode in Wheal Emma, these have been followed up well but the length of the ground operated on is but little more than 100 fathoms and the workings have reached the 115 fm level without meeting more than occasional spots in the lode of good paying ground and I fear that this lode is not to be depended on in Wheal Josiah or Wheal Emma for any large yield of ore.[20]

Paull felt that the way forward was to drive both north and south in search of untried lodes, Alford's crosscut being a case in point. He also felt that it would repay to "thoroughly" work older sections of the mine. Paull concurred with Captain Richards' report to the May meeting regarding the desirability of trying the postulated eastern extension of the West Maria and Fortescue Main Lode in Wheal Maria.[21] Interestingly Paull makes no mention of the Eastern Ground in his recommendations for future exploration.

Paull's gloomy outlook was shared by the market. The value of Devon Great Consols shares had dropped as low as £70 in August. The price had rallied slightly by the end of the year at which time shares were being quoted at between £90 and £100.[22]

1871

The diminution in the mine's prosperity and fall in share price was a cause of great concern to many of the shareholders who had come to regard their dividend as a dependable and regular portion of their incomes. On April 4th 1871 a small group of disgruntled shareholders, including Joseph Wills, R. B. E. Gill and J. A. Page, held a meeting in Plymouth which resolved that the Company's deed of settlement should be "undergo revision", the London Offices should be discontinued and the Directors' qualifications should be reduced. They also resolved that an approach be made to the Directors to convene an Extraordinary General Meeting to carry their resolutions.[23]

The Company's response to the Plymouth meeting was fairly rapid. The Company Secretary, Alexander Allen, alluded to the meeting's resolutions in his notice of the forthcoming Annual General Meeting. Allen, presumably speaking on behalf and with the full knowledge of the Directors stated that the resolutions made by the Plymouth meeting would "diminish the respect and high estimation which the company has pre-eminently commanded since its formation".[24] Allen also made the point that the Plymouth group only represented fifty six shares out of at total of 1,024.[25]

The 1871 Annual General Meeting was held on Tuesday the 9th May. The Company report quoted in the *Mining Journal* noted that "the directors have the mortification to show a serious diminution of dividends".[26] During the preceding year £8,192 had been divided amongst the shareholders.[27] In any other mine this would have been a cause for celebration; not so at Devon Great Consols. The ongoing decline in the mines fortunes was ascribed to the price of copper dropping to £5 12s 10d, an all time low in the history of the mine to date. In consequence the mine's management had agreed to a reduction in their pay: The Resident Director consented to a 50% reduction, Captain James Richards to 20% and the other Agents 11%. On a slightly more positive note the sale of arsenic were looking more positive having realised £12,132 4s 8d albeit at a cost of £4,261.[28]

Richards reported that New South Lode was proving rich in the bottom of Railway Shaft. He felt that this justified the expenditure of the previous two years and that it was now time to turn attention to the north lode leading from West Maria and Fortescue. Beyond that Richards noted that Cole's winze on the 130 on South Lode had opened up some "splendid ground".[29]

The 1871 Annual General Meeting was directly followed by an Extraordinary General Meeting called to consider the resolution and amendment of the Company's deed of settlement. Expenditure on the London office, Directors residences and Agents remunerations were called into serious question by Mr Gill, one of the "Plymouth Group". In spite of Mr Gill's no doubt eloquent criticism of the mine's management and operation, the meeting was not inclined to make radical changes to the Company settlement. The only resolution adopted was that clause 27 be amended, and the number of shareholders wishing to convene a meeting must posses not less than 350 shares in the capital of the company, instead of 650 as before, all other resolutions were abandoned.[30]

Although the criticisms aired at the May 1871 Extraordinary General Meeting only represented the views of what might be described as a small splinter group, the implication that the mine's management and agents were overpaid touched a nerve. In direct response to the "accusations" made at the Extraordinary General Meeting the management issued a Supplement to the Directors Annual report. The supplement, reproduced in *Mining Journal*, outlined both work being undertaken at the mine and the numerous and varied duties of the Agents. As a rebuttal of the Plymouth Group's accusations it is a convincing document; it also provides a fascinating snapshot of the mine that date.[31]

Whilst the management and shareholders of the mine were busy squabbling and point scoring amongst themselves, a fatal accident occurred on the mine on May 24th, 1871. Thomas Hinch fell in Anna Maria Engine Shaft whilst fetching some clay, his workmates found him dead at the bottom of the shaft with two broken arms, lying in about two feet of water. The inquest returned a verdict of accidental death. Whilst death or injury underground was not uncommon what makes the case of Thomas Hinch noteworthy was his age: at the time of his death he was just fourteen years old.[32,33]

With regard to operations on the mine no exploratory work was carried out on the Main Lode at Wheal Maria, Wheal Fanny, Wheal Anna Maria and Wheal Josiah. At Wheal Maria and Wheal Fanny a few tributers were employed extracting ore. "Several men" were working at Anna Maria. However most of the ore reserves from the wide, productive, sections of the lode had been exhausted. Captain Richards estimated that poorer ground might yield a further 2,000 tons.[34]

On South Lode the winze below the 130 had been continued during 1871, communicating with the 144. Unfortunately the ground cut in 1871 was not as good as the first seven fathoms. The 144 west of Agnes Shaft was driven throughout the year, generally in very poor ground. In spite of this the intention was to continue driving to explore a further seventy fathoms that lay between the current end and the 130 east of Hitchins' Shaft.[35]

At Wheal Emma little exploratory work had been carried out on the Main Lode, the only end to have driven was the 216 west of Thomas' Shaft, the deepest point on the mine. Paull felt that the work being undertaken here "may be considered a conclusive trial" of the lode at depth. He felt that if the exploratory work on the 216 did not yield and discoveries of copper ore the company could "hardly be expected to assay on further trials at this or greater depths in search of that mineral".[36]

New South Lode at Wheal Josiah continued to receive some attention. The 115 west had been driven throughout the year, for the most part in ground of "considerable promise" although since September the lode proved unproductive. Engine Shaft had reached the 130 and by December work had commenced on a 130 fathom level. Until August a crosscut from Agnes Shaft had been driven on the 144 with the intention of intersecting New South Lode at depth. Whilst Josiah Paull felt it desirable that the 144 crosscut should reach New South Lode he questioned whether the company would feel it worthwhile spending the £1,000 it would cost to drive the remaining seventy fathoms given that the lode had proved barren down to the 130.[37]

As in 1870 most exploratory effort was focussed on New South Lode at Wheal Emma. West of Railway Shaft work was carried out on the 100, 115 and 130 fathom levels none of which cut good ground. Throughout most of the year Railway Shaft was sunk by nine men and had, by December 1871, reached nine fathoms below the 130. Down to the 130 the lode in the shaft "contained a fine course of ore worth 12 tons per fathom". Between Railway Shaft and New Shaft two winzes had been sunk below the 115 in good ground. New Shaft reached the 115 in February and the 130 by the end of the year. East of New Shaft the 75, 100 and 115 had been driven during the year, the 75 and 100 in poor ground, the 115 in better ground.[38]

In his annual report of December 1871 Josiah Paull was dismissive of New South Lode:

> The New South Lode has been worked with a good deal of spirit both in

> Wheal Josiah and Wheal Emma. At times the prospects on it have been cheering but they do not hold to the end of the year. And the workings in both sections of the sett having reached the depth at which the Main Lode fell off in general in productiveness. I cannot hold out much hope of a lode so inferior to it in every element of richness as the New South Lode undoubtedly is proving very profitable in future at an increased depth.[39]

East of Thomas' Shaft a crosscut known as Bell's was driven southwards from Main Lode in order to intersect New South Lode. Josiah Paull considered that this crosscut was had strategic significance:

> The intersection of the lode at this point is an object of considerable importance it being parallel to the eastern deposit of ore on the Main Lode and 200 fathoms farther east than any working yet prosecuted on this.[40]

Some limited work had been carried out on minor lodes, including No.2 South Lode which lies between Main Lode west of Inclined Shaft and New South Lode.[41]

To Paull's satisfaction work had restarted on Alford's crosscut on the 100. By December 1871 the crosscut was forty six fathoms south of New South Lode and was being driven at a rate of "rather more than" three fathoms a month. Paull commented that if work continued at this rate there was some prospect of meeting "another south lode by the end of another year"; presumably he was gifted with second sight.[42]

During the year new exploratory works were carried out on two lodes lying to the north of the Main Lode at Wheal Maria. In June work started on clearing old workings on the eastward extension of the West Maria and Fortescue Main Lode (see Note 2). The lode in the 12 and 24 fathom levels east of the "boundary shaft" was reported as being large and mixed with ore including a great deal of mundic. However in the 40 the lode was considered poor. It was noted that little good ore had been found in West Maria and Fortescue at such shallow depths.[43]

Two hundred fathoms north of the West Maria and Fortescue Lode the company had started prospecting "a very promising lode". By December 1871 the lode had been explored for a length of 200 fathoms by a number of pits sunk from surface. Given these so promising indications, both the directors and agents were keen to develop this new lode. Interestingly the intention was to do so with fresh capital; a new company would be set up to work this portion of the sett without being a financial drain on the parent company. By the end of the year the Duke of Bedford had consented to allow the existing Devon Great Consols sett to be divided so as to allow this lode, which would be known as either as North Lode or New North Lode, to be worked separately.[44] Although this was an interesting idea it did not get much further, events in 1872 taking a significantly different course.

In 1871 the mine produced 30 tons of copper precipitate. They also sold 120 tons, leaving a stock of "rather more than" 100 tons.[45] Due a combination of a £4 per share dividend in May and a £6 per share dividend in November and an increase in the price of copper the mine's share price rallied during the year ending at between £135 to £140 a share.[46]

1872

A brief item to the effect that the five week month had been abolished on the mine appeared in the *Tavistock Gazette* of 9th February 1872. The five week month meant that the miners were paid twelve times a year, the move to a four week month meant that there were thirteen paydays a year. Whilst this seems a small change, the issue of the five week month would have significant implications latter in the decade. In June 1872, W. A. Thomas commented that he felt that the abolition of the five week month was "a very unwise step", noting that the agents were dissatisfied with the 12 months pay whilst the men were receiving 13 months pay.[47]

The 1872 Annual General Meeting was held at Gresham House on Tuesday, May 14th, 1872. W. A. Thomas was able to announce a slight improvement on the financial condition of the company. Mr John Blackwell was elected to the board of Directors in the place of the recently deceased Mr Blakeway.[48,49]

As in previous years, Captain Richards' report was presented to the meeting. There was good news regarding South Lode which was proving rich down to the 144. At Wheal Emma the Main Lode was looking promising on the 216. On New South Lode Railway Shaft was down to the 145. Richards also advocated re-starting work at Wheal Thomas which had not been worked for eight years. Work had started on the New North Lode at Wheal Maria which had been proved at shallow depth for 280 fathoms, comprising of fine gossan, "strong capel and a quantity of mundic, containing at this shallow depth both tin and copper ores".[50]

Interesting as these developments were, they were merely a *hors d'oeuvre* to

Richards' main course: the sinking of Richards' Shaft in the hope of finding tin at depth in emulation of Dolcoath.[51] The mines had been inspected on behalf of the Duke of Bedford (see Note 3) by Captains Josiah Thomas of Dolcoath and John Simmons, the Duchy of Cornwall's mineral agent, who reported that they were "forcibly struck with the similarity of the lode to the main lode which passes through Carn Brea, Tincroft, Cook's Kitchen and Dolcoath, and which in those mines is at the present time producing immense quantities of tin at depths varying from 200 to 330 fathoms".[52] Referring to this report Richards argued that Richards' Shaft should be sunk from its current depth of 230 fathoms to a depth of 300 fathoms in the hope of laying open the tin ground postulated to lie below the copper.[53]

The Richards' Shaft project would be a massive undertaking which, if expert opinion was to be believed, would prove to be the salvation of the mine. The magnitude of the proposal was such that an Extraordinary General Meeting, chaired by W. A. Thomas, was convened directly after the Annual General Meeting. The sinking of Richards' Shaft would be a hugely capital intensive project, the cost being estimated at £30,000. Thomas told the meeting that a number of schemes had been considered to raise the capital for carrying out the project, including approaching the Duke of Bedford regarding a reduction in dues.[54]

Negotiations with the Estate were in part successful; the meeting was informed that an agreement had been reached with the Duke on April 15th 1872. It was agreed that Wheal Maria should be deepened and extended; Richards' Shaft to be sunk to the 300-fathom level (65 fathoms below the present bottom), the lode to be cut through at each twenty fathoms, and carefully assayed for copper and tin; the levels below the 235 fathom level to be driven 100 fathoms at least, and the lode to be assayed as before once a month at least; the eastern part of Wheal Emma also to be deepened. Once these conditions had been met the Duke would grant a new lease for 21 years at 1-12th dues, reducible half-yearly as follows: tin ore, 1-22nd dues; arsenic and mundic, 1-18th dues; copper and lead ores from the north lodes, 1-15th dues; copper and lead ores from the main lode and all other lodes to the south of it, 1-18th dues.[55]

Having considered a number of proposals the meeting concluded that the best course of action would be to register and reconstitute the Company as a Limited Liability Company of 40,000 shares of £5 each. Of the 40,000 shares 9,280 would be offered to the general public at £3 whilst the remainder would be allotted to existing shareholders with a liability of £5. Thomas did not envisage that there would be a call on the existing shareholders as the capital raised from the sale of £3 to the general public would cover projected costs.[56]

The meeting resolved to adopt the proposals and empowered the directors to carry them out. The directors were also appointed directors of the new company.[57] On June 6th 1872 a further Extraordinary General Meeting was held with W. A. Thomas in the chair. The meeting confirmed the resolutions made on the 14th May.[59]

Another meeting was held on August 9th 1872 at which the Articles of Association of the new Limited Liability Company were agreed and were confirmed at a meeting held on August 24th which also resolved that the Memorandum and Articles of Association be registered with the Registrar of the Court of the Vice Warden of the Stannaries to be registered pursuant with the Joint Stock Companies Acts of 1862 and 1867.[60,61] The new company, Devon Great Consols Company Limited, was registered on August 31st 1872.[62]

William Russell, the 8th Duke of Bedford, died on May 27th 1872 aged 62. He was succeeded by Francis Russell, the 9th Duke, who would hold the title until his death in January 1891.[58]

Whilst the new Company was brought into being in the Company's board rooms, work continued apace on the mine. By the end of May 1872 operations had commenced at Richards' Shaft. In his report dated 30th May, Paull reported that the shaft was being drained below the 144 and by the beginning of July water levels had been dropped to the 170.[63]

The drainage of Richards' Shaft put considerable strain on pumping capacity in this part of the mine. Prior to the start of work at Richards' Shaft water in this section of this mine had been pumped to surface at Hitchins' Shaft. By pumping water to surface it was possible to cycle the water into a series of reservoirs as part of the mine's integrated water management system (see Note 4). In order to free up pumping capacity for the draining and deepening of Richards' Shaft it was decided to pump both Anna Maria and Wheal Josiah to the Blanchdown Adit, thus saving lifts of 60 fathoms from deep adit to surface. During June nine men were engaged in "stoping and cutting down the various lodes and crosscuts in order to bring the water through". Paull observed that this "very useful work" was "one of the most spirited I have known the company set about for many years past". The work was completed by the end of December 1872.[64]

By the end of July the water in Richards' Shaft was down to the 190 however work was hindered by the lack of "trained shaftmen" to carry out the work. At the end of August nine men were attending to the drainage of the shaft which had reached the 200 fathom level.[65] By mid October 1872 the shaft had been fully reconditioned from the 140 to the 212; pitwork and ladder ways being replaced. It was hoped to have the shaft drained to the 235 within the month, at which point sinking would commence.[66] The 235 was drained shortly after however a breakage of the capstan put the work back considerably, the water rising back up to the 224 at the end of the month. The problems were overcome and by late November 1872 sinking below the 235 had started. In the words of Josiah Paull:

> The search for tin may therefore be said to be now in hand.[67]

In October 1872 work was underway on New North Lode at Wheal Maria. Two perpendicular shafts, 65 fathoms apart, had been started, the intention being to intersect the lode fifty fathoms below surface. A run of flat rods 263 fathoms in length was under construction linking the new shafts with the "Old Wheal Maria water wheel". Work had also started on an adit both to drain the workings and to prove the lode.[68] The use of the Wheal Maria water wheel appears to have been intended as a temporary measure; by late November the mine had acquired "a 40 in cylinder steam engine". Captain Richards informed Paull that the engine "would soon be erected for the purpose of coping with the unusual quantity of water met with in this part of the sett". In December it was noted that "the weather has been too wet to admit any preparation for the engine house yet, which is to be erected on this part of the sett" (see Note 5).[69]

Limited work was also carried out on the West Maria and Fortescue Main Lode, although work was suspended on the 15 fathom level east of Boundary Shaft in June 1872.[70]

Preparatory to trying Wheal Maria in depth, as stipulated in the Company's agreement with the Estate, Gard's Shaft was being reconditioned, work starting in September 1872.[71] In October 1872 the old pitwork was being replaced with new and it was estimated that it would take two months to lower the water from the 60 where it was currently standing to the 100 fathom level at which depth exploratory drives would be commenced.[72] However the drainage of this section of the mine proved "a very difficult matter" and, at the end of the year, the water was "a little below" the 60. The delays were due, in part, to "the great accumulation of slime and sediment" in the shaft.[73]

The first meeting of the newly constituted Devon Great Consols Limited was held on Thursday 28th November, 1872 (see Note 6); as usual W. A. Thomas was in the chair. Reviewing past performance it was noted that the Devonshire Great Consolidated Copper Mining Co. had raised 569,824 tons of copper at a cost of £1,738,544 19s 11d realising £3,087,570 12s 4d; £242,537 14s 4d was paid to Duke in dues whilst £1,192,960 divided amongst shareholders; a truly impressive achievement. The directors informed the meeting that they were working hard to bring the development work outlined in the April 1872 agreement with the Estate to a point where the rebates would commence in line with the agreement. Captain Richards brought the shareholders up to date with progress on the mine. Importantly work had commenced on the sinking of Richards' Shaft below the 235, James Richards commenting that "depth only is required to meet with good results". Richards brought the meeting up to date regarding progress elsewhere on the mine including Gard's Shaft and New North Lode. In addition to work on the Devon Great Consols sett work was also being carried out at Colcharton, the lode on the 20 and 30 fathom levels comprising large quantities of mundic with "black oxide of copper".[74]

1872 ended on a sad note; on Tuesday 17th December a young boy was involved in a fatal accident underground in Wheal Josiah:

> A lad about 12 years of age, named Harvey, was engaged with his father kibble filling at the 144-fathom level in Hitchins' Shaft, and on moving aside to escape from some falling rubbish, he fell backwards into the shaft a distance of 26 fathoms. One of his thighs was broken, and the other injuries were of such a nature as to result in death on the following day.[75]

1873

The problems facing the mining industry in the south west manifested themselves in shortages of skilled labour, the lack of skilled shaft men available for Richards' Shaft already having been noted. In April 1873 Josiah Paull observed:

> The departure of Miners for the Iron and Coal districts of the North is on the increase and together with the emigration of large numbers to America is telling very seriously against the development of His Grace's Mines as well as those farther west. Those who remain are the worst of their class and it is found all but impossible to carry out works where experience and skill is required. It cannot be wondered at that men should be leaving us as they are doing when it is known that even those of the second class can earn

£2 per week in Lancashire, Cumberland and Durham.[76]

By the time of the next half yearly meeting held on Tuesday 27th May 1873 the company's financial situation was causing some concern. This was partly due to falling copper prices and partly due to the heavy expenditure on the preliminary works at Richards' Shaft. The company report notes that sinkers were being paid £80 per fathom if they sunk more than six fathoms a month or £75 if they sunk less than six fathoms; the shaft being opened out to twelve feet by six feet. This situation was exacerbated by the attitude of the Estate. The directors felt that the preliminary works they had undertaken entitled them to the rebates outlined in the April 1872 agreement. Unfortunately this was not the view of the Estate who, according to the directors, did not "put the same interpretation as the Directors upon a certain clause in the memorandum of April 15th 1872, specifying work to be accomplished, so as to entitle the company to the rebate in question". It would not be until November 1874 that the question of this rebate was resolved. From the shareholders point of view there was positive news, the cost of the work at Richard's Shaft was being met out of profits so a call would not be required.[77]

In spite of the assurances given at the May meeting a call on the shareholders proved unavoidable, a call of ten shillings per share being made in July 1873. This might be considered a pivotal moment in the history of the mine, it being the first call made since the initial £1 call of 1844.[78]

The draining of Gard's Shaft had been hindered during the early part of the year due to dry weather and consequent shortage of water at surface. In June 1873 it was reported that:

> the present water supply is quite unequal to the task of driving the pumping wheel so as to drain below the 80. A steam engine is purchased and it is intended to erect it in the Engine House already built by Gard's Shaft.[79]

By July there was enough water to drive the Wheal Maria wheel and Gard's Shaft was in fork to the 95 fathom level. During the latter part of the year there was sufficient water available to drive the pumps and keep this section of the mine drained. By the end of the year the 95 west had been driven nearly six fathoms, and the 95 west a little over eight fathoms. At the end of December Josiah Paull noted that a pumping engine was to be erected on Gard's Shaft.[80]

New North Lode received attention during the year. Initial progress at Eastern

Shaft was slow due to problems with water. In June a portable steam engine was in use driving pumps in the shaft. Eastern Shaft reached the 25 fathom level in August; however progress proved slow due to frequent breakages in the pumping gear. By the end of the year a line of flat rods was being constructed to connect Eastern Shaft "to the pumping engine at Gard's Shaft". No work was carried out on Western Shaft due to pumping problems. To the west of Western Shaft an air shaft had been sunk, communicating with the adit in December. Four men drove the adit a distance of one hundred fathoms during the year, the end of the adit level being about one hundred fathoms east of Eastern Shaft by the end of December.[81]

The sinking of Richards' Shaft continued during 1873. Progress was slow during the first half of the year only five fathoms were sunk. This was due to a combination of hard ground and a lack of skilled men being available for the job. Initially nine men were working in the shaft, however by May Captain Richards had increased the number to twelve. In spite of this "unusually large force" progress continued to be slow. In June the lode, which was composed mainly of capel and mundic, was tested for tin; only small traces being found. The sinking of Richards' Shaft was bedevilled by ongoing shortages of skilled labour.[82] In August Josiah Paull wrote:

> The men employed at Richards' Shaft and other deep workings are almost unmanageable, the news of from £2 to £3 per week being easily earned by very inferior miners who have gone to the coal and iron fields of the north is quite sufficient to make our men throw up their bargains, which they are doing daily".[83]

The ground in Richards' Shaft continued hard and intractable. By October the price being paid had increased to £87 10s per fathom. Even with an increase in price it was still proving extremely difficult to get men to work in Richards' Shaft; Josiah Paull commenting that "men are only got to sink it at all by humouring them in the way of giving premiums on their labour".[84]

In order to speed up work in Richards' Shaft, Captain Richards arranged for a trial of a boring machine on the mine during October 1873:

> a boring machine..... has been put to a fair test at surface in the hope that it might be used in the aid of the more speedy sinking of this shaft, every facility has been afforded to the agent of the patentee who conducted the

> trials but the result is that on rocks of the same hardness as that in Richards' Shaft, hardly any impression could be made, the boring tools blunted and broke against the stone.[85]

The second half yearly meeting of 1873 was held at the company's London offices on Tuesday 26th November. W. A. Thomas presented a dispiriting directors report noting that the progress of the "experimental works" had been unsatisfactory due to the hardness of the ground and the scarcity of skilled labour. The shareholders were informed that the mines were working on a much reduced scale due to low copper prices, high wage and material costs. When asked about Colcharton, Thomas replied that "it was purchased some time since ..... because there were certain lodes which were supposed to be identical with those of Devon Great Consols, but they had not found them yet to be so".[86]

Whilst the meeting must have been rather depressing for all concerned, Captain James Richards' report presented to the meeting at least demonstrated that progress was being made. At Wheal Maria the Main Lode had been cut on the 95 and was being opened out by easterly and westerly drives. On New North Lode the adit had reached the Great Crosscourse whilst Engine Shaft was 28 fathoms deep and a cross cut was being driven to intersect the lode. At Wheal Josiah, Richards' Shaft had reached 10 fathoms below the 235. Wheal Emma was still proving productive at depth: The Main Lode on the 216 was in good ground whilst on New South Lode good ground was being laid open and the 115 and 130 were productive.[87]

At the time of this meeting the directors were engaged in negotiation with the Estate regarding the renewal of mining, railway and arsenic leases.[88]

By the end of the year the ground in Richards' Shaft had improved, the ground containing much more quartz which, whilst hard, had better cleavage allowing faster progress. Paull noted that the mine had latterly managed to employ "a better class" of miner in the shaft. In addition the shaft had also been closely partitioned which improved ventilation and reduced temperature. Whilst both progress and working conditions had improved little if any tin was being discovered.[89] In his Annual Report, Paull commented that:

> The appearances of the Lode are by no means unfavourable for tin, but the assays made once a month or oftener do not show that it is on the increase, traces of tin are found in all but nothing more.[90]

The arsenic works was in full production, producing between 180 and 200 tons of arsenic a month during 1873.[91]

During the year a new precipitation works was laid out "in Blanchdown Wood just above the pumping wheels" at the mouth of the Blanchdown Adit. The need for a precipitation works here was occasioned by the change in the pumping regime resulting from the deepening of Richards' Shaft. Water which had been pumped to the shallow adit system at Hitchins' and other shafts was now being discharged through the deep adit.[92]

In consequence of the Company's increasing financial problems the share price had collapsed catastrophically during the year: At the end of 1872 a 10,240th share was valued at £72, at the end of December 1873 the same share was worth £1 15s. In one year £58,850 had been wiped off the share value of the mine. The market valued the mine at £17,920 which, as Josiah Paull noted, was "much less than the value of the plant and materials as they stand".

1874.

1874 would prove to be little better for the company than the preceding year. At the May 1874 half yearly meeting the state of the finances was so poor that the customary education grant was reduced to 30 guineas of which 20 guineas was for the school, attended at that time by between 50 and 60 pupils, and 10 guineas for the dispensary. W. A. Thomas informed the meeting that the new leases still had not been agreed and were in the hands of the Estate. Thomas felt that the "lease contained certain clauses incompatible with the interests of the company, and which he as chairman could not sign". To add to the, now customary, gloomy mood of the meeting it was announced that extremely hard ground had been encountered in Richards' Shaft, slowing progress to only three to four feet a month.[93] To compound the gloom a call of eight shillings was made in July 1874.[94]

By August 1874 the Company and the Estate had managed to conclude their lease negotiations, presumably in a mutually satisfactory manner. Three leases were signed by the Duke of Bedford, William Alexander Thomas and Thomas Morris. The three leases comprised the mining lease, the railway lease and the arsenic lease. Whilst the leases were dated 7th August 1874, they ran from Lady Day 1872 and were due to expire on Lady Day 1893.[95]

The mining lease made provision for and formalised the various exploratory works being undertaken, reflecting the April 1872 agreement. These included driving the 95 fathom level at Gard's Shaft forty fathoms east and forty fathoms west of the shaft at a rate of nine fathoms a year. Once this work was completed the company would be relieved of any further obligation to prosecute Wheal Maria or Wheal Fanny. Moving eastwards; the company was obliged to sink Richards' Shaft to the 300 at a rate of nine fathoms a year. One hundred fathoms of levels were required to be driven below the 260, fifty of which had to be driven on the 300 fathom level at a rate of twelve fathoms a year. At Wheal Emma the lease required the company to continue deep exploration; this included driving two levels east from Thomas' Shaft: one being the 216, the other being the company's choice. In return for carrying out these works to the satisfaction of the Estate a rebate in dues from 1/12th to 1/16th would be payable.[96,97]

Not surprisingly the companies exploratory activities during the year focused on carrying out the work laid down in the lease. The programme of deep driving in Wheal Maria was proving inconclusive, the lode, such as it was, being poor. Work was hindered by pumping problems. In June the water supply to the pumping wheel dried up as a result of a drought. In order to keep work progressing the Gard's Shaft pitwork was attached to the Gard's Shaft engine, and, it was hoped, that this would ameliorate further problems.[98]

At Richards' Shaft the ground continued hard; in November for example it was reported that the lode was of extreme hardness and "very little progress could be made". By April the shaft had reached fourteen fathoms below the 235 and work was in hand cutting a plat and carrying out work in connection with the installation of permanent pitwork. In his April report Paull observed that the lode in the shaft was promising, albeit for mundic, rather than tin. In August the lode in the shaft was assayed at between two to three pounds of tin per ton (see Note 7). Twelve men continued sinking Richards' Shaft through the year.[99] At the November 1874 meeting it was reported that progress in Richards' Shaft was still painfully slow, the shaft having reached a depth of 252 fathoms. Captain Richards had recently introduced dynamite which, it was hoped would speed progress.[100]

At Wheal Emma the Company chose, in addition to the 216, to drive east of Thomas' Shaft on the 145 fathom level. Both levels typically were driven by six men. The 145 was driven on lode, which was composed of iron carbonate and mundic. The 216 was not driven on lode, but a little to the north, presumably to speed up driving.[101]

If the May meeting was gloomy the mood was no better at the November half yearly meeting, held on Thursday 26th 1874; not even a rise in the copper standard could lift W. A. Thomas' spirits. A shareholder asked if they should stop searching for tin now that copper prices were picking up Thomas replied that "if they stopped the explorations they might as well close the mine, because the copper ore alone could not be supposed to last them very long".[102] On being asked if there were any real future prospects with regard to tin, Thomas hit a new low, replying that:

> he must confess that he was not very sanguine himself, but Mr Josiah Hitchins, who was a very great expert in his way, was satisfied that tin would be found, and certainly there was very great encouragement, because the lode at the deepest level was very promising, and sanguine people would, of course think they were going to have a very great thing.[103]

Work progressed rapidly in the Wheal Maria "deep workings". By the end of 1874 the 95 east of Gard's Shaft had advanced twenty four fathoms, whilst the 95 west had been driven twenty five fathoms. In his annual report for 1874 Josiah Paull noted that the lode was small and ill defined, yielding a little mundic and copper.[104]

By the end of December the Richards' Shaft had reached a depth of 22½ fathoms below the 235. The intention was to sink a further 2½ fathoms and then start driving the 260 fathom level.[105]

In December 1874 the 145 at Wheal Emma had cut a crosscourse which displaced the lode northwards. By the year's end the lode on the 145 was said to be large and promising yielding a good deal of mundic and a little copper. The 216 was also looking promising, having cut the lode at the end of the year, Paull reporting that "indications are better than at any time since the level was started".[106]

The exploratory works were, of course, tied in to rebates against dues paid. In November 1874 the company was paid the rebate for the period July 1872 to December 1873. This was paid in spite of the fact that the requisite work at Wheal Maria had not been carried out in accordance with the April 1872 agreement then in place. As noted the Company argued that they had put all their efforts into draining Richards' Shaft preparatory to sinking. The Estate accepted that the Company's case was just and consented to pay the rebate. During the first half of 1874 Company had failed to sink Richards' Shaft the necessary four and a half fathoms. However during the second half of the year the shaft was sunk six and a half fathoms which compensated for the poor performance in the earlier half. Satisfactory completion of the prescribed work meant that a full rebate would be paid for 1874 as soon as the Estate had received the final instalment of the 1874 dues which were due to be paid in April 1875.[107]

In addition to the compulsory works, exploration continued at a number of other points on the mine throughout the year: several levels on New South Lode had been advanced east of Railway Shaft including the 145 and the 160. Similarly the 115, 130 and 145 fathom levels had been driven east of New Shaft. New Shaft was sunk six fathoms below the 145 between August and December 1874. Considerable amounts of good ore were discovered during the year.[108]

Some exploration was carried out on Northey's Lode, a minor structure lying between No.2 South Lode and New South Lode, reached by Alford's crosscut on the 100 fathom level and a further crosscut on the 145. The lode was noted as being small and of little value.[109] Alford's crosscut continued to be driven southwards in the hope of cutting new lodes between New South Lode and Thomas'. On New North Lode Eastern Engine Shaft was sunk from the 25 to the 42 fathom level. Both the 28 and the adit had been extended during 1874. Unfortunately the ground here was proving to be poor.

1875

The compulsory exploratory works continued during 1875; in February four men were driving the 95 west in Wheal Maria where the lode was described as being "badly defined and quite unproductive" From the end of 95 east a further four men were driving a crosscut in a north east direction. The crosscut intersected two small branches "of no value".[110]

The ground at Richards' Shaft had improved, allowing more rapid sinking. By April the Shaft had reached the 260 and twelve men were engaged cutting a plat preparatory to driving a level.[111] Steady progress was being made on both the 145 and 216 fathom levels at Wheal Emma.

In contrast to recent half yearly meetings the directors were able to present a more positive outlook the May 1875 meeting. At Wheal Emma it was reported that valuable discoveries had been made on New South Lode. Things were also looking positive at Richards' Shaft where the ground had improved to such an extent that it was possible to sink eight fathoms a month rather then the paltry three feet six inches.[112] On the basis of the positive report presented to the meeting no call was made of shareholders.[113]

Arsenic was making a significant contribution to the financial position on the mine. In January Josiah Paull commented that "the arsenic is eagerly sought for by buyers" whilst the make fell "far short of the demand". To meet demand plans were in hand to increase output. In April preparations were underway to erect "five more revolving furnaces".[114]

At Richards' Shaft the lode on the 260 had been cut by June 1875. Unfortunately initial indications were not encouraging, the lode being "about four feet wide, composed of hard capel without any mundic of consequence and also without copper or tin ore".[115] By August the 260 east had been driven about six feet, the lode being about two feet wide. In order to establish the width of the lode at this depth crosscuts were driven both north and south, opening up ten fathoms of ground without finding any extension of the lode. Whilst the initial exploratory work was being undertaken on the 260 twelve men continued deepening the shaft, in easier ground than of late. However by October the poverty of the lode and the hardness of the ground led Captain Richards to decide to postpone work on the 260 and to carry out the compulsory drivcage of 50 fathoms on a deeper level.[116] By 21st October 1875 the Richards' Shaft was five fathoms below the 260 fathom level.[117]

Work on the 95 at Wheal Maria continued until 9th October 1875, forty fathoms having been driven both east and west of Gard's Shaft with little result. Paull considered that the lode in the ends of both drives was looking better than it had since work started. He was of the opinion that if the company was more prosperous the work should be continued, however he recognised that the current state of the Company's finances precluded further exploration at this point. Cessation of exploratory work meant that the Company would not need to operate the steam engine to keep this section of the mine drained which was costing £100 a month. Likewise they could save the miners wages at £4 a month each.[118,119]

At the November 1875 meeting the directors were pleased to announce that New South Lode was proving productive, enabling the "experimental works" to continue without a call. In addition to meeting exploration costs the Company had managed to make a small profit, not enough to justify a dividend, but a profit none the less. Shareholders were informed that there were only forty fathoms to go in Richards' Shaft before the 300 was reached. The meeting was also informed that, with the agreement of the Duke of Bedford's agents, exploration in Wheal Maria had ceased, further work being considered useless.[120]

No time was lost abandoning the deep levels in Wheal Maria. During December the pitwork was drawn from Gard's Shaft.[121] By the end of the year Richards' Shaft had reached the 270, the intention being to sink to the 280 before engaging in further lateral development. Whilst significant progress had been made in the shaft the lode had proved poor all year "containing but little mundic, only a few spots of copper ore here and there and scarcely a trace of tin".

The compulsory works in Wheal Emma made steady, if unspectacular progress during the year. On the 145 not much more than the obligatory fifteen fathoms was driven due to hard ground and poor ventilation. The 216 was driven slightly further, an advance of twenty fathoms, four feet being made during the year. Given the depth the men were working at and the hardness of the ground Josiah Paull, in his annual report, considered this to be "a very good piece of work".[122]

During 1875 the Company successfully completed the compulsory works, which qualified them for another rebate of dues.[123] The exploration of New South Lode at Wheal Emma continued during 1875, considerable development taking place in the deeper levels from the 115 to the 160. The 130 and the 145 in particular proving productive In June New Shaft reached the 160 and sinking continued through promising ground to the end of the year.[124,125,126]

Limited work was carried out on Northey's Lode, little ore ground being discovered.[127] In December the 145 north crosscut from New Shaft communicated with the corresponding level from the Main Lode west of Inclined Shaft. Work also continued in Alford's crosscut, by December it was 120 fathoms south of New South Lode, albeit without cutting anything of significance. Exploration continued on New North Lode during 1875. Unfortunately there was no improvement, the lode being uniformly poor. Apart from the 28 fathom level east all other works were suspended on 7th December.

The extensive remodelling of the arsenic works during the year led to a reduction in output. That said 1,212 tons of refined arsenic were sold during the year realising £10,503. A *Tavistock Gazette* correspondent writing in January 1875 estimated that one month's make of arsenic would be sufficient to kill 500,000,000 people![128]

In his annual report of December 1875 Josiah Paull noted that the make of copper precipitate was much reduced. Paull ascribed this to a reduction of copper sulphate in the water discharged from the mine.[129]

In his December 1875 annual report Josiah Paull highlighted the anomalous situation the mine was in. In Paull's opinion in pure financial terms it would be far better for the company to realise its assets rather than carry on working:

> On their 30,000 tons or ore, 4,000 tons of refined arsenic, the plant and leasehold houses the Company could discontinue explorations, extract and return the ore and arsenic, sell the plant and houses and then wind up the concern with profit on the transaction of from £60,000 to £70.[130]

However Paull realised that there were other factors concerned in addition to profit:

> For the sake of good mining and also for the sake of the 629 people employed in the mine I hope they will go on as they are doing, but I don't think many other mining companies would practice the same self denial.[131]

1876

The mid 1870s saw a general industrial depression, the copper trade being no exception. Between July 1875 and July 1876 the price of copper ore fell by £1 per ton. On the 20th July the price fell by 7s 6d. With regard to Devon Great Consols Josiah Paull observed that, at current prices, copper would barely repay the cost of working.[132]

In spite of straitened times the expensive exploratory work continued. By August 1876 Richards' Shaft had reached the 280 and work hard started on driving both east and west of the shaft; two fathoms having been driven in either direction by 23rd August. By mid October a crosscut to the south was underway. At this depth the lode proved to be "quite unproductive of copper and tin"; indeed it was difficult to distinguish the lode from the surrounding rock except for the fact that the lode was harder, being composed of capel. By the end of the year the shaft had reached a depth of four fathoms below the 280, nine fathoms having been sunk during 1876.[133] The Shareholders, at the November 1876 meeting, were informed that Richards' Shaft would reach 300 fathoms within twelve months.[134]

The deep workings on the Main Lode at Wheal Emma progressed steadily during

1876. The 145 had been driven a distance of twenty fathoms, one foot and eight inches during the year. At years end the lode was described as "a kind very congenial for ore". The 216 had been advanced twenty eight fathoms, one foot and five inches. The lode at this depth was poor, largely being composed of hard capel and mundic with no signs of copper ore.[135]

As in previous years the work undertaken in Richards' Shaft and the Wheal Emma deep workings was sufficient to qualify the Company for the rebate of dues.[136] Limited exploratory work was being undertaken on the Main Lode. At Wheal Maria the 50 fathom level had been driven throughout the year with little result. A small amount of work had also been carried out on and in the vicinity of the 60 fathom level west of Field Shaft in Wheal Josiah. On South Lode a winze between the 130 and 144 fathom levels opened up a good deal of paying ground.

Considerable effort was expended in searching for the postulated western end of New South Lode. Between February and April Jeffry's crosscut on the 80 fathom level was advanced.[137] In April the crosscut intersected a rather poor lode believed to be part of New South Lode. At the November half yearly meeting held on Tuesday 28th November 1876 the shareholders were informed of the discovery of a promising lode at Anna Maria.[138] The lode was thought to be the western extension of New South Lode. Work started in August and involved the sinking of a new shaft known as Blackwell's. Initial indications were good, the shaft being sunk through a very rich gossan. By the end of the year the shaft had reached a depth of ten fathoms, unfortunately due to the wet nature of the ground, work was suspended. In Wheal Josiah, Drew's crosscut on the 144 was driven south towards new South Lode. In December 1876 it was hoped that New South Lode would be intersected within three months.[139]

East of Railway Shaft work on development of New South Lode at depth continued. New Shaft reached the 175 at the end of August and the 175 east cutting good ground. In order to development the eastern extremities of New South Lode a crosscut from the 137 east of Inclined Shaft was started in July 1876. In December it was noted that the 137 crosscut would intersect New South Lode in about eighteen months time.[140]

A small amount of work was carried out on the 100 and 115 fathom levels on Northey's Lode. Josiah Paull noted that "just sufficient ore is met with to cover the costs of working".[141] South of New South Lode, Alford's crosscut was driven throughout the year by two men who advanced the level fifteen fathoms. By

December the crosscut was one hundred and thirty fathoms south of New South Lode without intersecting anything of real interest. Limited exploration was carried out on the 28 fathom level on New North Lode during 1876. The lode was reported as "very small and poor".

It was evident at the November 1876 meeting that W. A. Thomas was in a poor state. He informed the meeting that he had suffered so badly from bronchitis during the preceding ten months that he was not able to give the company the attention it deserved and asked the meeting either to accept his resignation or excuse him from such constant attendance. The shareholders expressed the hope that Thomas would not resign and re-elected him chairman, however it must have been fairly obvious to all concerned that Thomas' days were numbered.[142]

At the end of a poor year Josiah Paull was able to find a silver lining:

> In these dull times when miners and mine labourers know not where to turn for employment it is satisfactory to know that these mines are giving employment to 754 persons as against 624 persons at the end of 1875.[143]

1877

The early part of 1877 must have felt very similar to 1876. At the first half yearly meeting of the year held on May 22nd 1877, the shareholders were told that the exploratory work required by the Duke had progressed continuously to the satisfaction of his agent, but as yet without any profitable result. New machinery for dressing halvans had now been completed and the company hoped to increase the returns to 1000 tons per month.[144]

In spite of low copper prices and the expensive cost of exploratory work the Company found itself able to pay a dividend of five shillings per share in May.[145]

If the May meeting was dull the November meeting was anything but: The meeting marked a hugely significant change in the mine's management. Although W. A. Thomas was nominally still chairman, he was very ill and had had to go to the Continent for the sake of his health. The meeting was chaired by one Peter Watson (see Note 8) a larger than life character who would rapidly become the most important figure in the life of the mine. The changes did not stop here, it was announced that Captain James Richards (see Note 9) was too ill to continue in his post. After thirty three years service he was replaced by his brother, Captain Isaac Richards, who had already more than proved his ability during his

Figure 25.
Peter Watson, Chairman of Devon Great Consols from the late 1870s.

PETER WATSON, ESQ., C.C.

long service in charge of the mine's ore dressing department. This catalogue of ill health was completed by the absence of Thomas Morris, the Resident Director who, it was reported, was too ill to attend the meeting. The cynic might suggest that this represented a wholesale cull of the "old guard" and its replacement with a more dynamic team.[146]

Changes in management aside, of prime importance was the announcement that Richards' Shaft had reached the 300 after five years of hard, expensive and often frustrating work. Whilst this was good news tin had not been found; something that was felt was imperative for the continuing survival of the mine. The urgent need to reinvent the mine as a tin producer was driven by the long term decline in the price that the mine was getting for its copper ore. The figures presented to the meeting were unequivocal. In 1847 the mine was receiving £6 15s a ton, dropping to £6 in 1857, £5 in 1867 by the time of the November 1877 meeting this figure had plummeted to £2 5s 5d a ton. To speed up exploratory driving on the 300 the meeting was informed that the directors had decided to adopt rock drills, the "boring machine" being a then emergent technology. Watson quoted the success of such machines at both Dolcoath and Carn Brea and he noted that in the "Western mines" the Lords had contributed to the expense of boring machines. To this end the directors had consulted the Duke of Bedford's Agent however, Watson reported, "so far the Duke had not consented to contribute anything, but he still hoped that the Duke would see his way to give the company a helping hand in the purchase of these machines". On the subject of cost Watson stated that it "would depend on the number of drills, they might go to an expense of £500, £1,000, £2,000". He felt that the expense would be more than justified as "there was no doubt that these drills would be a great assistance at Richards' Shaft".[147]

In December 1877 four men were engaged in driving the 280 at Richards' Shaft, the ground being noted as hard, the lode being composed of "capel and a little mundic". By the end of the year the 280 had been driven sixteen fathoms in hard, unyielding ground.[148]

Work had also started on the 300 fathom level, the lode being "very hard capel.... without mineral of any kind". In his annual report Josiah Paull was sorry to note that the lode on the 300 contained "no copper, tin or mundic ore in it". Interestingly Josiah Paull corresponded with Josiah Thomas, the Dolcoath Mine Captain, on the subject in November. Thomas, on whose advice the deep exploration had been undertaken, was of the opinion that Richards' Shaft had not gone deep enough:

> Devon Consols has been rather disappointing in depth, if the granite could only be touched I have no doubt that the desired result would have been attained and the lode have changed into a tin producing one. If the trial had not been made there would always have been a feeling that it ought to have been.[149]

With regard to the other compulsory works the 145 in Wheal Emma had been driven twenty six fathoms during the year, reaching the "eastern crosscourse or slide". This completed the Company's obligation to advance this level. The 216 had been driven twenty three fathoms on an unproductive lode.[150]

The Company had exceeded its compulsory obligations during the year qualifying them for the rebate.[151] Beyond the compulsory works at Richards' Shaft and the deep driveages in Wheal Emma considerable exploratory work had been carried out on New South Lode. To facilitate drainage at Blackwell's Shaft an adit was driven during the first half of the year communicating with the shaft in June 1877. Blackwell's Shaft cut the lode at a depth of thirty three fathoms, the lode proving poor. No further work was carried out from September.[152]

Drives were advanced both east and west of Jeffery's crosscut on the 80 fathom level, but the lode proved poor in both directions.[153] The 144 fathom level crosscut communicated with New South Lode Engine Shaft (Counthouse Shaft) on the 130 fathom level. The lode at this level was said to be promising, albeit yielding little ore. The shaft itself had been sunk eleven fathoms below the 130.

The deep exploration of New South Lode at Wheal Emma continued apace, the 100, 115, 130, 145, 160 and 175 fathom level having been advanced during 1877. Progress on the 137 crosscut east of Inclined Shaft had been driven throughout the year and it was estimated that the lode would be reached within three months. Two further deep crosscuts had been started during 1877, one on the 145 and the other on the 190. Both were driven from the Main Lode with the intention of rapidly opening up the eastern end of New South Lode.[154] Work on Alford's crosscut was suspended at the end of August, a victim of the depressed state of the copper trade. During 1877 an average of 106 "picked men" were engaged in exploratory work.[155]

There had been a very considerable production of arsenic during 1877, amounting to nearly 2,400 tons. Unfortunately Mr Drayton, who had contracted to buy the year's output, went bankrupt and no further sales or shipments were made after September. In spite of this the company continued to produce arsenic and, by the end of the year, five thousand barrels of arsenic, containing 870 tons, were stockpiled either at the works or at Morwellham. In his December 1877 annual report Paull noted that a new purchaser, Dr. Emmens (see Note 10), had been found for both the remainder of the 1877 and the 1878 make. Paull was not overly impressed with Dr Emmens' *bona fides*, noting that "the arsenic trade seems to have got into some very disreputable hands":

> Dr Emmens is a person who has made himself in connection with such rotten concerns as Holmbush Mine Callington and Greenhill Brick and Arsenic Works, two speculations which have absorbed over £300,000 of capital to no one's good but Dr. Emmens so far as is known and he is himself at the present time a bankrupt without certificate.[156]

Evidently the Company came to share Paull's less than glowing opinion of Dr Emmens. Emmens had failed to provide the Company with satisfactory security for payments against the arsenic and the Company was refusing to deliver the arsenic.[157] In October a case of arsenic poisoning was reported. Fumes from the arsenic works had been blown over Hele Farm to the north, the farmer, Mr Fuge, losing three bullocks.

1878

The problems that the mine was experiencing with Dr Emmens continued into the new year. On 25th January 1878 Paull report that "no shipments of arsenic have yet been made." Issues appear to have been temporarily resolved as the mine despatched three hundred tons of arsenic during February. However on 22nd March Paull observed that the contractor was reported to be bankrupt and it was certain that he would not be able to "carry out his engagement to take away the 300 tons per month agreed or any portion of it". The failure to secure a reliable contract for their make of arsenic put the Company in a very difficult position. Over the preceding decade the mine had become increasingly dependent of the arsenic revenue as copper prices slumped. Without the sale of arsenic the mine simply could not meet its costs. At the end of March it was noted that "the monthly sales of copper are not meeting costs by more than £1,000". The only way the mine could continue to function was to make significant reductions in cost.[158]

In an attempt to save money peripheral, and not so peripheral, exploratory work was suspended in the spring of 1878. For example in February work on the 80 west of Jeffrey's crosscut stopped. Similarly in Wheal Josiah the 130 on New South Lode East of Engine (Counthouse) Shaft was suspended on March 25th, as were driveages on the 144.[159] More significantly the *Mining Journal* of March 30th noted that work on the 300 at Richards' Shaft had been suspended "due to the depressed state of mining" (see Note 11).[160]

In addition to scaling back exploration miner's pay was also cut. By mid April two pay cuts had been imposed.[161]

A tragic accident occurred on the mine on Tuesday 6th March 1878:

> On crossing the railway of these mines on Tuesday afternoon last, a little boy about 5 years of age, son of Captain Mitchell, was run over by a truck which was on its way to the loading place, and was so seriously injured that he died the same evening. A little sister was with him but she saw the truck approaching and fortunately escaped. Yesterday an inquest was held before Mr. R. R. Rodd, district coroner, when a verdict of "accidental death" was returned.[162]

In April 1878 labour unrest manifested itself on the mine. The first hints of a problem started to appear early in April, for example a *Mining Journal* correspondent writing under the name "XYZ" wrote:

> A great deal of excitement and surprise exist in the neighbourhood of Tavistock and Gunnislake on its becoming intimated that in all probability within the next few weeks the old and much dreaded five week month is to be revived in the system of paying miners wages at some of the principal mines in the district.[163]

It has already been noted that Devon Great Consols had abolished the five week month in 1872. The resurrection of the idea meant a reduction from thirteen to twelve pay days a year. In theory this should not have an impact on the amount miners were earning, however in practice the system could lead to real hardship amongst the miners and their families (see Note 12).

One assumes that the re-introduction of the five week month was part and parcel of the programme of financial stringency resulting from the Company's failure to sell its arsenic. It seems that the introduction of the five week month was considered by many miners as the last straw. The local vicar D. P. Alford summed the situation up succinctly commenting that the miners had already suffered the two recent pay cuts which they had accepted "with great cheerfulness and patience" given the state of trade and the belief that the management would treat them fairly. In Alford's words the miners felt that the five week month was unfair, regarding its introduction "as an indirect and underhand way of still further reducing their wages".[164]

In an attempt to get the directors to reconsider the imposition of the five week month, 448 Devon Great Consols miners signed and submitted a petition reproduced in the *Mining Journal* of 13th April, 1878:

> To the Directors of the Devon Great Consols
>
> We, the undersigned working men employed in these mines having been informed that the five week system is again to be adopted in these mines, humbly beg that there be no alteration from our present system, nor any further reduction in wages, seeing that the late reductions have placed us on a very meagre sustenance. We humbly beg that the directors will reconsider the matter, and not enforce a system that is in itself unjust and opposed to the wishes and feelings of the country at large. - Your petitioners will ever pray.
>
> Signed by 448 miners of the Devon Great Consols Mine.[165]

The miners petition was accompanied by a similar petition signed by "171 of the gentlemen and trades people of Tavistock".[166]

Thomas Morris and the Agents were also opposed to the re-introduction of the five week month.[167] Feeling in the district was almost uniformly against the re introduction of the five week month, however strength of local opinion had little impact on "two or three of the non resident directors" who, in the words of the *Mining Journal* had "but imperfect knowledge of the serious consequences such as step might involve".[168] Chief amongst this group of non resident directors was Peter Watson.

Josiah Paull, in his capacity as the Estate's Mining Agent noted that:

> the men have a real grievance, and always had when this obnoxious month was in force.[169]

On April 17th the Directors issued the following resolution described by the *Mining Journal* as "an undisguised return to the five week month":[170,171]

> The directors having further considered the alterations in the time and work of payment to be made they consider that for the future there should be for the 52 weeks work 12 monthly payments, six months of four weeks, and four months of five weeks, making in all 52 weeks.[172]

Saturday 20th April 1878 was the bi-monthly setting day at Wheal Emma. Prior to the setting the Directors resolution had been read out to the assembled men. A spokesman for the miners stated that the men had held a meeting that morning and had unanimously decided not to accept bargains under the five week system. Thomas Morris, presumably having recovered from his recent illness, and Isaac Richards, then proceeded with the setting; true to the miners word not a single pitch was let and, as reported in the *Mining Journal*, "the whole of the men throughout the mines (in number about 600) struck there and then.[173] That the whole of the mine struck at this stage is an overestimation as, initially, only the Wheal Emma men "came out, the men in the western sections of the mine were still working on the old system, their setting day not being until the following month. A closer estimate would be that 150 Wheal Emma miners withdrew their labour, with the additional consequence that a further 150 surface workers, dependent on Wheal Emma ore, also stopped work.[174]

From the outset the strike had strong local support, local tradesmen offering to raise subscriptions for the striking men and offering them credit.[175]

The unrest regarding the introduction of the five week month was not confined to Devon Great Consols. For example Wheal Crebor also attempted to introduce the five week month with a similar result to that at Devon Great Consols. Other mines in the area were more pragmatic. The management at Gunnislake Clitters rejected the five week month feeling that "the interest of the company will be best served by the continuance of the present system of payment of wages every four weeks". It was noted that Clitters had local management which, it was felt, was more responsive to local circumstances than to the London based management of Devon Great Consols.[176]

As April drew to a close the Wheal Emma men stood firm. On Saturday 27th April three hundred men assembled on the mine to see if any response had been received from the Directors. No response was forthcoming and, in consequence the gathered miners adopted the following resolution:

> The machinery and engines to work on without any obstruction; the shaftmen to watch or look after such machinery as might be required of them until the next meeting. That thanks be tendered to the press and to the ministers of all denominations for the sympathy and support they have shown in the struggle.[177]

By this time the strike was confined to Devon Great Consols, the Wheal Crebor men having returned to work by early May, the attempt to introduce the five week month having been abandoned there. In the wider mining world it was felt that the Devon Great Consols Directors' stubborn adherence to the five week month was ill judged and ill informed. The strike was already starting to have potential long term impacts on the viability of the mine. The best and most skilled miners were starting to seek employment elsewhere; a commentator in the *Mining Journal* noting the several of the best men had already "returned their powder cans and materials and do not intend to return".[178]

The strikers were united in their position buoyed by the knowledge of strong local support. In early May a public meeting, chaired by the Portreeve of Tavistock, was held in Tavistock. Apart from expressing support for the miners the meeting took the practical step of establishing a relief fund for the striking miners and their families.[179]

In an attempt to resolve the situation Thomas Morris went to London and met with Watson and his fellow Directors. Morris returned with the power to assure the tutworkers that if they adopted the five week month they would be paid for the extra week with no advantage taken in the price of the setting. Incidentally the Directors offer failed to address the issues faced by the "monthly account men". On Saturday 11th May Morris and Captain Richards met with the striking miners at the Wheal Josiah count house. It soon became evident to Morris that only about 100 of the 150 striking men were present. On asking why, Morris was informed that they represented the whole of the strikers, the absentees either having found work elsewhere or had left the district. Morris proceed to convey the Directors offer to the remaining men who rejected the proposal, stating that whilst they had every confidence in Morris and the mine's Agents they had no confidence in the non resident Directors "whom they did not care to trust".[180] The failure to resolve the dispute at this stage was nothing short of disastrous. To date the strike had only involved the "Eastern Division" of the mine, in other words Wheal Emma. However the following Saturday would be the setting day for the "Western Division" of the mine, that is to say Wheal Maria, Wheal Fanny, Wheal Anna Maria and Wheal Josiah.

The "Western Division" bargains expired on Saturday 18th May and on that day "the whole of the hands, even the women and children, unanimously refused to continue on the five week month"; production had ceased.[181]

The Company's half yearly meeting, chaired by Peter Watson, was held on 29th May, 1878 at Gresham House. The meeting determined to stand by the resolution of 17th April and not to compromise with the strikers. Discussion amongst the shareholders was heated and served only to intensify feelings which were already running high.[182,183] Josiah Paull reported that:

> Mr Watson appeared to have the majority of the share interest with (him), and would take no advice and brook no opposition, Mr Morris could do nothing.[184]

The no compromise stance taken by the shareholders at the 29th May meeting was, realistically speaking, untenable. Watson's performance at the meeting appears, to some degree, to have been bluster and bravado. After the meeting he suggested a compromise. Peter Watson proposed that the five week month should be introduced "but at the end of four weeks in such months the back months' wages should be paid with an advance of one week from the current five weeks month". At a meeting held at noon on 31st May the proposal was put before the men who unanimously rejected it.[185]

Josiah Paull was greatly impressed by the manner in which the Devon Great Consols men conducted themselves at the 31st May meeting:

> they refused to have anything to do with the obnoxious five week month in any shape or form. They readily agreed to watch the pumping Engine and wheels at surface and the pumps underground so long as the struggle lasts, all besides is at a stand still including the arsenic works. I am thankful to say the men could not behave with more respect than they did today to Mr Morris and the Agents. I have never seen such self control exercised by nearly 400 men.[186]

By the beginning of June there were rumours that a solution might be emerging.[187] A meeting had been arranged between Peter Watson and the miners' representatives by the end of the first week in June, this may have been occasioned by reports that the Duke of Bedford was rapidly loosing patience with the situation.[188] It was widely felt in the district that should the mine come to an absolute halt the Duke of Bedford would revoke the company's lease.[189] Unfortunately by 15th June the negotiations had not reached any conclusion; neither side being prepared to give in on what it saw as matters of principle.[190] The *Mining Journal*, reflecting the wider view, was becoming increasing exasperated by the situation commenting that:

> nothing would surprise us less to see this once grand undertaking being thrown utterly idle, to the far more serious loss of the shareholders than a judicious continuation of working under the old system, which admitted of the reduction of wages to any desirable extent, would have involved.[191]

Prolonging the strike was in nobody's interest and eventually common sense prevailed, the strike ending on June 17th 1878. In an attempt to break the deadlock Watson had come down from London on 15th June, meeting the strikers' representatives at the Bedford Hotel in Tavistock on the 17th June. During the four hour meeting during which time Watson made, in the words of the *Tavistock Gazette*, "several futile efforts..... to obtain the consent of the men to different schemes". The miners stood firm and eventually Watson did what he should have done weeks earlier, finally backing down and abandoning the five week month. In return the miners agreed to a reduction in wages amounting to one thirteenth, something that they had been prepared to do from the outset.[192,193]

Watson informed the miners' representatives that there was a large quantity of arsenic on the mine which could not be sold until prices picked up, the market being at a particularly low ebb. However Watson assured the men that when the arsenic was sold there would be a rise in wages.[194] The compromise achieved at

Devon Great Consols was more or less identical to that reached at Wheal Crebor, the key difference being that Watson's stubbornness delayed a resolution by several costly weeks. Watson's about turn cannot have been easy for a man of his character and it is to his credit that he did back down as, in doing so, he may well have saved the mine. The miners certainly appreciated Watson's gesture as they voted him a special award of thanks. The strike over, the whole district was able to breathe a collective sigh of relief as the failure of the mine would have been a disaster, the company spending between £40,000 and £45,000 in the district annually. To mark the end of the strike the church bells in Tavistock were rung in celebration.[195]

The May 1878 half yearly meeting, in addition to discussing the all pervading issue of the five week month, also resolved to employ boring machines.[196] To this end the following advertisement appeared in the *Mining Journal* of 1st June 1878:

> Rock-Boring Machinery required.
>
> The directors of Devon Great Consols Company (Limited) solicit full particulars from manufacturers of rock boring machinery &c., for sinking, driving or stoping at the company's mines.
>
> The particulars to be sent to Alexander Allen Esq., Secretary, The Devon Great Consols Company (Limited), 134 Gresham House, Old Broad Street, London E.C.

Although the strike was over the company still faced significant problems. Foremost was the Company's ongoing failure to dispose of its make of arsenic. By the end of July the company had run out of storage space. Stocks of refined arsenic amounted to 10,600 barrels or "rather more than 1,850 tons". If the arsenic had been sold at the contract price the Company would have realised £15,000. Josiah Paull put the blame squarely at Peter Watson's door:

> This lamentable block in the business is entirely owing to Mr Watson having entered into a contract for the years' make with Dr Emmens, who was bankrupt when the agreement was come to and who is still in the same difficulty.[197]

It was not until August 1878 that the Company managed to find a new buyer and started selling arsenic again, about twenty three tons being sold during the month. During September one hundred and seventy six tons were sold at around £7 per ton, thirty shillings a ton less than the contract with Dr Emmens.[198,199]

The difficulties associated with the arsenic trade, the loss of production during the strike and the fall in the price of copper, which had reached a low point of £1 18s 6d a ton in November 1878,[200] put the company in a very difficult financial position. To avoid a call on the shareholders the company had taken out a loan of £7,000 to carry the mine through the hard times it was experiencing. By the time of the November meeting £1,000 of the loan had been repaid.[201]

With the exception of the strike, exploratory work at Richards' Shaft continued throughout 1878. On the 280 thirteen fathoms had been driven, the lode typically being from four to five feet wide and comprised of "capel & chlorite and some arsenical mundic". During November and December the lode took on "a very Tinny appearance" occasioning Josiah Paull to forward a sample to the Dolcoath assayers "in the hope that a proper assay being made it would be found to produce Tin". Unfortunately no traces of tin were found. The drive on the 300 fathom level east had been advanced fifteen fathoms, the lode containing some arsenical mundic but no tin or copper. The 300 west had been driven two fathoms in similarly unrewarding ground. The results were so poor that by the end of December the Company had applied to the Estate to abandon deep exploration in Richards' Shaft.[202]

Whilst the compulsory work on the 145 at Wheal Emma had been completed in 1877 driving continued during the first four months of the year, work presumably ending with the strike. After the strike work did not recommence. The 216 was still a compulsory working and it was driven twenty one fathoms "on a poor lode". As with Richards' Shaft the Company applied to the Estate to abandon exploratory work on the 216.[203]

New South Lode was proving productive on the 100, 175 and 190 levels.[204] During the remainder of 1878 New Shaft was sunk below the 190.[205,206,207] By October 1878 work had also started at Railway Shaft which was being sunk below the 160, the intention being to open up New South Lode on the 175 and 190 levels west of New Shaft.[208] Friend's crosscut on the 137 from Inclined Shaft cut New South Lode in July.[209,210]

Although New South Lode was productive exploration was being scaled back as part of the general cost cutting. For example promising levels such as the 145 east of New Shaft were not driven at all during 1878 whilst other driveages such as the 100 east, the 160 east and the 175 east were suspended during the year as were the 145 and 190 crosscuts from Main Lode.[211]

Given the depressed state of both copper and arsenic trade a reduction in expensive exploratory work would appear, at least in the short term, to be a reasonable strategy. However the Estate was more interested in the long term development of the mine rather than the short term financial expediency that had driven Watson during 1878. Josiah Paull noting:

> exploratory work is absolutely necessary for its working many years longer and knowing that there is much unexplored ground not only on New South Lode but on Wheal Thomas as well, possibly on other south lodes as well I much regret that the desire of the Agents as well as myself to see more doing in the way of explorations is thwarted by the Directors of whom Mr Watson is the ruling spirit.[212]

In his first year at the helm Watson had made a poor impression on the Estate, Josiah Paull being particularly unimpressed:

> the control of the mine having got into Peter Watson's hands.... (has) been much against its proper development and I feel that the last hindrance (the strike) is one which I fear we shall experience until it suits Mr Watson to go out of the concern as a shareholder in the same way as he came into it.[213]

1879

Whilst Watson had managed to quell labour unrest for the time being its consequences were felt into the New Year. The *Tavistock Gazette* of 21st February 1879 carried a report that Thomas Morris had submitted his resignation as Resident Director, feeling "compelled by the force of circumstances".

The actuality the resignation was not as dramatic as it first appears; in practice it meant that Morris would not seek re-election at the May 1879 meeting, although he would continue as Resident Director until then. Morris attended the May meeting where Peter Watson, on behalf of the company, thanked him for his service, a vote of thanks being passed by the meeting. Lord Claud Hamilton replaced Morris as one of the board of directors.[214] Morris' resignation may be seen at the culmination of the cull of the old guard. Supporting the miners in the five week month dispute seems to have been the last straw for the predominantly London based shareholders who appear to have become exasperated with Morris who, they felt, had "gone native". At an E. G. M. held on 13th June 1879 H. C. Stewart, one of the more vocal shareholders, commented that the office of Resident Director had become nothing more than a "warm and genial and comfortable sinecure".[215] It was resolved at the May 1879 half yearly meeting to alter articles 55 and 60 of the company's articles and memorandum, this effectively removed both the position of the Secretary and the Resident Director, their responsibilities being taken over by the Managing Director.[216] This move marked a significant shift in the balance of power within the company, concentrating power in Watson's hands whilst diminishing the role of the local management whose loyalty, in the eyes of the directors, had, in the recent strike, been misplaced. The *Tavistock Gazette* was less than impressed with the situation, commenting:

> .....underlying the whole is an evident desire on the part of the "Dictator" to take into his sole hands the entire reins.... from the .....alterations in the articles of association, we should not be surprised to find that Mr Peter Watson is proposed as "lock, stock and barrel", with a country residence at Tamar View.[217]

Away from the boardroom news from the mine was mixed. In a report dated 10th January, 1879 Isaac Richards noted that there was no improvement on either the 280 or 300 at Richards' Shaft.[218] On 23rd January he reported that work on both levels would cease within the week.[219] In fact work was scheduled to stop on Saturday 25th February.[220] In his annual report Josiah Paull wrote:

> It should be stated here as a record for those to come that in the 280 fm level at the time of its stoppage there was a large lode (5 feet wide) 2½ feet of which was composed of capel and also mundic, the remaining 2½ feet was chiefly clay slate.
>
> At the 300fm level east the lode at the time of stoppage was 2½ feet wide composed of capel and sulphurous mundic. The ground in this level was very hard.
>
> In both levels the lode may be said to have been of such character as would have warranted further prosecution had there been any view or prospect of a change of rock such as from clay slate to granite before us, but there was not and with a poor mine generally I don't think further outlay here could have been pressed for.[221]

The compulsory work on the 216 east of Thomas' Shaft was also abandoned on 25th January 1879. The 216 had been driven 138 fathoms from Thomas' Shaft without discovering any ore.[222,223]

With the abandonment of the compulsory works, exploratory work was confined to New South Lode. "Very fair progress" was being made at New Shaft which was sinking below the 190, whilst Railway Shaft was being sunk below the 160.[224,225] By 13th February New shaft was 11 fathoms below the 190; the intention being to continue sinking and to drive the 202 level. Richards envisaged that New Shaft would reach the 202 in about two months.[226] By 3rd April work had started driving a crosscut on the 205, Richards noting that the ground was mineralised.[227] Railway Shaft was also making good progress having reached nine fathoms, four feet and seven inches below the 160 by 27th March.[228]

By the end of March the mine was again bedevilled by strike action. The roots of the dispute lay in the resolution of the "Five week month strike" of the previous year. It will be recalled that Peter Watson had assured the miners that once the mine's stocks of arsenic had been sold they would receive a pay rise. Between the end of the strike and before the arsenic was sold the miners suffered "two or three subsequent reductions" in pay. Early in 1879 the mine managed to sell its arsenic at what the *Tavistock Gazette* described as "a very favourable price" to Messrs. Field of Cornwall. Not unnaturally the miners expected Peter Watson to honour the promise that he made to them in June 1878. However instead of the promised rise Watson cut the miner's pay by a further 10% reducing the miners average monthly pay to £2 10s a month or 12s 6d a week. The directors had decided to use the proceeds of the arsenic sale to pay of the remaining balance of £6,000 of the 1878 £7,000 loan, rather than increasing the miners pay. Arguably it was somewhat disingenuous of Watson to offer the miners a pay rise on the back of the sale of the arsenic, given that the proceeds must have been earmarked for paying off the loan.[229,230,231]

On the 25th January setting day the miners agreed to accept the cut and to take bargains under the terms offered. However on the same day the miners held a meeting and one of their number was deputed to write to Watson outlining the dire straits the miners were in and the hardship that the pay cuts were causing. In reply Watson dug his heels in, insisting on the 10% reduction. In addition to the pay cut, the directors also announced that the miners would also be expected to work on "Maze Monday". "Maze Monday" was the Monday following pay day which the miners traditionally used to repair their tools and make the "necessary preparations" prior to commencing a new bargain (see Note 13). Saturday 22nd March 1879 was the Wheal Emma setting day by which time the men were aware both of Watson's refusal to reconsider the reduction in their pay and also the abolition of "Maze Monday". Two of the Wheal Emma men, acting as spokesmen, highlighted their grievances stating that the conduct of the Directors was shameful and, in consequence, refused to work. Both Isaac Richards and Thomas Morris attempted to persuade the men to return to work citing the depressed state of trade. The men asked permission to withdraw to discuss the matter. After a short discussion the men returned and announced that they would not return to work, their spokesman stated that "it was not so much the simple question of "Maze Monday" as the succession of "cuts" which they had to submit to, and it was the last feather that broke the camel's back".[232,233]

Late on the following Monday Peter Watson sent a telegram stating that the directors would not alter their position and that the miners were told that whenever they chose to accept their bargains they would be let. In the light of the Directors position Captain Richards, who was well respected by the men, persuaded them to return to work, making the point that it was folly to reject work when they had nothing to fall back on and "starvation staring them in the face". Reluctantly the Wheal Emma men saw the truth in what Richards was saying and returned to work on the Wednesday.[234]

Whilst the Wheal Emma strike was short lived the issues which caused it had not gone away. The setting for the western section of the mine was held on 19th April 1879 at which the men resolutely refused to take any of the bargains offered and the setting was abandoned. On Tuesday the men quietly returned to work having observed "Maze Monday" and announcing that they would follow this course of action at every setting (see Note 14).[235,236]

In an attempt to resolve matters the Wheal Emma men addressed a "memorial" outlining their grievances to Captain Richards, the "memorial" being dated 6th May 1879, this was forwarded by Richards to Peter Watson. Watson's reply dated 11th May 1879, was unequivocal, Watson and his fellow directors were not prepared to concede ground.[237]

Matters had not resolved themselves by the time of the half yearly meeting held on Tuesday 27th May 1879. The miners frustrated by a lack of progress decided to sidestep the directors and address a "memorial" outlining their grievances to the shareholders. The memorial, reproduced in full in the pages of the *Tavistock Gazette*, outlined the miners view on the abolition of traditional holidays such as midsummer, low wages and "Maze Monday".[238] At the insistence and to the credit of Peter Watson the memorial was read at the meeting, albeit with sarcastic interjections from a Mr John Stuart who appears to have fancied himself as something of a wit, a view shared by many of his fellow shareholders who greeted his comments with laughter. Watson explained to the meeting that he was unable to accede to the miners demands particularly in regard to the question of wages Even though the directors had managed to clear the £7,000 loan the financial climate was such that the Company really had very little room for manoeuvre. The average price the mine was receiving for its copper ore had dropped as low as £1 10s 1d per ton. The company had again, unsuccessfully, approached the Duke of Bedford with a view to reducing his royalty.[239]

Towards the close of the meeting John Stuart proposed the following resolution which was seconded by a Mr Fitzpatrick:

> That this meeting, after having considered the unsigned appeal of the miners and others, regret that they cannot comply with the request contained therein, and that the whole matter of wages and holidays be left to the chairman and directors.[240]

The next setting day was held on Saturday 14th June. Unlike previous setting days when bargains were set for either the Eastern of Western parts of the mine the whole of the mine was being set. On the previous setting day the Eastern men (see Note 15) had only accepted a bargain of a month, as opposed to the usual two months, so that their bargains would expire at the same time as those taken by Western men. On the previous Thursday a notice had been posted around the mine to the effect that the miners were not to leave work until the Saturday afternoon and were to return to work on the Monday morning. Traditionally the miners left work on Friday afternoon during the last week of their bargain and, as discussed, did not start work again until the following Tuesday. In the words of the *Tavistock Gazette* "against this a great many of the men exhibited signs of resistance and threatened to strike".[241] During the Saturday morning a large crowd gathered in front of the count house. Shortly after one o'clock Captain Isaac Richards came onto the balcony and informed the gathered crowd that Peter Watson, who was visiting the mine, would address them. The miners suggested that they would prefer to send in a deputation to discuss the matter, accordingly ten miners, five of whom were, what might be termed, militant and five moderate held a meeting with Watson.

Shortly after the meeting concluded Watson, Richards and several of the mine's agents appeared on the balcony in much the same manner as W. A. Thomas had in February 1866. A small minority in the crowd jeered Watson, however he was more than equal to the situation: he waited a few moments and then remarked that "if they did not think it worth their while to listen to him, he could not help it".[242] This seemed to quell the rowdy elements; Watson was master of the situation. In his usual bluff manner he explained to the men that the company had made serious losses during the previous year and that in an attempt to control cost all he was doing was bringing practices at Devon Great Consols into line with those of the larger Cornish mines. Given the financial situation all pitches would be arranged on a basis of fourteen shillings a week as before, however if the price of copper rose matters might change. He did offer a degree of concession; miners would have to come into work on the Monday after a setting, however not until 2 o'clock and then only to arrange the return of tools. Having finished Watson thanked the men for listening; Richards then proceeded with the setting. Evidently the men must have been satisfied by Watson's speech as all the bargains were taken.[243] Watson must have been pleased with his performance; he had diffused a potentially explosive situation whilst keeping concessions to a minimum.

Watson's appearance on the count house balcony marked the high water mark of the dispute. Isolated incidents apart, the dispute was over. For example on Midsummer Day a number of Devon Great Consols miners took an unofficial holiday in direct opposition to the contracts they had agreed to at the last setting. Apparently many spent the day "standing at the corners of streets with their hands in their pockets, smoking pipes". On returning to work on the Wednesday a number of miners who had absented themselves on the previous day were "spaled" (fined) 2s 6d whilst some tutwork bargains were "altogether put an end to".[244]

MOSES BAWDEN, ESQ.

Figure 26.
Moses Bawden, Purser of Devon Great Consols.

On Midsummer Day 1879 the directors made a significant appointment to the management of the mine, employing Moses Bawden as Purser.[245] Bawden (see Note 16) was a man of real talent; Peter Watson, in typical style, stated: "I think he is the best man in Cornwall but myself".[246] In effect the appointment of Bawden was recognition on the part of the Directors that they needed a man of experience on the ground filling much the same role as Thomas Morris had as resident director. Unlike Thomas, Moses Bawden did not benefit from the tenancy of Abbotsfield, which had been sold for £3,000 by the time of the November 1879 meeting. Bawden resided, until his death in January 1916, at Tamar View, a property owned by the mine for which he paid an annual rent of £50.[247]

The November half yearly report notes that Peter Watson had been appointed Managing Director, a role that he had effectively been carrying out since he took over the role of Chairman in W. A. Thomas' stead in 1877. [248]

At the mine New South Lode continued to prove productive and development work was continuing apace. By 9th October 1879 Railway Shaft had reached the 175 and nine men were engaged in sinking deeper, meanwhile New Shaft had been cased, divided and ladder and skip roads had been installed between the 190 and the 205.[249] Shareholders at the November 1879 meeting were informed that work would restart at Wheal Thomas, although in true style the mine would be renamed "Watson's Mine".[250]

Whilst the directors had been seriously considering the introduction of rock drills it was not until the autumn of 1879 that matters became more concrete, the *Tavistock Gazette* reporting that "no less than three have been ordered and their relative merits will be tested.[251] By the time of the November half yearly meeting held on Wednesday 26th November 1879, £130 had been paid "on account of rock boring machinery". On the ground a compressor was being erected and "tubing" being put in. The directors commented on the lack of assistance they had received from the Duke of Bedford in relation to the introduction of rock drills and they expressed the hoped that the introduction of drills might influence the Duke to reconsider his position.[252,253] As the year and indeed the decade drew to a close Isaac Richards reported on 11th December 1879 that men were installing air pipes for the rock drills in New Shaft; the shape of things to come.[254] Josiah Paull noted that the mine had introduced both Barrow and Eclipse machines.[255]

The 1870s had been a traumatic decade for the mine: the collapse of copper prices, reconstruction as a limited liability company, the search for tin and the failure to find it, the wholesale change in the mine's management and the labour disputes. Only the ongoing prosperity of New South Lode and the increasing importance of arsenic had kept the mine's head above water. Certainly the *Tavistock Gazette* felt that they were witnessing the last days of the once great mine:

> Devon Great Consols – Then and now
> ..... one is reminded....when..... Devon Great Consols was in the full swing of its prosperity, now some thirty years ago. As compared with its present condition, it was what the jolly, robust, well fed, well to do farmer is to the half starved, whining cowering pauper, whose wasting substance seems ever on the verge of total disappearance. Ah! Those were right royal days, days when ore was piled in mountain heaps, when demand watched supply with hungering eyes, and still cried "give us more", when dividends poured into the laps of lucky shareholders with bewildering frequency, and royalties were so ample as to make a substantial addition to even Ducal incomes. In those days everything was large and free – we may say magnificent – about Devon Great Consols. It built railways, created townships, and supported an army of cheerful prosperous workers. Everybody coming into the neighbourhood went to see the great mine, and the officials showed its wonder to the enquiring stranger with quite an air of proud elation as of those who should say, "we have the honour to belong to this vast, well managed, wealth producing enterprise". And indeed the bodily presence of many of the officials worthily represented the greatness around them. They were tall bulky men, Sauls, not occasionally, but always among the *profits*, and as being head and shoulders above their fellows. And the hospitality of the mine was quite in keeping with its general grandeur. Good dinners, numerous and joyful guests, most excellent vocal music, and, oh reader! *Such punch*. The skill that went into the composition of this beverage, nectar fit for the gods! was the result of many years of studious, loving thought and patient experiment, animated by the hope of attaining a grand success. It was wonderful drink, and men who partook of it recalled their first sip with the emotion that only great critical events of life can inspire. Punch is to be had at the mine now we suppose on pay days, but we reckon that the gatherings round the board, are only the ghosts of what they were, and with the absence of the old prosperity, and many of the old familiar faces, the mysteriously compounded liquor is drunk

> as by men at a funeral gathering, not for exhilaration but for oblivion of painful memories. And this change over the spirit of their dream, and the dream of their spirit, has all be wrought by the change in the value of copper. Should the fickle metal rise again, departed glories may return, and Capital and Labour be once more united in the enjoyment of a common prosperity.[256]

With the benefit of hindsight the *Tavistock Gazette's* "obituary" for the mine was somewhat premature, the mine having another twenty years of productive life albeit predominantly as an arsenic producer. Indeed it would be more than a century later when the last mining related activity took place at Devon Great Consols.

Chapter 7 Notes

Note 1

The actual measurement from the top of the collar at New Shaft to the 100 is 86 fathoms.

Note 2

This lode would latterly be known as Capel Tor Lode.

Note 3

The search for tin at depth was driven by the Bedford Estate. Writing in 1902 and with the benefit of hindsight Josiah Paull observed:

> By the year 1872 it had become evident that the lodes were getting barren of Copper ore in length, the Eastern ground was quite unproductive and years before the lodes had failed in depth, they lasted productive to between 100 and 140 fathoms from surface when sterile lodes became the general rule. It was an analogous experience to mining in Cornwall, there all the great lodes had been rich in Copper poverty followed, but not despairing of finding mineral deep sinking was persevered with and Tin was found in great quantities. Mine after mine became rich for Tin as depth was reached. Feeling as we did a like search should be made for it under the Copper deposits of Devon Great Consols we asked for an inspection of the mine by the most experienced Tin miners of Cornwall (Messrs. Josiah Thomas of Dolcoath and John Simmons Mineral Agent for the Duchy of Cornwall) they made a careful examination of the mine which occupied three days in

doing. Their report was an encouraging one, this laid before the Directors of the Mine and their local manager brought about an arrangement ...... to sink Richards' Shaft...[257]

Note 4

The fundamental problem of water management at Devon Great Consols was maintaining a sufficient supply of water to the dressing floors and plant at the top of the hill. Solutions included pumping water from the Tamar up to the top of the mine and the management of small surface streams. A further solution involved the development of a series of shallow adits tens of fathoms above the Blanchdown Deep Adit. Wheal Anna Maria, Wheal Josiah and Wheal Emma all had shallow adits feeding extensive reservoir systems. Having a two tier adit system meant that the mine had the flexibility either be drained at the valley floor by the Blanchdown Adit or at the top of the hill by the shallow adits. The adit drainage strategy would have been something of a juggling act: Pumping to Deep Adit would have required less energy than pumping to the shallow adit system. However water pumped to Deep Adit would not be available for dressing whereas the water pumped to the shallow adits would.

Note 5

It is possible that this 40″ engine was originally intended to be erected on New North Lode. As matters transpired it was erected in the vicinity of Gard's Shaft. Josiah Paull in his annual report for 1873 noted that "a steam engine will soon be erected at Gard's Shaft for the drainage here as well as for that of the North. In the same report it was noted that New North Lode Engine Shaft was being connected to "the pumping engine at Gard's Shaft" during December 1873. By June 1874 the Gard's Shaft pitwork was connected to the engine.[258]

Note 6

Unlike the joint stock company, which held an annual meeting, the new limited liability company was obliged to hold two shareholders meetings a year, an A.G.M. in May and an O.G.M. in November.

Note 7

Writing in December 1874 Josiah Paull commented that:

> The quantity of Tin required to pay working expenses would be 25 lbs per ton of stuff. 35 lbs per ton would be required to give profits, this is about the average yield at Dolcoath.[259]

Note 8

Peter Watson, a share dealer, mining commentator and speculator had interests in a large number of mines including Devon Great Consols, Great Laxey, Roman Gravels and Lead Hills. As chairman of the Directors and the majority shareholder at Devon Great Consols he would go on to steer the mine through the thin years of the late 1870s, 1880s and 1890s; a task which he carried out in his own unique style. Watson was what might be described as a pillar of his community; he was also an Officer of the Honourable Artillery Company. An active Freemason, Watson was a Past Master of the Fitzroy Lodge which he had joined in 1852. He was extensively involved in local politics being a member of the Heston and Isleworth Local Board where he chaired the finance committee; he also sat as a County Councillor for Middlesex.[260]

He was a man of huge energy and charisma, a talented man of business whose larger than life, sometimes bombastic, character served to hide a keen, occasionally ruthless, intelligence. Watson's 1911 obituary in the *Mining Magazine* paints a broad picture of a man much respected by his peers:

> By the death of Peter Watson the Cornish mining industry loses one of its chief supporters. At one time he was able to boast of being a shareholder in 250 different mines in the old county. As chairman of Devon Great Consols he was connected officially with an historic enterprise in the adjoining county, but his interest did not cease here, for he was connected with mines in Wales, Scotland, and the Isle of Man. He was conspicuously just to employees and enjoyed general regard among mine captains by reason of his loyalty to them as long as they did their duty. His invincible optimism cheered them and served to stimulate Cornish mining in its dark days. In personal relations he was lovable, adding frankness to gentleness, and courtesy to a high order of intelligence. He attained the age of 81 and died rather suddenly, but peacefully, after a career full of generous effort and useful endeavour.[261]

Note 9

Captain James Richards was appointed Captain at Wheal Maria in "about September 1844" by Josiah Hitchins, becoming Chief Agent after Hitchins' resignation in 1850. His illness, noted at the May 1878 meeting would eventually lead to his death on the 5th August 1878. Richards' funeral was recorded as being the largest seen in Tavistock for many years. By way of an epitaph the *Tavistock Gazette* commented that: "he lived an exemplary life, and

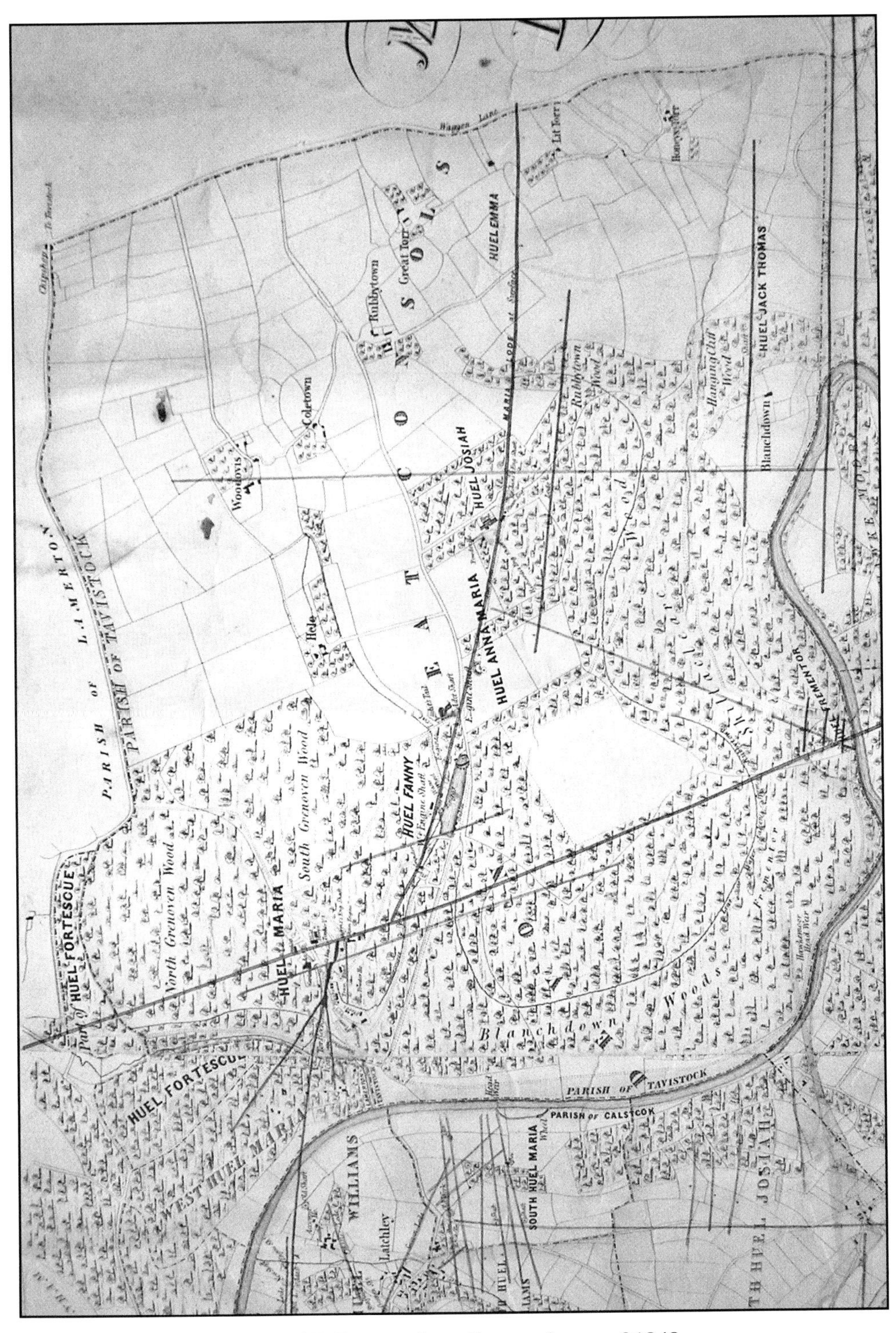

Figure 27. Extract from Symons' map of 1848.

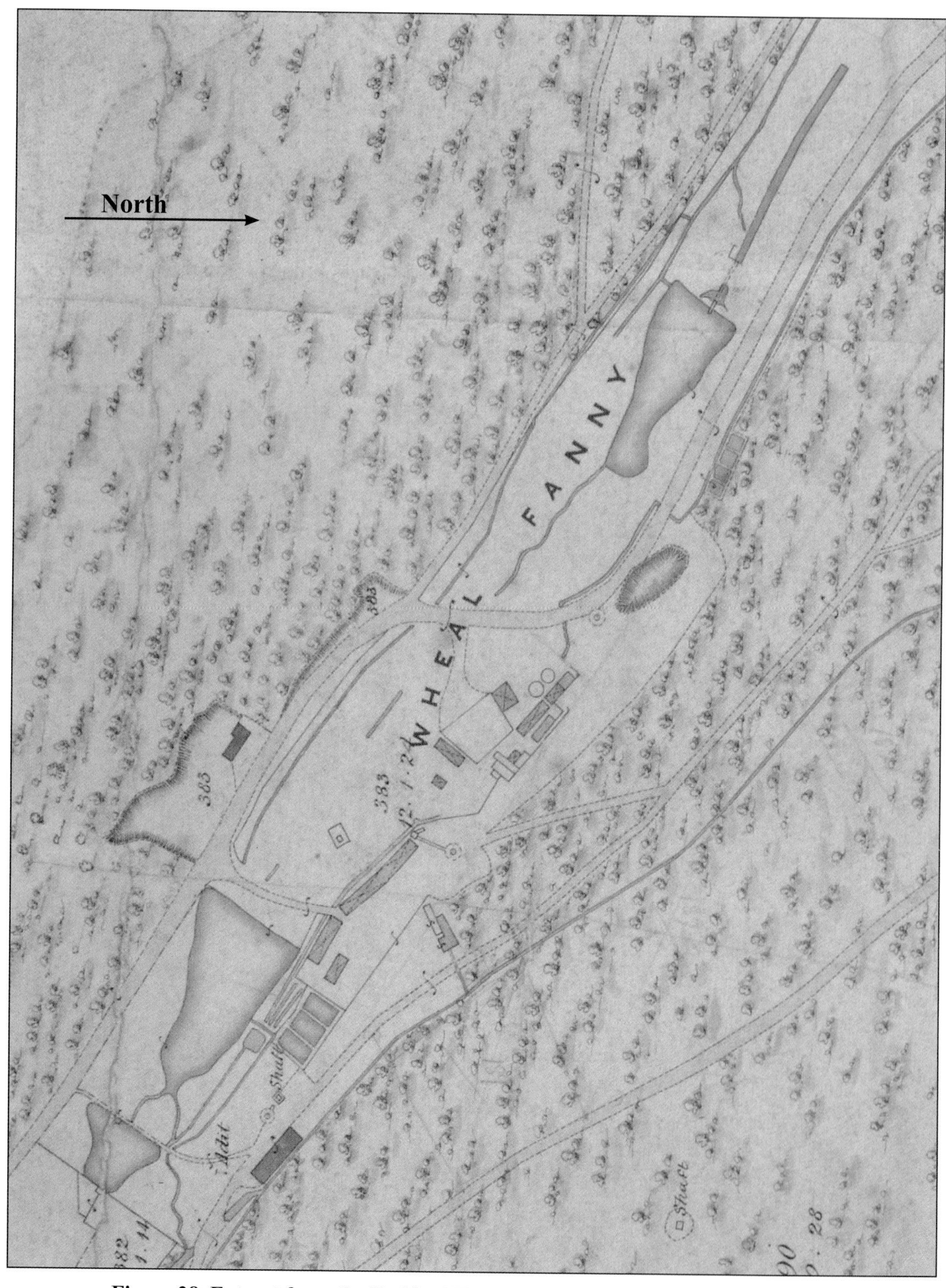

Figure 28. Extract from the Bedford Estate map 1867 showing Wheal Fanny.

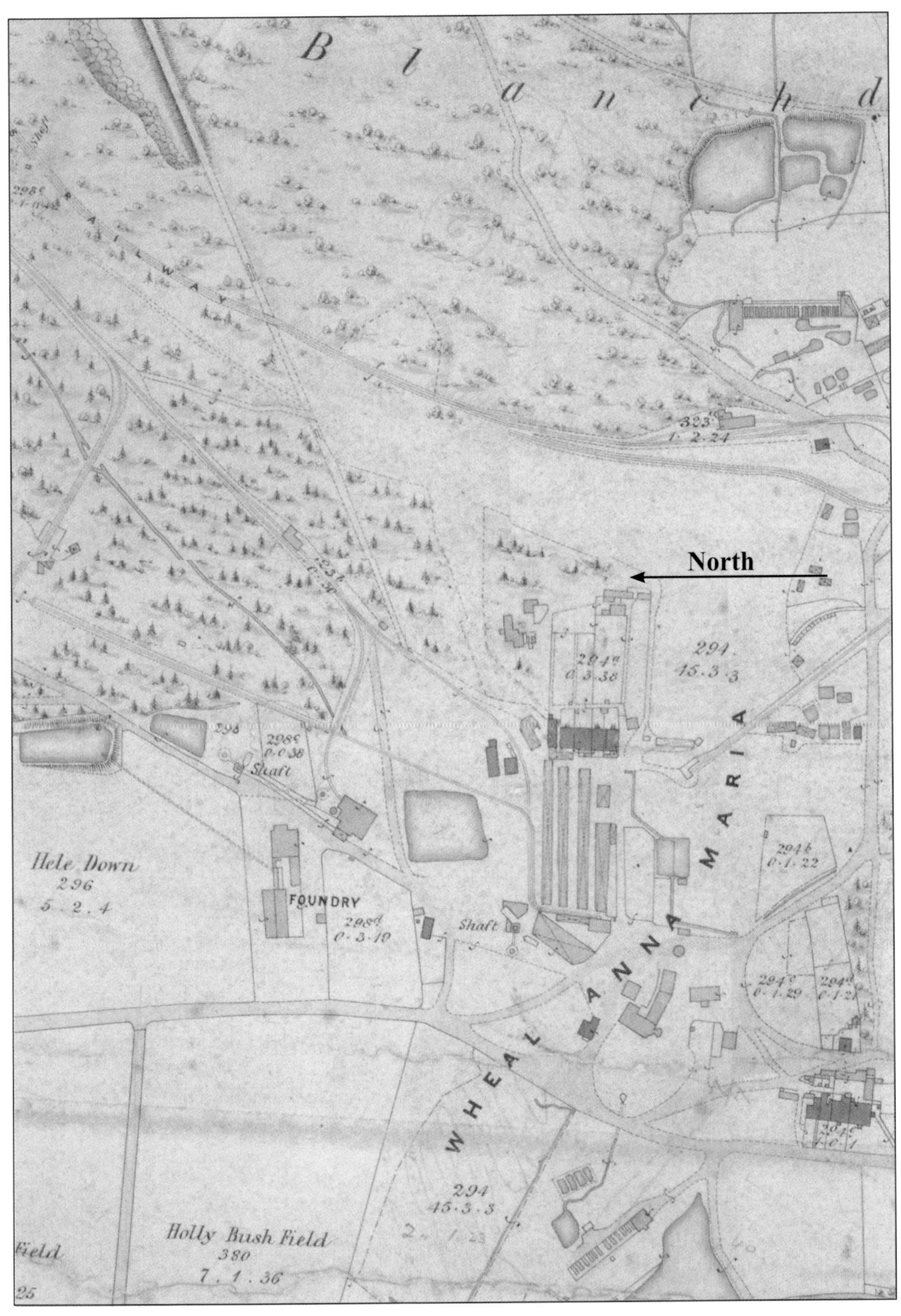

Figure 29. Extract from the Bedford Estate map 1867, showing the upper dressing floors at Wheal Anna Maria.

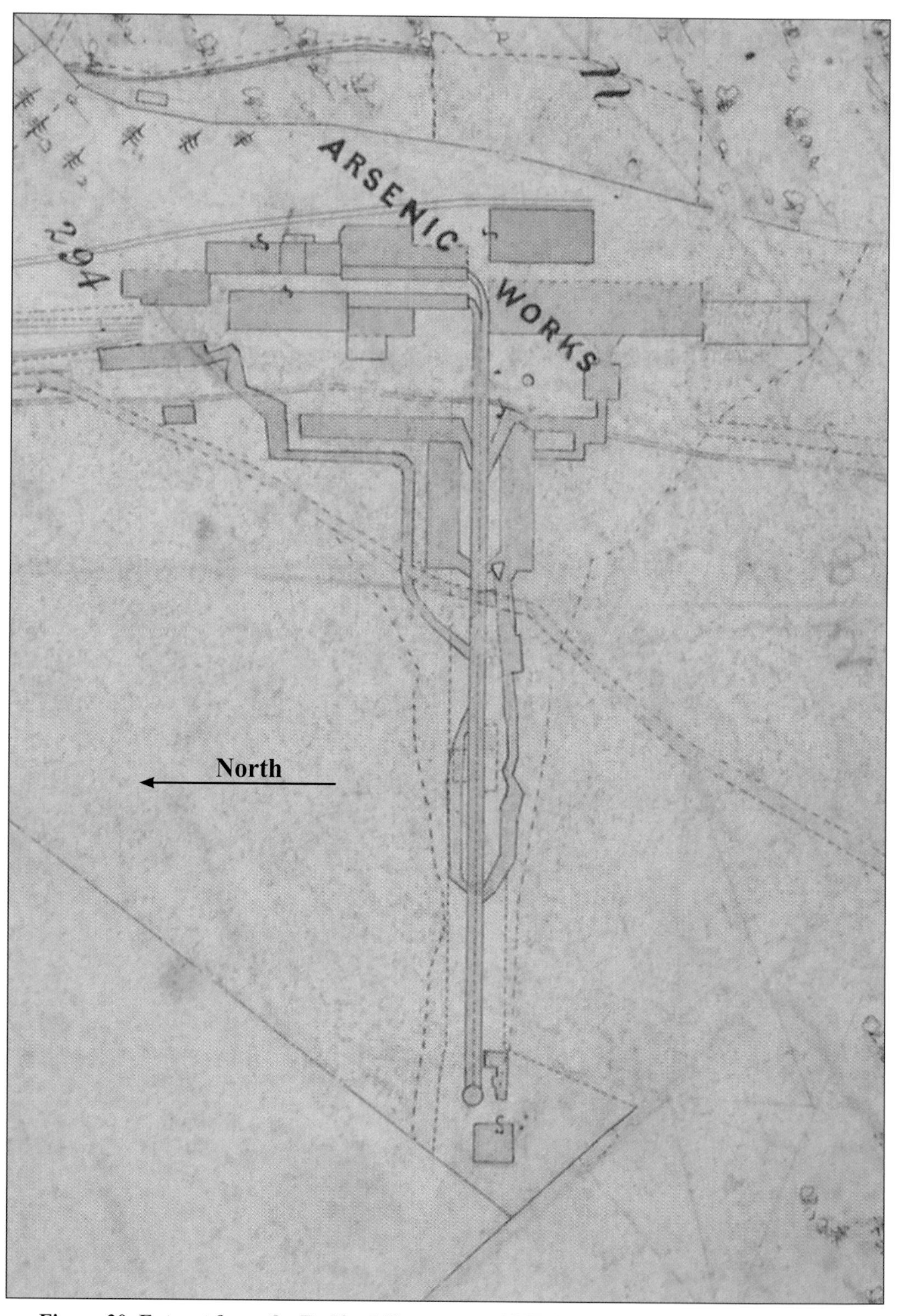

Figure 30. Extract from the Bedford Estate map 1867 showing the Wheal Anna Maria arsenic works.

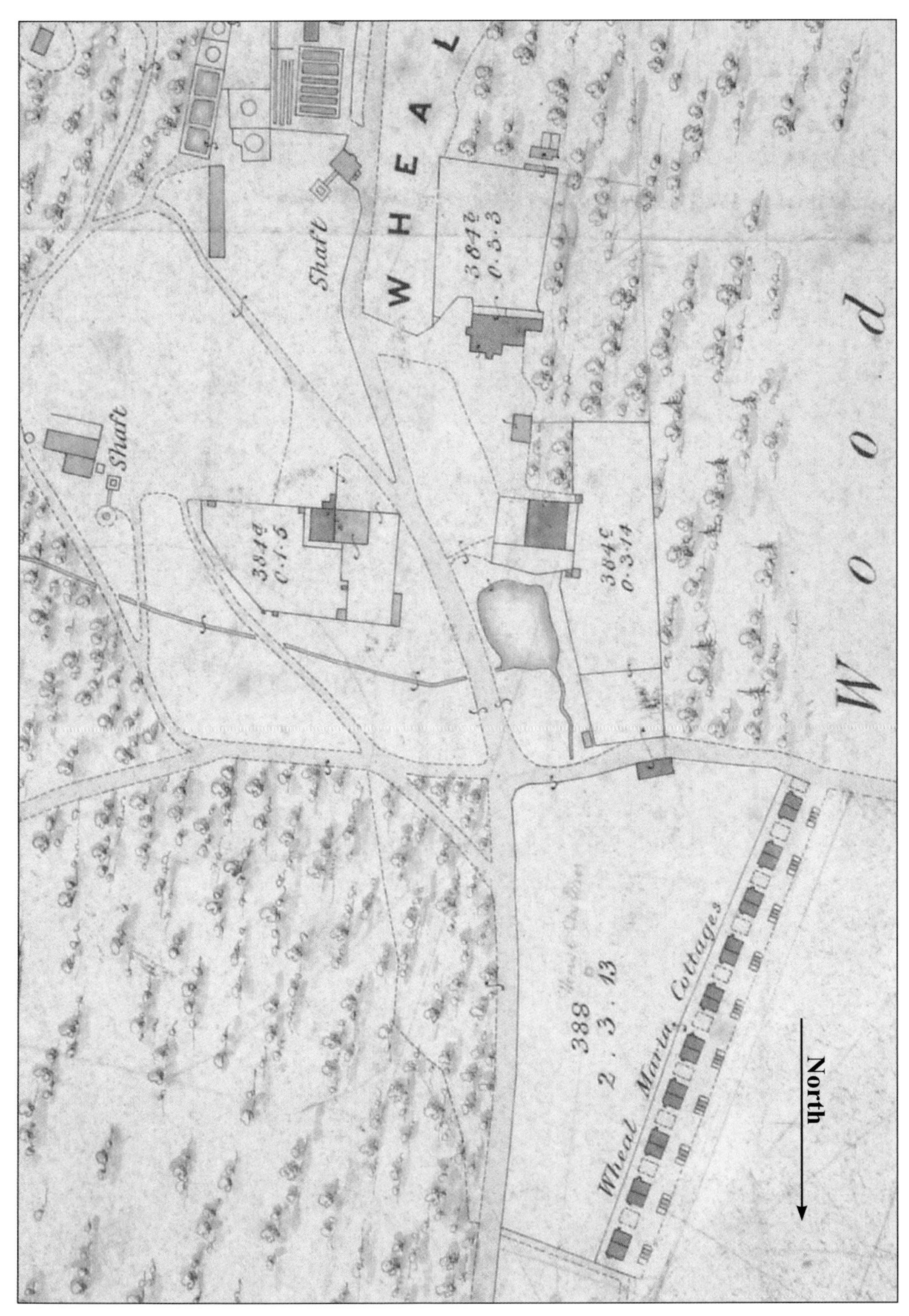

Figure 31. Extract from the Bedford Estate map 1867 showing the Wheal Maria Cottages, Morris' and Gard's shafts and part of the dressing floors.

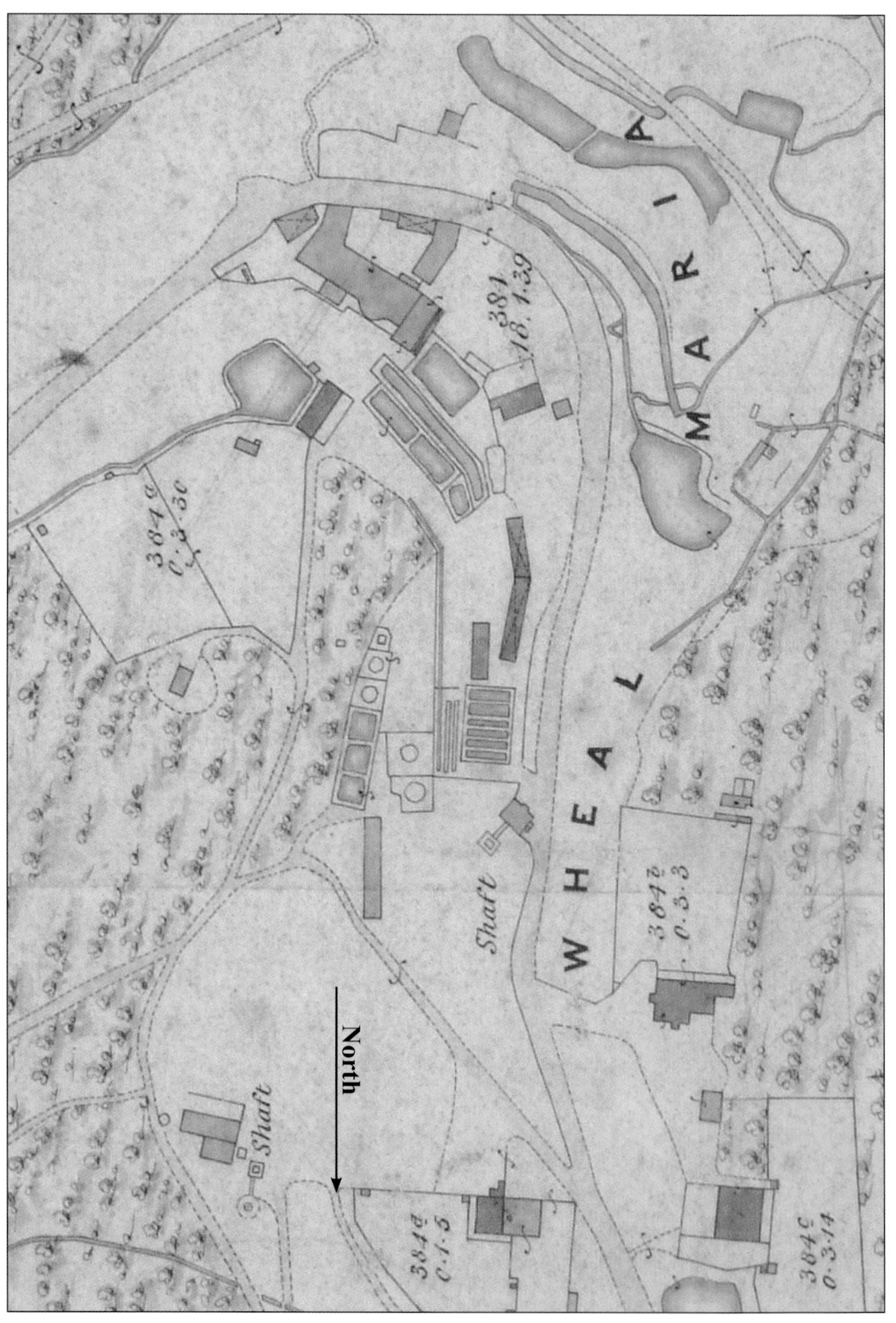

Figure 32. Extract from the Bedford Estate map 1867 showing the Wheal Maria dressing floors and foundry.

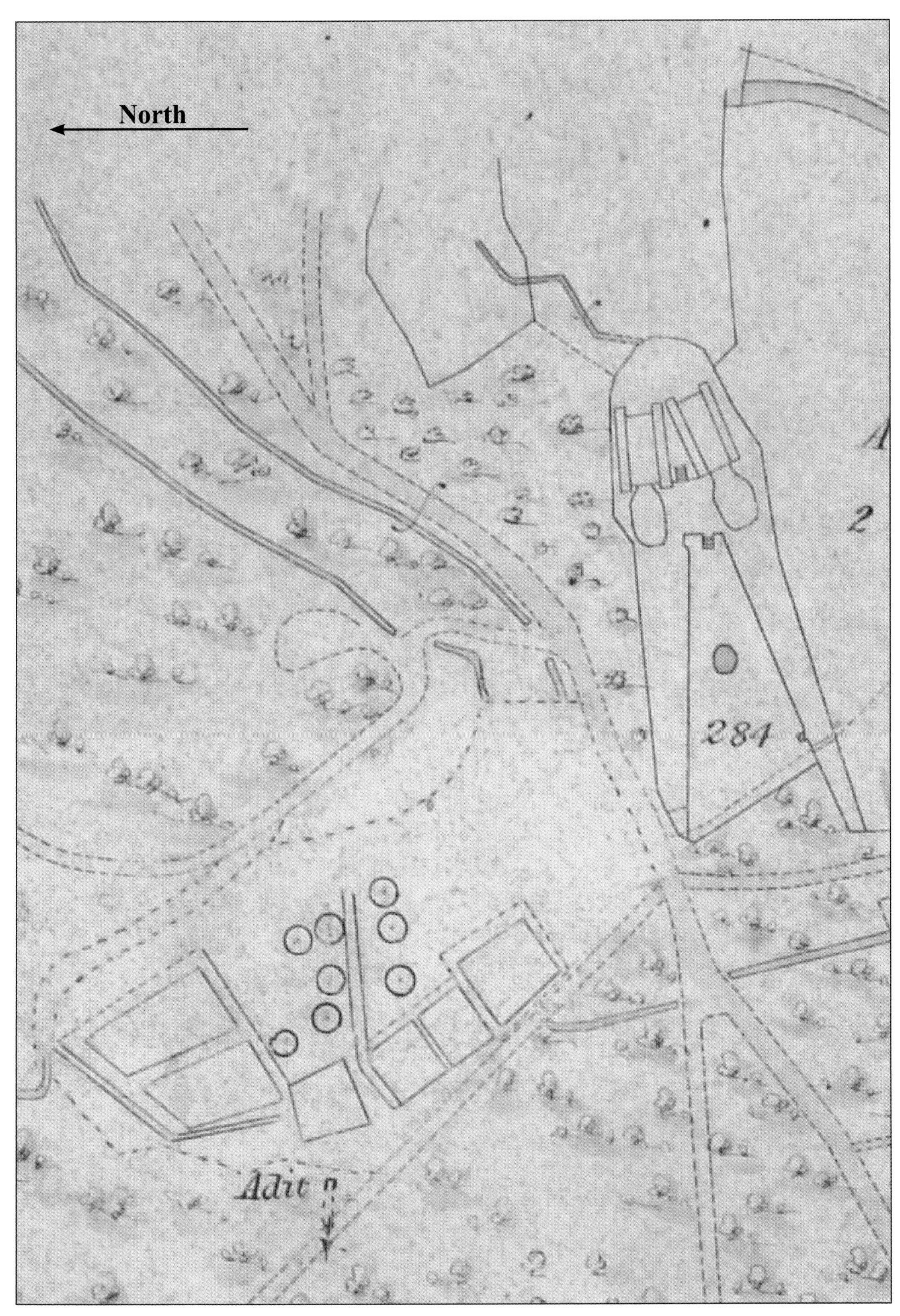

Figure 33. Extract from the Bedford Estate map 1867 showing the Blanchdown Great Wheels, precipitation works and the Blanchdown adit.

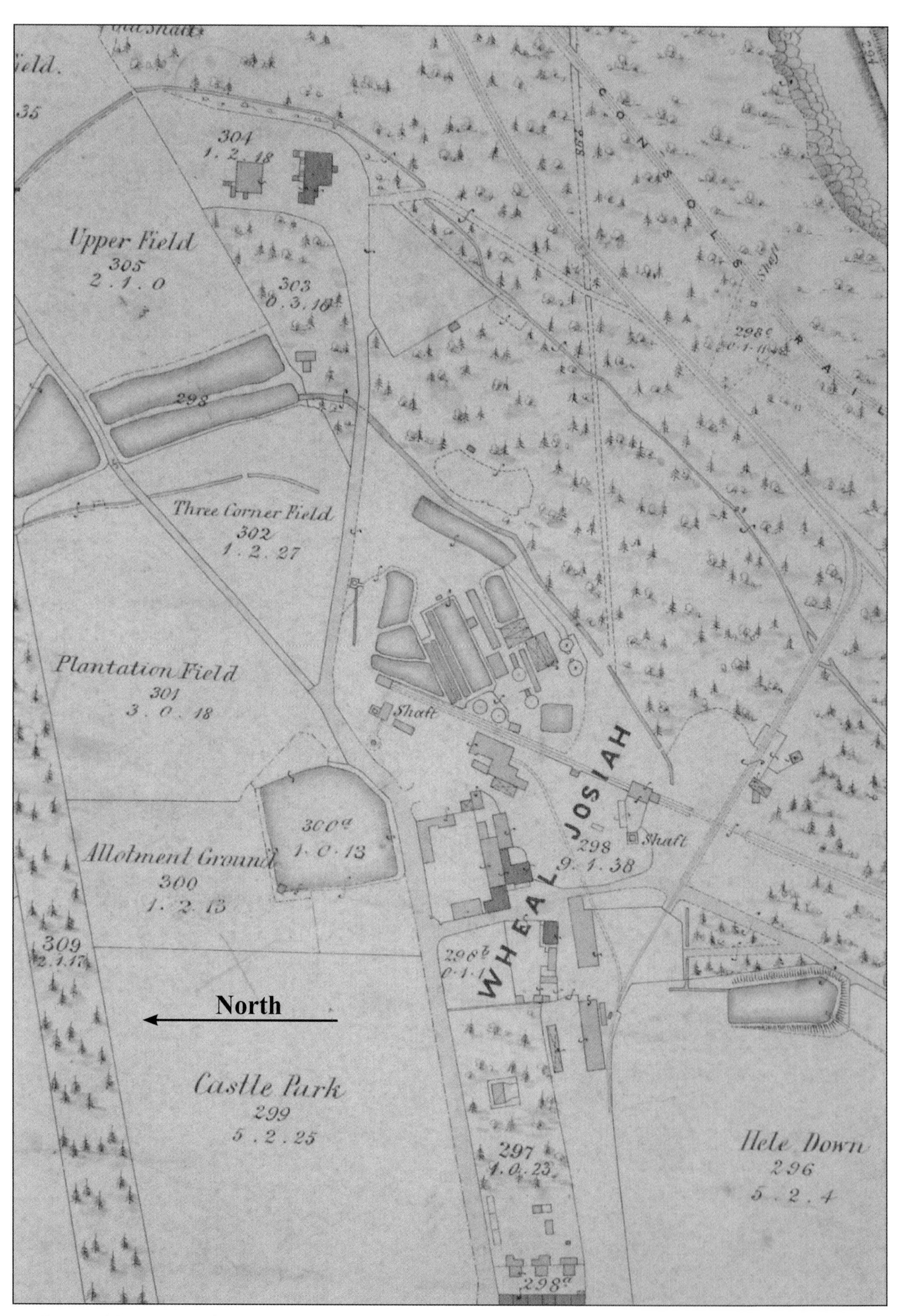

Figure 34. Extract from the Bedford Estate map 1867 showing the Wheal Josiah dressing floors.

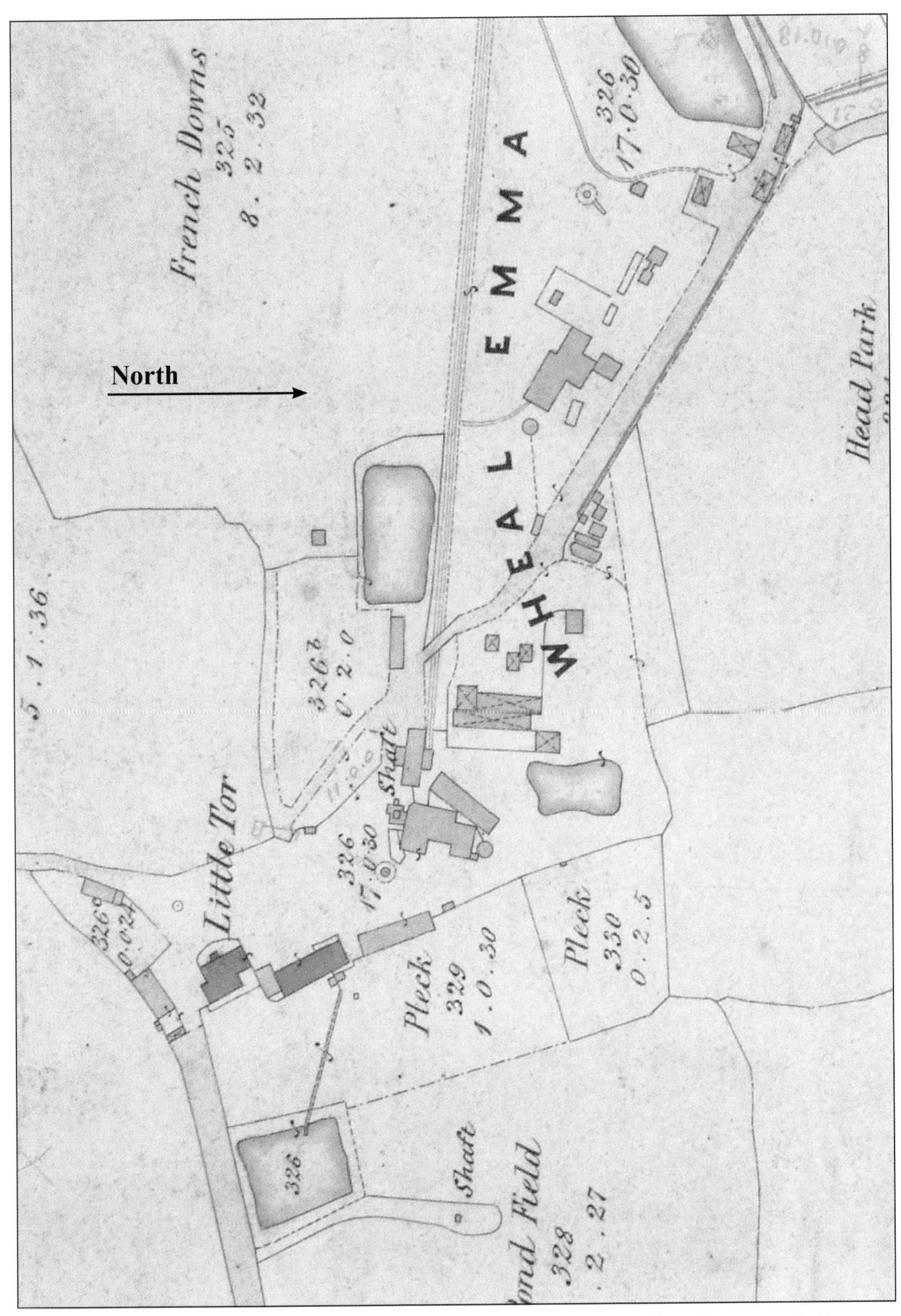

Figure 35. Extract from the Bedford Estate map 1867 showing the east end of Wheal Emma.

Figure 36. Watercolour of Abbotsfield painted on 6th June 1879 by Mary Josephine Browne (1845 – 1927).

Figure 37. Impressive post medieval /early modern tin work on the caunter lode to the north of Wheal Frementor.

Figure 38. Shaft marker erected in October 1904 during clearance and abandonment of the mine. The shaft it is marking is Gard's Shaft.

Figure 39. Flat rod tower in Rubbytown Bottom.

Figure 40. The 1920s Brunton Calciners.

Figure 41. A 2011 view of the 1920s arsenic works.

Figure 42. The ball mill from the 1970s tin recovery operation remains in situ.

Figure 43. Mechnical jigger above the 1970s tin mill.

Figure 44. Early 1970s view of the mill.

Figure 45. The 1970s mill building in its final form.

Figure 46. The 15 fathom level at Wheal Fanny.

Figure 47. Stull on the 15 fathom level at Wheal Fanny.

Figure 48. Ore chute at the 25-fathom level in Wheal Fanny.

Figure 49. Launder in Wheal Josiah Shallow Adit.

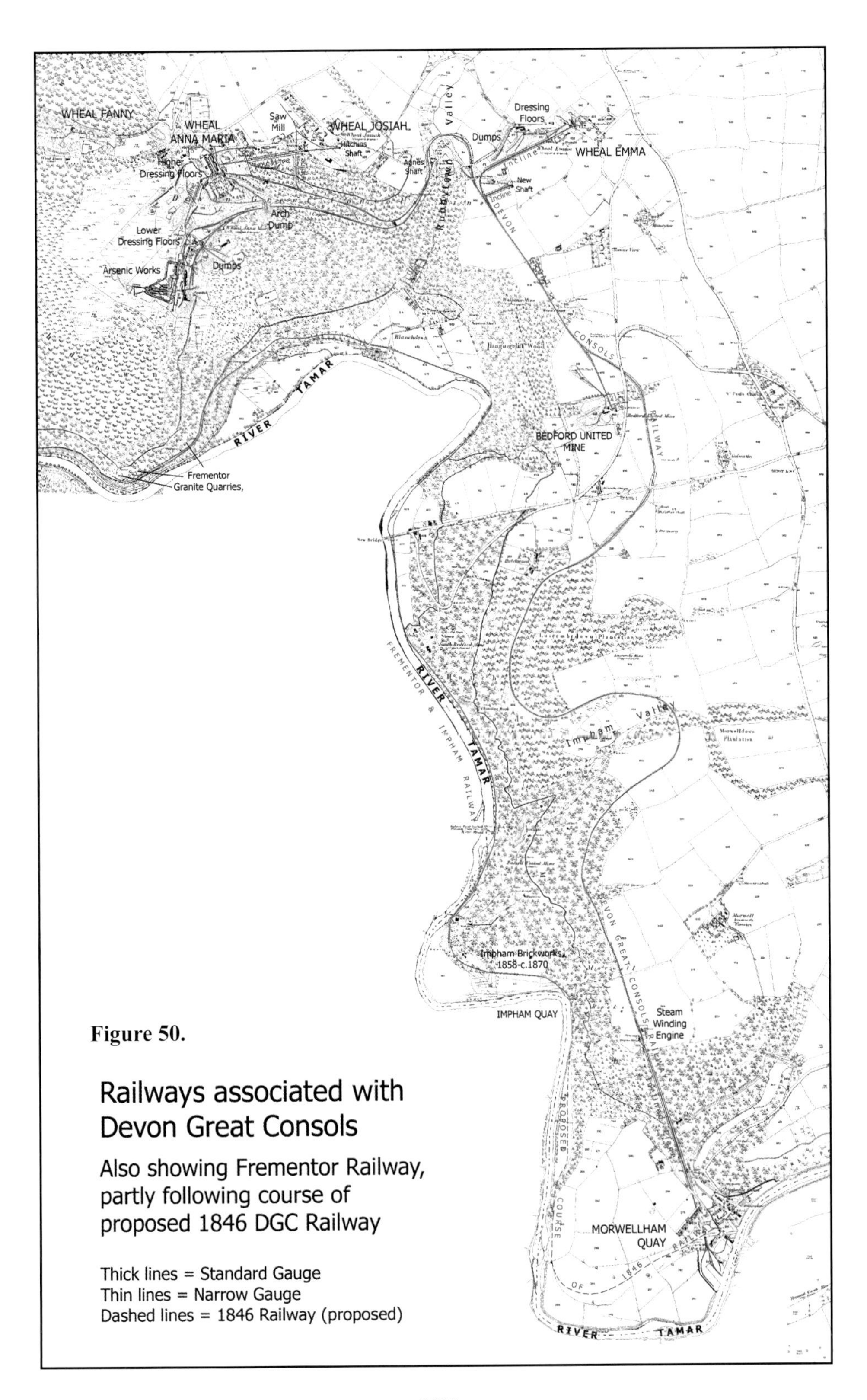

Figure 50.

Railways associated with Devon Great Consols

Also showing Frementor Railway, partly following course of proposed 1846 DGC Railway

Thick lines = Standard Gauge
Thin lines = Narrow Gauge
Dashed lines = 1846 Railway (proposed)

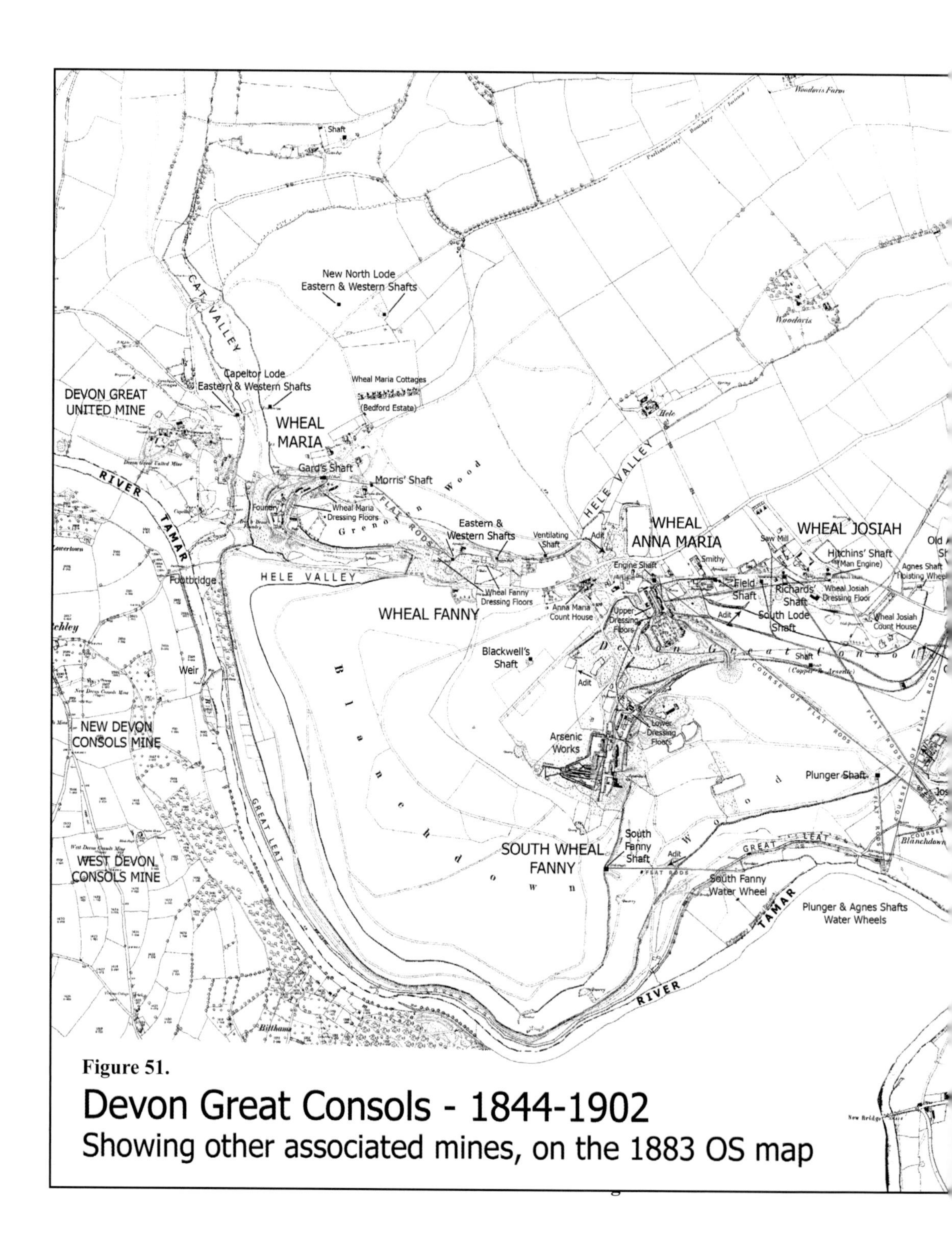

Figure 51.

Devon Great Consols - 1844-1902

Showing other associated mines, on the 1883 OS map

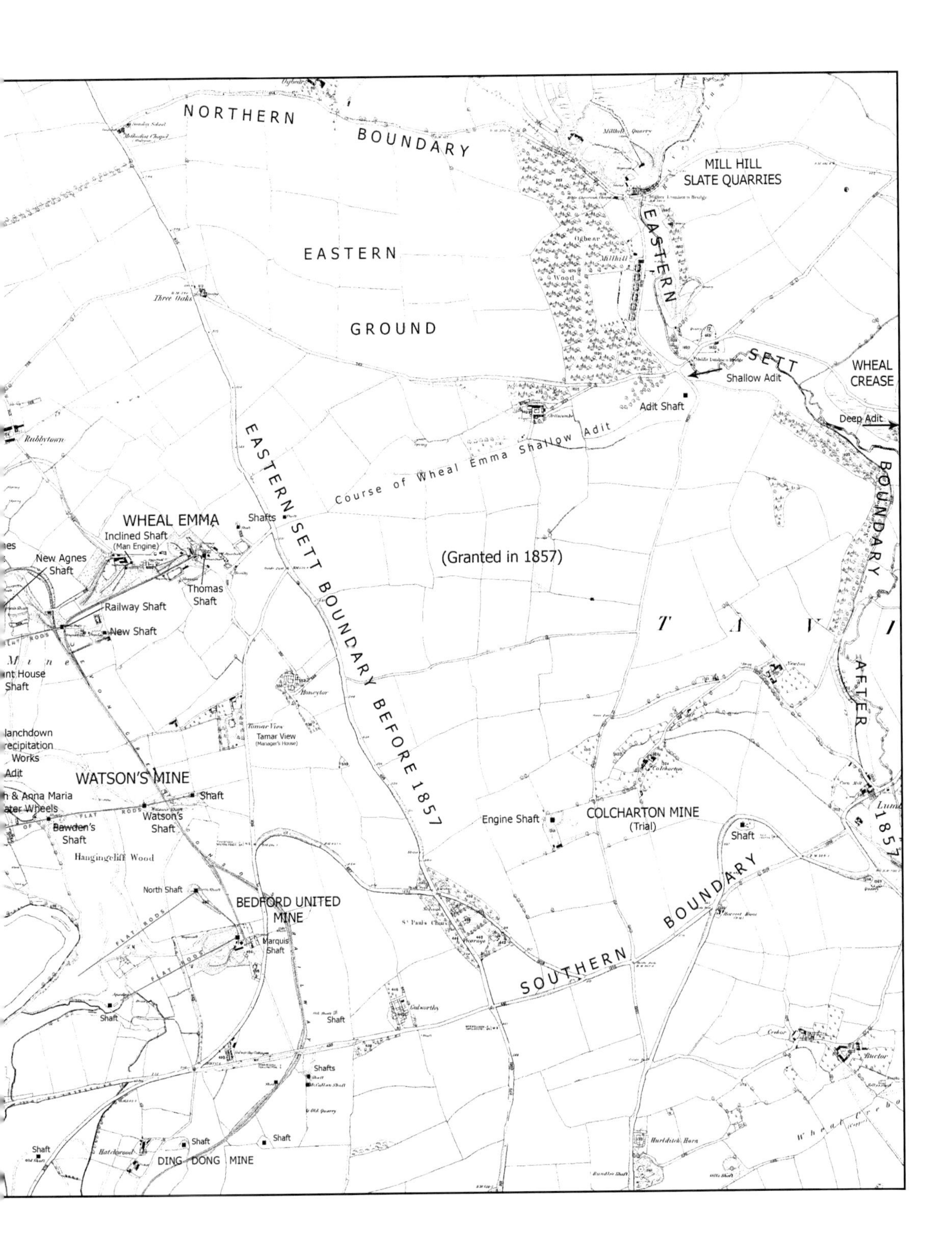
NORTHERN BOUNDARY
MILL HILL
SLATE QUARRIES
EASTERN
GROUND
EASTERN SETT
WHEAL
CREASE
Shallow Adit
Adit Shaft
Deep Adit
Course of Wheal Emma Shallow Adit
BOUNDARY AFTER 1857
EASTERN SETT BOUNDARY BEFORE 1857
WHEAL EMMA
Shafts
Inclined Shaft
(Man Engine)
New Agnes
Shaft
Thomas
Shaft
Railway Shaft
New Shaft
(Granted in 1857)
Tamar View
(Manager's House)
WATSON'S MINE
Shaft
Watson's
Shaft
Bawden's
Shaft
Hangingcliff Wood
Engine Shaft
COLCHARTON MINE
(Trial)
Shaft
North Shaft
BEDFORD UNITED
MINE
Marquis
Shaft
SOUTHERN BOUNDARY
Shaft
Shafts
Shaft
Shaft
Shaft
DING DONG MINE

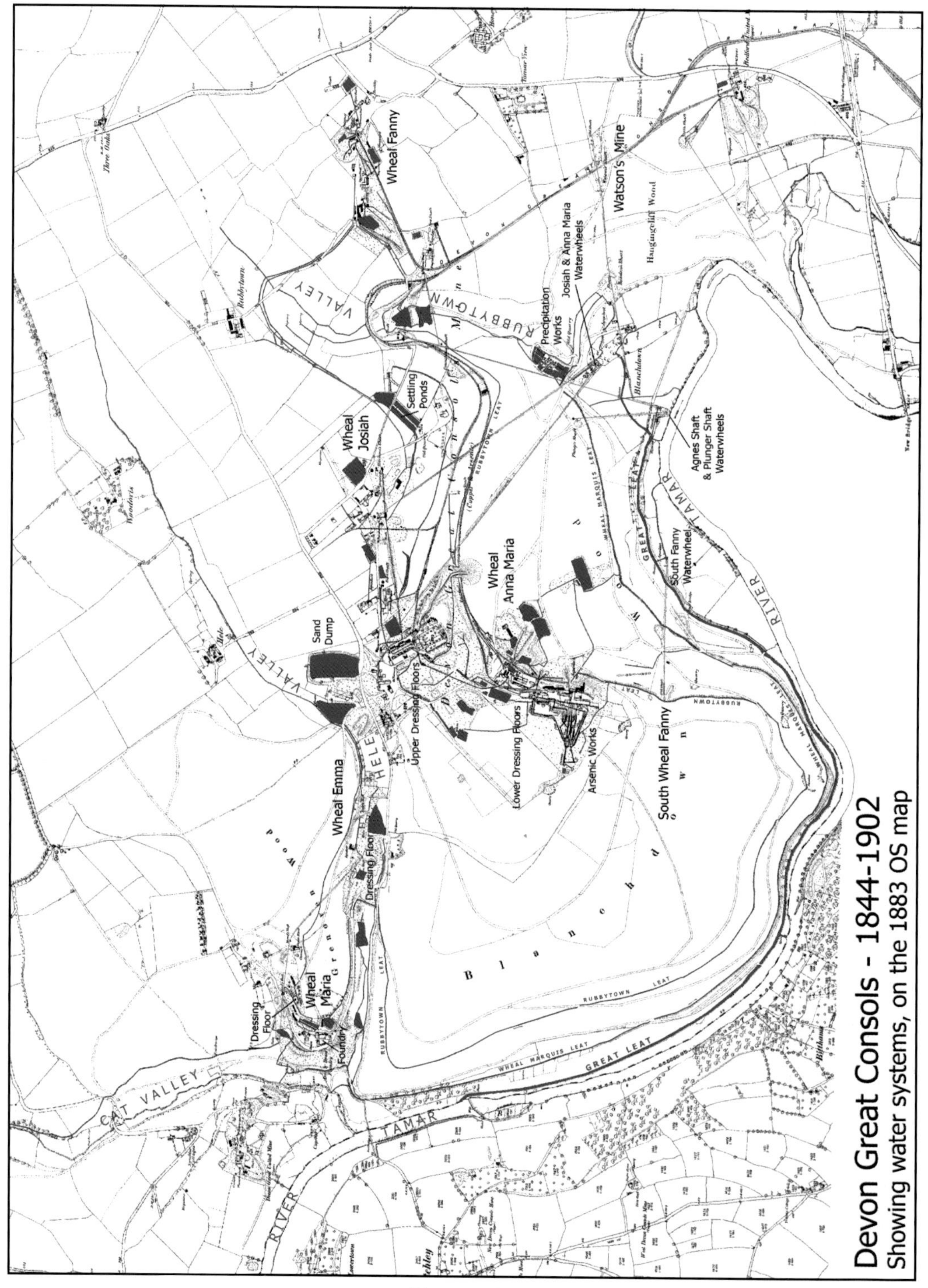
Devon Great Consols - 1844-1902
Showing water systems, on the 1883 OS map
Wheal Fanny
Watson's Mine
Josiah & Anna Maria Waterwheels
Precipitation Works
Settling Ponds
Wheal Josiah
Agnes Shaft & Plunger Shaft Waterwheels
Wheal Anna Maria
South Fanny Waterwheel
Sand Dump
Upper Dressing Floors
Lower Dressing Floors
Arsenic Works
South Wheal Fanny
Wheal Emma
Dressing Floor
Wheal Maria
Dressing Floor
Foundry
HELE
VALLEY
CAT VALLEY
RIVER TAMAR
RUBBYTOWN
GREAT LEAT
WHEAL MARQUIS LEAT
RUBBYTOWN LEAT

Figure 52.

died universally regretted".[262]

Note 10

Dr Stephen Henry Emmens was an intriguing if somewhat elusive character (thus warranting a longer footnote than is really justified). Born in 1844 in Lewisham, he studied chemistry at Kings College, London. In January 1865 he was elected an Associate of the Institute of Actuaries. Emmens was a partner in the eminently respectable Church of England Life and Fire Assurance Institution succeeding his father as chairman in 1868. Something of a polymath Emmens published books as diverse as his *Treatise of Logic: Pure and Applied* (1865) and the *Philosophy and Practice of Punctuation* (1868), he was also involved in patents for improved railway equipment. In the mid 1870s Emmens branched out as a financier under the name Emmens Brothers, a concern which led to Emmens' bankruptcy in May 1875. During the mid 1870s Emmens was associating with sometime diplomat and fraudster Horatio Nelson Lay who had interests in Tamar Valley mining. It was around this time that Emmens started taking an interest in Tamar Valley mining; his interests including Wheal Newton, Holmbush and Greenhill Brick and Arsenic Works. In 1875 the Emmens Process (hydrometallurgical extraction by roasting and chloridisation) was being used to treat low grade Holmbush copper ores. In September 1876 Emmens was chairman of the Profit Union Ltd which went into liquidation. In October 1884, when the Electric Light Company was wound up, Emmens was Chairman. Sometime after this date Emmens emigrated to the United States where he became a member of numerous learned societies. He developed a process for treating nickel matte. By December 1890 Emmens had developed an explosive known as Emmensite. By the end of the decade he had developed a secret process by means of which he claimed to be able to turn Mexican silver dollars into gold via an intermediate stage known as argentaurum. Some have suggested that Emmens' process was a cover for fencing stolen gold; others have called him the last alchemist.

Note 11

The suspension of work at Richards' Shaft must have been very short lived; no mention being made of it in the Estate Mining Reports. The Company would have had to have had the Duke's permission to suspend work before the compulsory fifteen fathoms had been driven.

Note 12

Josiah Paull sums up the issues succinctly:

you are aware the old system – abolished about 7 years ago consisted of 8 months of four weeks and four months of five weeks – for the year of 52 weeks. In practice it resulted in the monthly men getting no more in the five weeks than they did in the four, and to men on contract work the price was always lowered when a five week month was to be provided for in the lettings so that contract men did not and were not allowed to earn five weeks wages in the month of that duration.[263]

Note 13

A description of Maze Monday from the miners' viewpoint was included in a "memorial" presented to the May 1879 meeting by the miners of Devon Great Consols:

> There remains for us now to deal with the question of Mondays after the setting. There seems to be a great deal of misunderstanding in reference to this question, and it requires a great amount of explanation to set the matter clear before the shareholders. They have been told no doubt, that the miners keep holiday on this day, when the truth is it is the most busy day in the whole taking. On this day he has to return his tools and take them out and fit them for use; take out his materials and return that which he has left from the former taking, for at the end of every taking half of the pares are broken up, some joining other pares, some going abroad, and some to other mines; other men are coming from adjoining mines to this, but before they can do so they have to settle their affairs at the mines they are about to leave, then bring their clothes to this mine, in most instances five or six miles. It may not be known to the shareholders that the men of this mine have upwards of five tons of steel charged to their account, the value of it might be about £100, which if we do not return on this day, you exact a fine besides the value of the steel, of £400 and we have paid at that rate for every pound of steel that has been lost or mislaid from the commencement of these mines, and when it is known that this weight of steel means one thousand drills, all stamped wit the marks of different pares, it will readily be seen that to look over the whole for lost drills entails a great deal of time and trouble and as to performing any other work on this day it is out of the question. And the miners look upon it as absurd and up to this time have always treated it with ridicule. We have endeavoured to the best of our abilities to place the case before the shareholders as clear as possible.[264]

Note 14

In analysing the "Maze Monday" dispute the *Tavistock Gazette* suggested that the issue went deeper than the current round of cuts and the abolition of "Maze Monday". The *Gazette* offered the opinion that the Devon Great Consols men had common cause and saw parallels with the men who were striking at Great Laxey on the Isle of Man. The Great Laxey strike was a response to a new code of regulations introduced by the Directors one of whom was none other than Peter Watson. The new code of regulations compelled the Great Laxey men to give up all holidays except Christmas and Easter. In addition the miners were obliged to work six eight hour shifts a week, this being in addition to the time spent getting to and from their work places which could be up to a couple of hours.[265] The *Isle of Man Times* reported that:

> Next step an iron cage was introduced, into which the poor "slaves" were thrust and there kept under the "taskmasters" eye and "lock secure" until the "shift" to the very last second was completed; and no matter how wet and shivering they might be in the cold, windy level, they must stand until the iron bars are withdrawn under penalties of fines or dismissal.[266]

Note 15

The *Tavistock Gazette* of 20th June, 1879 states that it was the western men who took a one month setting in May to bring them in line with the Wheal Emma. This is out of sync with previous reports of settings: Eastern settings on odd months, Western on even.

Note 16

Moses Bawden 1834-1916 was a mining man to the core of his being. Born in Camborne in 1834, his father was an agent at Tincroft Mine. He started work at the age of ten at Godolphin Mine as an assistant to both the surface and underground Agents. He subsequently served an apprenticeship in South Wales with Messrs. Williams, Foster and Co., the copper smelters which, in later years, must have given him a fairly unique perspective on the copper mining industry. After his apprenticeship he became Purser and assistant Agent at Redmoor Mine near Callington, giving him an early taste of the Tamar Valley mining field. Bawden also became an Agent at Tincroft. Following his spell at Tincroft Bawden took up an appointment at the San Fernando Lead Mine at Linares in Spain. On his return to Cornwall he took up purserships at various Redruth mines including Wheal Busy, Boscawen, Hallenbeagle, Great North Downs and Great Briggan.[267] Bawden left Redruth to take up the Purser's job at Devon Great Consols in 1879.

During the following decades Bawden became a key figure in the local mining scene, becoming involved in numerous local mines including Drakewalls, Wheal Crebor and various Dartmoor mines such as Hexworthy. Whilst Bawden's interests focussed on the Tamar Valley his view was anything but parochial, for example in 1884 he was appointed UK agent for Rio Tinto mines and also the Quedraba Railway, Land and Copper Company.[268] Beyond his mining interests Moses Bawden was a Cornish J.P., sitting, on occasion, at Callington. In 1914 Bawden presented a paper, largely dealing with Devon Great Consols, to the Devonshire Association. During his later years he lived quietly at Tamar View (the house he had rented from Devon Great Consols during his pursurship) where he died on Wednesday, January 19th 1916 in his 83rd year.[269]

Chapter 7 references

1. *Tavistock Gazette* 10 June 1870
2. *Tavistock Gazette* 10 June 1870
3. *Tavistock Gazette* 10 June 1870
4. *Tavistock Gazette* 10 June 1870
5. Devon Record Office document 8187 Bedford Estate Mine Reports 1867-1872
6. Devon Record Office document 8187 Bedford Estate Mine Reports 1867-1872
7. Devon Record Office document 8187 Bedford Estate Mine Reports 1867-1872
8. *Tavistock Gazette* 10 June 1870
9. Devon Record Office document 8187 Bedford Estate Mine Reports 1867-1872
10. Devon Record Office document 8187 Bedford Estate Mine Reports 1867-1872
11. *Tavistock Gazette* 10 June 1870
12. Devon Record Office document 8187 Bedford Estate Mine Reports 1867-1872
13. Devon Record Office document 8187 Bedford Estate Mine Reports 1867-1872
14. Devon Record Office document 8187 Bedford Estate Mine Reports 1867-1872
15. *Tavistock Gazette* 10 June 1870
16. Devon Record Office document 8187 Bedford Estate Mine Reports 1867-1872
17. Devon Record Office document 8187 Bedford Estate Mine Reports 1867-1872
18. Devon Record Office document 8187 Bedford Estate Mine Reports 1867-1872
19. Devon Record Office document 8187 Bedford Estate Mine Reports 1867-1872
20. Devon Record Office document 8187 Bedford Estate Mine Reports 1867-1872
21. Devon Record Office document 8187 Bedford Estate Mine Reports 1867-1872
22. Devon Record Office document 8187 Bedford Estate Mine Reports 1867-1872
23. *Tavistock Gazette* 21 April 1871
24. *Tavistock Gazette* 28 April 1871
25. *Tavistock Gazette* 28 April 1871

26. *Mining Journal* 13 May 1871
27. *Tavistock Gazette* 5 May 1871
28. *Mining Journal* 13 May 1871
29. *Mining Journal* 13 May 1871
30. *Tavistock Gazette* 12 May 1871
31. *Mining Journal* 27 May 1871
32. *Tavistock Gazette* 26 May 1871
33. *Tavistock Gazette* 2 June 1871
34. Devon Record Office document 8187 Bedford Estate Mine Reports 1867-1872
35. Devon Record Office document 8187 Bedford Estate Mine Reports 1867-1872
36. Devon Record Office document 8187 Bedford Estate Mine Reports 1867-1872
37. Devon Record Office document 8187 Bedford Estate Mine Reports 1867-1872
38. Devon Record Office document 8187 Bedford Estate Mine Reports 1867-1872
39. Devon Record Office document 8187 Bedford Estate Mine Reports 1867-1872
40. Devon Record Office document 8187 Bedford Estate Mine Reports 1867-1872
41. Devon Record Office document 8187 Bedford Estate Mine Reports 1867-1872
42. Devon Record Office document 8187 Bedford Estate Mine Reports 1867-1872
43. Devon Record Office document 8187 Bedford Estate Mine Reports 1867-1872
44. Devon Record Office document 8187 Bedford Estate Mine Reports 1867-1872
45. Devon Record Office document 8187 Bedford Estate Mine Reports 1867-1872
46. Devon Record Office document 8187 Bedford Estate Mine Reports 1867-1872
47. *Mining Journal* 8 June 1872
48. *Tavistock Gazette* 17 May 1872
49. *Tavistock Gazette* 24 May 1872
50. *Tavistock Gazette* 24 May 1872
51. *Tavistock Gazette* 24 May 1872
52. *Mining Journal* 4 January 1873
53. *Tavistock Gazette* 24 May 1872
54. *Tavistock Gazette* 24 May 1872
55. *Tavistock Gazette* 24 May 1872
56. *Tavistock Gazette* 24 May 1872
57. *Tavistock Gazette* 24 May 1872
58. Wikipedia: William Russell, 8th Duke of Bedford & Francis Russell, 9th Duke of Bedford
59. *Mining Journal* 8 June 1872
60. *Mining Journal* 10 August 1872
61. *Mining Journal* 31 August 1872
62. *Mining Journal* 30 November 1872
63. Devon Record Office document 8187 Bedford Estate Mine Reports 1867-1872

64. Devon Record Office document 8187 Bedford Estate Mine Reports 1867-1872
65. Devon Record Office document 8187 Bedford Estate Mine Reports 1867-1872
66. *Tavistock Gazette* 18 October 1872
67. Devon Record Office document 8187 Bedford Estate Mine Reports 1867-1872
68. *Tavistock Gazette* 18 October 1872
69. Devon Record Office document 8187 Bedford Estate Mine Reports 1867-1872
70. Devon Record Office document 8187 Bedford Estate Mine Reports 1867-1872
71. Devon Record Office document 8187 Bedford Estate Mine Reports 1867-1872
72. *Tavistock Gazette* 18 October 1872
73. Devon Record Office document 8187 Bedford Estate Mine Reports 1867-1872
74. *Mining Journal* 30 November 1872
75. *Tavistock Gazette* 20 December 1872
76. Devon Record Office document 8187 Bedford Estate Mine Reports 1873-1878
77. *Mining Journal* 31 May 1873
78. Devon Record Office document 8187 Bedford Estate Mine Reports 1873-1878
79. Devon Record Office document 8187 Bedford Estate Mine Reports 1873-1878
80. Devon Record Office document 8187 Bedford Estate Mine Reports 1873-1878
81. Devon Record Office document 8187 Bedford Estate Mine Reports 1873-1878
82. Devon Record Office document 8187 Bedford Estate Mine Reports 1873-1878
83. Devon Record Office document 8187 Bedford Estate Mine Reports 1873-1878
84. Devon Record Office document 8187 Bedford Estate Mine Reports 1873-1878
85. Devon Record Office document 8187 Bedford Estate Mine Reports 1873-1878
86. *Mining Journal* 29 November 1873
87. *Mining Journal* 29 November 1873
88. *Mining Journal* 29 November 1873
89. Devon Record Office document 8187 Bedford Estate Mine Reports 1873-1878
90. Devon Record Office document 8187 Bedford Estate Mine Reports 1873-1878
91. *Mining Journal* 31 May 1873
92. Devon Record Office document 8187 Bedford Estate Mine Reports 1873-1878
93. *Mining Journal* 30 May 1874
94. Devon Record Office document 8187 Bedford Estate Mine Reports 1873-1878
95. Devon Record Office document L1258 DGC Leases 1874
96. Devon Record Office document 8187 Bedford Estate Mine Reports 1873-1878
97. Devon Record Office document L1258 Mining Lease 1874
98. Devon Record Office document 8187 Bedford Estate Mine Reports 1873-1878
99. Devon Record Office document 8187 Bedford Estate Mine Reports 1873-1878
100. *Mining Journal* 28 November 1874
101. Devon Record Office document 8187 Bedford Estate Mine Reports 1873-1878

102. *Mining Journal* 28 November 1874
103. *Mining Journal* 28 November 1874
104. Devon Record Office document 8187 Bedford Estate Mine Reports 1873-1878
105. Devon Record Office document 8187 Bedford Estate Mine Reports 1873-1878
106. Devon Record Office document 8187 Bedford Estate Mine Reports 1873-1878
107. Devon Record Office document 8187 Bedford Estate Mine Reports 1873-1878
108. Devon Record Office document 8187 Bedford Estate Mine Reports 1873-1878
109. Devon Record Office document 8187 Bedford Estate Mine Reports 1873-1878
110. Devon Record Office document 8187 Bedford Estate Mine Reports 1873-1878
111. Devon Record Office document 8187 Bedford Estate Mine Reports 1873-1878
112. *Mining Journal* 29 May 1875
113. Devon Record Office document 8187 Bedford Estate Mine Reports 1873-1878
114. Devon Record Office document 8187 Bedford Estate Mine Reports 1873-1878
115. Devon Record Office document 8187 Bedford Estate Mine Reports 1873-1878
116. Devon Record Office document 8187 Bedford Estate Mine Reports 1873-1878
117. *Tavistock Gazette* 22 October 1875
118. Devon Record Office document 8187 Bedford Estate Mine Reports 1873-1878
119. *Mining Journal* 27 November 1875
120. *Mining Journal* 27 November 1875
121. Devon Record Office document 8187 Bedford Estate Mine Reports 1873-1878
122. Devon Record Office document 8187 Bedford Estate Mine Reports 1873-1878
123. Devon Record Office document 8187 Bedford Estate Mine Reports 1873-1878
124. Devon Record Office document 8187 Bedford Estate Mine Reports 1873-1878
125. *Tavistock Gazette* 1 October 1875
126. *Tavistock Gazette* 22 October 1875
127. Devon Record Office document 8187 Bedford Estate Mine Reports 1873-1878
128. *Tavistock Gazette* 15 January 1875
129. Devon Record Office document 8187 Bedford Estate Mine Reports 1873-1878
130. Devon Record Office document 8187 Bedford Estate Mine Reports 1873-1878
131. Devon Record Office document 8187 Bedford Estate Mine Reports 1873-1878
132. Devon Record Office document 8187 Bedford Estate Mine Reports 1873-1878
133. Devon Record Office document 8187 Bedford Estate Mine Reports 1873-1878
134. *Mining Journal* 2 December 1876
135. Devon Record Office document 8187 Bedford Estate Mine Reports 1873-1878
136. Devon Record Office document 8187 Bedford Estate Mine Reports 1873-1878
137. Devon Record Office document 8187 Bedford Estate Mine Reports 1873-1878
138. *Mining Journal* 2 December 1876
139. Devon Record Office document 8187 Bedford Estate Mine Reports 1873-1878

140. Devon Record Office document 8187 Bedford Estate Mine Reports 1873-1878
141. Devon Record Office document 8187 Bedford Estate Mine Reports 1873-1878
142. *Mining Journal* 2 December 1876
143. Devon Record Office document 8187 Bedford Estate Mine Reports 1873-1878
144. *Mining Journal* 26 May 1877
145. Devon Record Office document 8187 Bedford Estate Mine Reports 1873-1878
146. *Mining Journal* 1 December 1877
147. *Mining Journal* 1 December 1877
148. Devon Record Office document 8187 Bedford Estate Mine Reports 1873 - 1878
149. Josiah Thomas to Josiah Paull, November 13th 1877, *in* Devon Record Office document 8187 Bedford Estate Mine Reports 1873-1878
150. Devon Record Office document 8187 Bedford Estate Mine Reports 1873-1878
151. Devon Record Office document 8187 Bedford Estate Mine Reports 1873-1878
152. Devon Record Office document 8187 Bedford Estate Mine Reports 1873-1878
153. Devon Record Office document 8187 Bedford Estate Mine Reports 1873-1878
154. Devon Record Office document 8187 Bedford Estate Mine Reports 1873-1878
155. Devon Record Office document 8187 Bedford Estate Mine Reports 1873-1878
156. Devon Record Office document 8187 Bedford Estate Mine Reports 1873-1878
157. Devon Record Office document 8187 Bedford Estate Mine Reports 1873-1878
158. Devon Record Office document 8187 Bedford Estate Mine Reports 1873-1878
159. Devon Record Office document 8187 Bedford Estate Mine Reports 1873-1878
160. *Mining Journal* 30 March 1878
161. *Mining Journal* 13 April 1878
162. *Tavistock Gazette* 9 March 1878
163. *Mining Journal* 6 April 1878
164. *Mining Journal* 13 April 1878
165. *Mining Journal* 13 April 1878
166. *Mining Journal* 13 April 1878
167. Devon Record Office document 8187 Bedford Estate Mine Reports 1873-1878
168. *Mining Journal* 27 April 1878
169. Devon Record Office document 8187 Bedford Estate Mine Reports 1873-1878
170. *Mining Journal* 27 April 1878
171. *Mining Journal* 25 May 1878
172. *Mining Journal* 27 April 1878
173. *Mining Journal* 27 April 1878
174. *Mining Journal* 18 May 1878
175. *Mining Journal* 27 April 1878
176. *Mining Journal* 27 April 1878

177. *Mining Journal* 4 May 1878
178. *Mining Journal* 4 May 1878
179. *Mining Journal* 11 May 1878
180. *Mining Journal* 18 May 1878
181. *Mining Journal* 25 May 1878
182. *Mining Journal* 25 May 1878
183. *Mining Journal* 1 June 1878
184. Devon Record Office document 8187 Bedford Estate Mine Reports 1873-1878
185. Devon Record Office document 8187 Bedford Estate Mine Reports 1873-1878
186. Devon Record Office document 8187 Bedford Estate Mine Reports 1873-1878
187. *Mining Journal* 1 June 1878
188. *Mining Journal* 8 June 1878
189. *Mining Journal* 15 June 1878
190. *Mining Journal* 15 June 1878
191. *Mining Journal* 15 June 1878
192. Devon Record Office document 8187 Bedford Estate Mine Reports 1873-1878
193. *Mining Journal* 22 June 1878
194. *Tavistock Gazette* 28 March 1879
195. *Mining Journal* 22 June 1878
196. *Mining Journal* 25 May 1878
197. Devon Record Office document 8187 Bedford Estate Mine Reports 1873-1878
198. Devon Record Office document 8187 Bedford Estate Mine Reports 1873-1878
199. *Mining Journal* 22 June 1878
200. *Mining Journal* 30 November 1878
201. *Tavistock Gazette* 23 May 1879
202. Devon Record Office document 8187 Bedford Estate Mine Reports 1878-1886
203. Devon Record Office document 8187 Bedford Estate Mine Reports 1878-1886
204. *Mining Journal* 6 July 1878
205. *Mining Journal* 31 August 1878
206. *Mining Journal* 21 September 1878
207. *Mining Journal* 5 October 1878
208. *Mining Journal* 12 October 1878
209. Devon Record Office document 8187 Bedford Estate Mine Reports 1873-1878
210. *Mining Journal* 21 September 1878
211. Devon Record Office document 8187 Bedford Estate Mine Reports 1878-1886
212. Devon Record Office document 8187 Bedford Estate Mine Reports 1878-1886
213. Devon Record Office document 8187 Bedford Estate Mine Reports 1878-1886
214. *Tavistock Gazette* 30 May 1879

215. *Mining Journal* 14 June 1879
216. *Mining Journal* 14 June 1879
217. *Tavistock Gazette* 23 May 1879
218. *Mining Journal* 11 January 1879
219. *Mining Journal* 25 January 1879
220. Devon Record Office document 8187 Bedford Estate Mine Reports 1878-1886
221. Devon Record Office document 8187 Bedford Estate Mine Reports 1878-1886
222. Devon Record Office document 8187 Bedford Estate Mine Reports 1878-1886
223. *Mining Journal* 25 January 1879
224. *Mining Journal* 11 January 1879
225. *Mining Journal* 25 January 1879
226. *Mining Journal* 15 February 1879
227. *Mining Journal* 5 April 1878
228. *Mining Journal* 29 March 1879
229. *Mining Journal* 29 March 1879
230. *Tavistock Gazette* 28 March 1879
231. *Tavistock Gazette* 30 May 1879
232. *Mining Journal* 29 March 1879
233. *Tavistock Gazette* 28 March 1879
234. *Tavistock Gazette* 28 March 1879
235. *Mining Journal* 26 April 1879
236. *Tavistock Gazette* 25 April 1879
237. *Tavistock Gazette* 23 May 1879
238. *Tavistock Gazette* 30 May 1879
239. *Tavistock Gazette* 30 May 1879
240. *Tavistock Gazette* 30 May 1879
241. *Tavistock Gazette* 20 June 1879
242. *Tavistock Gazette* 20 June 1879
243. *Tavistock Gazette* 20 June 1879
244. *Tavistock Gazette* 27 June 1879
245. *Tavistock Gazette* 27 June 1879
246. *Mining Journal* 29 November 1879
247. *Tavistock Gazette* 21 November 1879
248. *Tavistock Gazette* 21 November 1879
249. *Mining Journal* 11 October 1879
250. *Tavistock Gazette* 21 November 1879
251. *Tavistock Gazette* 10 October 1879
252. *Mining Journal* 29 November 1879

253. *Tavistock Gazette* 21 November 1879
254. *Mining Journal* 13 December 1879
255. Devon Record Office document 8187 Bedford Estate Mining Reports 1899-1906
256. *Tavistock Gazette* 20 June 1879
257. Devon Record Office document 8187 Bedford Estate Mining Reports 1899-1906
258. Devon Record Office document 8187 Bedford Estate Mine Reports 1873-1878
259. Devon Record Office document 8187 Bedford Estate Mine Reports 1873-1878
260. *Mining Journal* 9 June 1892
261. Mining Magazine Vol. 4 p327
262. *Tavistock Gazette* 9 August 1878
263. Devon Record Office document 8187 Bedford Estate Mine Reports 1873-1878
264. *Tavistock Gazette* 30 May 1879
265. *Tavistock Gazette* 25 April 1879
266. *Tavistock Gazette* 25 April 1879
267. *Mining Journal* 30 July 1892
268. *Mining Journal* 30 July 1892
269. *Tavistock Gazette* 21 January 1916

Chapter 8

1880-1889: The search for tin resumed

"it would be cowardice if we did not sink this shaft, and try what is underneath us there" – Peter Watson, November 1883. (Mining Journal, November 24th, 1883)

1880

The 1880s started well for the mine; in January the *Tavistock Gazette* announced that news that the company had managed to sell not only all of its existing stocks of arsenic but also the make for the coming year, the purchaser being Mr Field. The sale of arsenic and the rumours of a good dividend resulted in a rush for Devon Great Consols shares, £70,000 being added to the market value of the mine within a week which, in the words of the *Tavistock Gazette*, was "a rise unprecedented in the history of mining".[1] On 22nd January Josiah Paull, somewhat cynically, commented:

> By puffing the mine on the share market more than the rise in copper and arsenic the value of the concern has been enormously increased in the present month.[2]

Speaking at an E. G. M. on Wednesday 28th January Peter Watson commented that the contract with Mr Field was the largest ever entered into by the Company.[3] The rumours of a dividend proved to be true, eight shillings being paid on each share in February 1880.[4]

Whilst the prospects and condition of the mine were discussed at the 28th January Extraordinary General Meeting this was not the purpose of the meeting. The recent death of W. A. Thomas and the resignation of Thomas Morris meant that the company was obliged to appoint two new trustees. The role of trustee was significant in that the trustees had the authority, amongst other things, to sign leases. It will be recalled that the leases currently held by the company had been signed by W. A. Thomas and Thomas Morris. Given the situation occasioned by Thomas' death and Morris' resignation the appointment of new trustees was somewhat urgent, the positions being filled by Peter Watson and Stanley Morris.[5] The matter was discussed at the May 1880 half yearly meeting, Watson informing the meeting that new leases had been drawn up and were awaiting signature at the Bedford Estate Office.[6] The 1874 leases were formally surrendered on 21st June 1880. The new leases were signed by Peter Watson and Hugh Stanley Morris on the 22nd June 1880; to run for twenty one years from the 25th March 1872 to the 25th March 1893.[7]

On the mine, rock drills had been successfully put to work by mid January 1880.[8] At the 28th January E.G.M. Watson informed shareholders that:

> The sum of £1,000 had been spent on boring machinery, and the expenditure was still going on; the apparatus was started about a month ago, and in the first month, even with the imperfect handling of the men, who had not been accustomed to work boring machinery, they had been enabled to drive, in hard rock at the rate of about 5 fms a month, or nearly three times more rapidly than if the same work had been performed by hand labour.[9]

By the end of March 1880 rock drills were in use in two locations on New South Lode: The 137 east, east of Friend's crosscut and Dawe's crosscut on the 190.[10] Commenting in the *Mining Journal* Isaac Richards noted that progress was twice that of hand drilling.[11] In the six months up to May 1880 the company had purchased "seven rock drills, one air compressor and receiver and a large quantity of piping".[12] Isaac Richards was able to report to the May 1880 half yearly meeting that rock drills were in use on the 137 and the 190 and that:

> Their progress at these points has considerably improved, and this desirable and effective power will be of great importance in the future development of the vast extent of mineral ground still remaining to be explored throughout the mines.[13]

In June it was noted that a Darlington machine was in use on the 137 east, a Barrow in the 190 crosscut and a further machine in the 190 west.[14]

The "imperfect handling" of rock drills by the Devon Great Consols men must have been something of an issue. In July 1880 Josiah Paull reported that the driveage of two of the "leading levels" in Wheal Emma had been let on contract to "the owners of Darlington's Boring Machine".[15]

In his September 1880 report Paull noted that an Eclipse machine was in use on the 137 east.[16]

Towards the end of April 1880 the *Mining Journal* reported further labour unrest on the mine. This appears to have had its origin "in a desire on the part of these men to keep holidays as accustomed". The extent of the difficulty appears to have been very limited, indeed the *Mining Journal* doubted that the action could "fairly be termed a strike".[17] Certainly it was all over by the end of April, the *Mining Journal* correspondent noting that "the difficulty at Devon Great Consols – it was never worth while calling it a strike – has come to an end almost as soon as it arose".[18]

The May 1880 half yearly meeting was held on Wednesday 26th May at the company offices at Austin Friars. Unlike recent meetings there was cause for celebration. The average price that the mine was receiving for its copper had risen to £2 14s a ton. The sale of arsenic during the six months to the 30th April 1880 had realised a very satisfactory £15, 517, this included one third of the money from the January 1880 contract. Peter Watson, in the chair, was at his bombastic best and with good reason, for the mine seemed to be pulling itself out of the hard times of the 1870s. In describing Devon Great Consols Watson uttered his oft quoted statement: "It is not an ordinary mine, but a mine of mines." The meeting shared Watson's optimism and enthusiasm voting him an increase in his managing directors remuneration from £250 to £500 per annum. The shareholders were not left out of this bonanza, after the meeting the directors met and resolved to pay a dividend of 10s a share.[19,20,21] Watson's credit was riding high, not only with his fellow shareholders but also in the wider mining world, the *Mining Journal* being in particularly eulogistic voice:

> just in the nick of time, when things were looking about as black as they possibly could, Mr Peter Watson was induced to undertake the administration of affairs, and under his energetic and judicious supervision, ably and loyally backed by the heads of departments at the mine, confidence has been restored to the shareholders, harmony and goodwill amongst the work people, and splendid quarterly dividends have been resumed...[22]

In Wheal Josiah some exploration was undertaken on New South Lode, the 115, 130 and 144 all receiving attention during the year. Unfortunately none of these levels cut good ore ground.[23]

At Wheal Emma work was ongoing during 1880 to develop New South Lode at depth, the 137 and the 190 having being mentioned in connection with the introduction of rock drills. In early December 1880 Isaac Richards reported that the lode on the 137 east was eight feet wide and of "very fine description, being composed of capel, quartz, peach, fluor, prian, and small quantities of copper and mundic ores".[24] By 16th June Dawe's crosscut on the 190 had cut New South Lode.[25] On July 1st Isaac Richards commented that the lode on the 190 was six feet wide.[26]

During the first half of 1880 Railway Shaft was being sunk below the 175, the 190 having been reached by the end of March.[27] The shaft between the 175 and the 190 had been divided, cased and equipped with both ladder and skip roads by June 16th.[28] The shareholders at the May 1880 meeting were informed that a 190 crosscut would be driven north from Railway Shaft to intersect New South Lode[29] and by June 12th Isaac Richards announced that work had started driving crosscuts both north and south.[30] By July 1st it was noted that the lode on the 190 was four feet wide and yielding three tons of copper and four of mundic per fathom.[31] The 190 east of Railway Shaft had communicated with the 190 west of New Shaft by September 9th 1880.[32] The communication between the two shafts having been made, attention turned to the 190 west of Railway Shaft, the *Mining Journal* noting in a report dated 16th September that men were engaged in installing air tubing prior to driving with a Barrow rock drill.[33]

The deepest level on New South Lode in 1880 was the 205. At the November 1880 meeting Isaac Richards reported to shareholders that the lode on the 205 west of New Shaft was large and of a very fine character.[34] Richards' intention to prosecute the 205 seriously was demonstrated in December when he announced that a Darlington drill was to be used on this level.[35]

Whilst New South Lode was the focus of a great deal of attention in 1880 it was not the only point of interest, considerable resources being committed to the reopening of Watson's Mine. By the early May pumping had commenced, Captain Richards commenting on 8th May that the pumping machinery had been completed and connected via a run of flat rods 280 fathoms long to "Richards' large waterwheel".[36,37] By the end of May the mine had been drained at least as far as the 52, Moses Bawden exhibiting specimens of ore from this level at the May meeting.[38] No time was lost in recommissioning the shaft; in a *Mining Journal* report dated 24th of June Richards wrote that work was progressing well, the hauling machine was nearly complete and would be ready for work in the next couple of days.[39] Isaac Richards was as good as his word, by the 1st July work had started sinking Watson's Engine Shaft below the 64.[40] Sinking proceeded rapidly, by 3rd September the shaft was 14 fathoms below the 64 and work was underway to case and divide the shaft and install a ladder road.[41] This had been completed by 16th September and the lode had been cut on what would become the 76 fathom level (see Note 1).[42] Shareholders at the November 1880 meeting were informed that the lode on the 76 was three feet wide and of a promising character.[43]

The November Half Yearly meeting was held on Wednesday 24th. The spirit of optimism was still very much in evidence bolstered by the income from the January sale of arsenic, the company receiving £15,390 10s 5d. The price of copper had fallen slightly, the mine receiving an average price of £2 12s 3d per ton. Shareholders were informed that the new leases had been signed. Of particular significance was a reduction in the dues demand by the Duke of Bedford, a reduction from 1/12th to 1/18th having been agreed.[44] An interesting discussion took place regarding the reasons why mechanical rock breakers had not been introduced in favour of hand labour for primary reduction. It was argued that although a rock breaker would pulverise ore at 1/12th price of hand labour, a machine had not the judgement of a hand picker. Moses Bawden made the point that this was very important in regard to low grade ores; he argued that "boys and girls exercise judgement in separating minerals" recovering between 5,000-6,000 tons of per annum with a value of £18,000, money that would be lost if the process were to be mechanised.[45]

A dividend of six shillings was announced at the meeting.[46]

During 1880 no exploratory work was carried out on either Main Lode or South Lode.[47]

Copper precipitation continued: during 1880 28 tons of precipitate was produced.[48]

The dividends of 1880 did much to bolster the Company's share price, share value increasing from £6 10s in December 1879 to £16 in December 1880, representing a £97,280 increase in the share value of the mine.[49]

1881

During the dying days of 1880 and the first weeks of 1881 the Tamar Valley experienced extremely bad weather. Around New Year heavy rain caused serious floods which "threatened to overpower" most of the mine's pumping waterwheels.[50] On 30th December 1880 Isaac Richards, in one of his regular *Mining Journal* reports, wrote that they were having problems keeping the Counthouse Shaft section of the mine drained due to the heavy rain.[51] Almost immediately as this crisis had been dealt with, snow and freezing weather set in; in a report dated 20th January Isaac Richards commented that surface operations had been brought to a standstill and that they were having great difficulties keeping the pumping wheels at work as a result of a snow storm.[52] It was not just the mine that was effected, the weather had paralysed the whole of the valley. Vessels were frozen in the tidal Tamar which meant that supplies of material, such as coal, could not reach the mine and, likewise, the mine could not dispatch its arsenic.[53] However by 27th January a rapid thaw had set in and operations returned to normal, all that remained was to repair the damage caused by the weather which came to £500.[54,55,56]

In a letter dated 23rd February 1881 Peter Watson wrote to the shareholders informing them that the directors were not in a position to pay a dividend, a key reason being that as yet they had not been able to secure a contract for the sale of that years' make of arsenic. Beyond the failure to sell the arsenic further reasons were cited in a letter to the directors and shareholders dated 17th February and signed by Isaac Richards and Moses Bawden. Richards and Bawden outlined a host of problems they were experiencing on the mine. Foremost on their list was the recent poor weather which had significantly interrupted operations. Whilst the weather was at the fore, of longer term importance was the worrying intelligence that some of the "exploratory levels have not been producing as much copper ore and mundic as heretofore", the 205 on New South Lode proving particularly disappointing. The mine's ageing infrastructure was also an issue, the railway had been extensively re-laid, Plunger Shaft required re-timbering at a cost of £400. The sum of £2,000 had been spent on the arsenic works; the flues and condensers had been rebuilt whilst the calciners and furnaces were repaired. However not all was doom and gloom, Richards and Bawden held out the promise of future prosperity, drawing attention to the 137 which was exploring the Eastern Ground at depth. The Eastern Ground, secured at great expense, was unworked and according to Richards and Bawden, a discovery here was imminent. They also felt that Watson's Mine, currently sinking below the 76, would prove rich at depth.[57] Whilst the agents needed to reassure both the directors and shareholders that future prospects were good they might have been well advised to recall the old miners' adage that "you can't see further than the end of your pick"!

In April 1881 it was noted that a Darlington boring machine was in use on the 115 east on New South Lode in Wheal Emma.[58]

The May 1881 half yearly meeting was held on Wednesday 25th at the Company's London Office; Peter Watson, as usual, presiding. To add to the company's problems the copper price had continued to fall dropping from an average of £2 12s 6d in the previous half year to £2 per ton. On a more positive note the mine had finally been able to make two sales of arsenic.[59] In spite of the sales of arsenic Watson informed shareholders that they would not, in consequence of the recent heavy expenditure, be receiving a dividend at the present time.[60]

A full report on the mine, prepared by Isaac Richards and dated 7th May 1881, was presented to the May meeting. Richards' report outlined work undertaken at surface, referring to Plunger Shaft, the railway and the arsenic works. As rebuilt the arsenic works comprised seven calciners, three refineries, 5,429 feet of flues and an arsenic mill driven by a steam engine.[61]

During the preceding six months a monthly average of 192 tutworkers had been employed on the mine earning an average of £3 10s 8d a month.[62]

At Wheal Josiah pumping at Counthouse Shaft had been proving problematical, as had been highlighted during the bad weather at New Year. Counthouse Shaft had been pumped by pitwork driven by the Agnes Shaft water wheel, which had proved inadequate.[63] In March 1881 the 144 east of Hitchins' Shaft had communicated with the 144 west of Agnes Shaft via Penrose's winze, apart from improving ventilation in this section of the mine it enabled some of the Agnes Shaft water to be diverted to Richards' Shaft to ease the burden on the Agnes Shaft wheel.[64] Evidently this had not proved a solution as by the time of Richards' report the Counthouse Shaft pitwork had been connected to the Richards' Shaft waterwheel via 41 fathoms of additional flat rods and an angle bob at a cost of £310. Richards reported that ore reserves in Wheal Josiah amounted to 1,363 tons.[65]

Wheal Emma was still very much the focus of development work with reserves of 19, 289 tons in May 1881. At Inclined Shaft the 190 had cut the lode at a number of points which although large was not very productive. The 137 east of Friend's crosscut was cutting good ore ground. At New Shaft the 205 had communicated with the 190 via Hockaday's winze improving ventilation, the 100 was also active. Railway Shaft was sinking and had reached ten fathoms below the 190. The 175 west and 160 west were said to be promising and Richards predicted that good ore would be encountered between the two levels.[66]

Peter Watson told the May 1881 meeting that they also intended to drive westwards at depth, back towards the Great Crosscourse, in the hope of picking up anything that had been previously missed.[67]

With regard to Watson's, Isaac Richards informed shareholders that work was concentrated on sinking the shaft; plats had been cut at 76 and a permanent plunger lift had been installed to the 40.[68]

Sinking Watson's Engine Shaft continued throughout the year, by 1st September the shaft had reached the 88 and work had started on casing and dividing the shaft down from the 76.[69] By the beginning of August a new shaft, known as Western Shaft, had been started at Watson's. The new shaft lay 140 fathoms west of Engine Shaft.[70] After sinking through loose ground Western Shaft the lode was cut at a depth of 12 fathoms and nine inches in early October. Isaac Richards described the lode as comprising of fine capel, quartz, peach "and a little of both copper ores of good quality".[71] Unfortunately work at Western Shaft was suspended at the end of October due to problems with water, to remedy the problem it was intended to install a lift pump driven by a line of flat rods attached to the Engine Shaft flat rod run.[72,73]

Work on and in the vicinity of Railway Shaft continued throughout the summer of 1881; by 14th July the shaft had reached the 205 and work had started on driving a crosscut north to insect the lode.[74] On 4th August the 205 crosscut, referred to as Bray's, was reported to be driving in highly mineralised ground.[75] In spite of such optimism work on the 205 north was suspended by October 6th, not having found the lode.[76]

Peter Watson had bad news for the shareholders at the November 1881 half yearly meeting, announcing to the meeting that the average price the of copper ore had continued to fall during the preceding six months, the mine now only receiving £1 14s 11d per ton. Returning to a recurring theme Watson put the blame firmly at the door of the South Wales smelters and the price they were paying at the Cornish ticketings. The shareholders were informed that the directors were exploring other avenues for the sale of their ore. On the mine the decision had been taken to speed up exploration and development work. To do this Moses Bawden proposed to suspend all hand labour development work, employing rock drills instead, Bawden commenting that rock drills were laying ground open twice as fast as hand labour. The men who were formerly engaged in hand development would be "put to breaking copper".[77,78]

1881 proved to be a very poor year in terms of discoveries. In December Josiah Paull expressed grave concerns regarding future prospects:

> Devon Great Consols has been fairly well worked taking into account the acceleration made by the use of Boring machines. The results have been very discouraging and I am led to fear that the New South Lode has reached the limit of productiveness in both length and depth. The only hope I venture now to indulge is that Watson's may turn out well if the explorations are persevered with. It is very much to be desired that a good discovery may be made in that direction and that very soon.[79]

1882

The shareholders received some encouraging news in January 1882; the company had entered into a contract to sell £25,000 worth of arsenic, the second largest contract it had made to date.[80,81]

At Watson's the emphasis was being placed on deepening the mine. By mid February 1882 Engine Shaft was sinking below the 88.[82,83] Pumping had been started on Western Shaft at the beginning of February and by the 22nd the shaft was in fork and sinking had recommenced, reaching a depth of seventeen fathoms by 23rd March.[84,85,86]

Work at Wheal Emma was also progressing; the 137 east which was noted as yielding some good copper ore after having passed through indifferent ground.[87] In the vicinity of New Shaft, Jones' rise in the back of the 145 was very productive; the lode, six feet wide at this point, was producing three tons of copper and six tons of mundic per fathom.[88] In May Moses Bawden informed shareholders that rock drills were in use at four locations including the 137 east at a point 213 fathoms east of the shaft, the 115 east, the 160 west and the 190 west.[89] Presumably the drives on the 160 and 190 constituted the deep western drives announced at the May 1881 meeting. By 21st June the drills on the 160 were being moved from the 160 to the 205.[90]

The May 1882 meeting was held on Thursday, 25th May. Peter Watson was able to inform the meeting that the average price the mine was receiving for its copper had increased to £2 2s 3d per ton. Watson was also able to announce that during the six months to 30th April the mine had made a profit of £1,730. The shareholders were informed that a rich find had been made in Watson's Engine Shaft during the previous few days and that the mine was being deepened with all speed. The ever optimistic Isaac Richards reported that "the extension of the workings westward into the great length of ground onto the Wheal Maria great crosscourse, and beyond it on to the Tamar" could "scarcely fail to be attended with success".[91,92]

In August Moses Bawden and Isaac Richards met with the Estate's representatives to discuss the possibility of drawing the pumps from sections of the mine which the Company had no intention of either working or exploring. In effect this would mean abandoning the main Lode below Deep Adit. Given the financial position of the Company this would seem to be a reasonable proposition. That said, Josiah Paull was of the opinion that the proposal was a pretext on the part of Moses Bawden to induce the Estate to grant a rebate in dues against the cost of the introduction of boring machines.[93]

Boring machines were being used in earnest in Wheal Emma. In October 1882 six men were driving the 160 west of Railway Shaft with an Eclipse, six men were driving the 190 west with a Barrow, six men were driving the 115 with a Darlington whilst three men were driving the 137 east with a Barrow machine.[94]

Neither Watson's bombast nor Richards' optimism could disguise the fact that this expensive exploratory work was not yielding results. It was becoming evident that the eastern end of New South lode was, to all intents, barren.

At the November 1882 meeting Richards was forced to admit that the trials undertaken were not as successful as could have been wished. However Richards did feel that the lode on the 137 east and 115 east was encouraging and he held out hopes that great discoveries would be made in the, largely unexplored, Eastern Ground. At Watson's Engine Shaft the lode was looking better on the 100 than it had on the 88.[95]

Whilst the mine's current prospects were fairly uninspiring, Richards was able to throw the meeting a lifeline:

> In looking forward to the future success of the mines it is important that another feature in connection with our general explorations should be specially mentioned. For some time past we have been meeting with specimens of tin ore, raisings from various points, especially on New South Lode, and although it has not been found in regular paying quantities several tons of good quality have been selected, and it is more than probable that having, as it has, all the characteristics of a tin bearing lode, large quantities of this mineral will be found on a further development of the workings at depth.[96]

Peter Watson confirmed this stating that "the directors have been down to the mine, and were pleased to find great blocks, I may say, of rich tinstuff on the mine at surface".[97] Watson told the meeting that he was convinced that with time Devon Great Consols would become a great tin mine and he recommended that they sink deep "to leave the copper bearing measures".[98]

As an interesting digression the Devon and Cornwall Central Railway Act 1882 authorised a railway which, if built, would have bisected the Devon Great Consols sett (see Note 2).[99]

1883

In mid January 1883 the Directors issued a circular announcing that they had sold the years' make of arsenic for £25,000.[100]

With regard to tin, prospects remained positive during 1883, "good specimens" of tin ore being found from the 170 to the 190 fathom levels. At the May 1883 meeting it was reported that "fine specimens of tin" were constantly being raised from Wheal Emma "strongly indicating the existence of this mineral at depth on this strong and masterly lode". It was announced at the May meeting that Captain Josiah Thomas of Dolcoath had been commissioned to report on future prospects for tin.[101] It will be remembered that it was on the strength of Josiah Thomas' 1872 report that the Company undertook the expensive and, ultimately, fruitless deepening of Richards' Shaft in search of tin. Thomas inspected the mine in July 1883, his report being submitted to the shareholders on 11th July. Thomas strongly recommended sinking below the 205 in the hope of finding payable quantities of tin at depth. The directors visited the mine in September at which time Isaac Richards presented them with his report which mirrored Thomas'. Both Thomas' and Richards' reports were endorsed by the agents on the mine.[102] By the beginning of October 1883 the decision had been taken to start sinking

Railway Shaft. On 4th October Isaac Richards wrote that "it is fully anticipated that discoveries of copper and tin ores will be made as depth is attained".[103] Sinking started on Saturday 24th November 1883, the work being let to six men at £18 per fathom. At the November half yearly meeting, held on Wednesday 28th, Watson stated that "it would be cowardice if we did not sink this shaft, and try what is underneath us there".[104] Initial progress at Railway Shaft was mixed, on 19th December Richards noted that "the ground is scarcely so favourable for exploring, some floors of quartz having come across the killas".[105] By year's end the shaft was three fathoms below the 205 and the ground was "proving tolerably favourable for progress".[106]

The resumption of the quest for tin at depth should not obscure the fact that the mine was being developed at several points during 1883. At Wheal Emma New Shaft on New South Lode, for example, rock drills were being installed on the 215 during August that year.[107] Between June and the end of the year a Darlington drill was in use on the 205.[108]

At Wheal Josiah, Counthouse Shaft was sinking below the 144.[109] Watson's Mine was still considered to be a strong prospect: In July 1883 a permanent "drawing lift" had been installed between the 88 and 100 fathom levels.[110] During August Richards noted that progress was being made in the 32 at Western Shaft.[111] In October work started on sinking Midway Shaft to improve ventilation on the 22 and 32.[112] During the later part of the year work recommenced on "the long neglected South lodes". In December it was noted that two men were driving a small level west from the "Fremator" (South Fanny) Adit.[113]

In December it was reported that three men were "rising and stoping" above the 137 east of New Shaft using a Darlington machine.[114] This appears to be the first use of a boring machine at Devon Great Consols for stoping rather than development work.

Christmas 1883 must have been somewhat marred by the death of 14 year old Thomas Johns on 13th December. Johns was employed as a fireman and shunter on the railway. He was run over by a train, having jumped off the wagon he was riding in before the train had come to a halt.[115]

1884

As emphasis on the mine switched from copper to arsenic, a fundamental event in the life of the mine became the securing of an annual contract to sell the mine's make of arsenic. In February 1884 the mine was able to enter into a contract for the following years' make, worth £24,000 to the company. In consequence new condensers were constructed at the arsenic works.[116]

The sinking of Railway Shaft aside various points were being pursued during the early months of 1884; examples include Watson's Midway Shaft had communicated with the 20 fathom level in January.[117] On 13th March Richards reported that the 112 at Watson's was yielding two tons of copper and mundic per fathom.[118]

In February Counthouse Shaft communicated with the 160 west of Railway Shaft.[119] In April it was noted that the 160 west was being driven parallel to the Main Lode in the hope of cutting undiscovered ore ground, arguably something of a forlorn hope.[120]

Work in the vicinity of the Frementor Adit continued through the early months of 1884, however work had been temporarily suspended by late March, Richards commenting that the lode was "small and unproductive".[121,122,123] Writing on June 12th Richards informed readers of the *Mining Journal* that work had recommenced on the adit.[124]

At the May 1884 meeting held on Wednesday 28th, the dominant theme was the future prosperity of the mine. As might be expected chief amongst the wonders on offer was the prospect that "a productive tin mine may be found in Devon Consols under the great copper ore deposits, as in the case of Dolcoath Mine".[125] Railway Shaft had reached a depth of fourteen fathoms below the 205, it was planned to sink a further two fathoms and then drive a crosscut to intersect the lode. Lesser wonders on show included Watson's which was confidently expected to cut rich at any moment. Richards announced to the meeting that he intended to explore the ground to the north west of Wheal Maria, the proposal being passed by the meeting The reasoning for this was based on experience at the neighbouring Devon Great United, formerly known as West Maria and Fortescue, which had largely similar management to Devon Great Consols. At Devon Great United Capel Tor lode was proving rich at depth and it was hoped this would carry on into Devon Great Consols where it had previously been worked in the early 1870s.[126]

By the end of June 1884 Railway Shaft had reached a depth of fifteen fathoms below the 205 and by August work had begun driving a 220 crosscut south. Progress of the crosscut, known as Bray's, was hampered by poor ground, Richards writing on 14th August that "the ground is of a rather troublesome nature for exploration, the killas being intermixed with floors of quartz".[127,128] By 11th December 1884 the crosscut had intersected the lode and Richards announced the intention to derive west on lode.[129] On 18th December Richards reported that the lode on the 220 west was composed of capel with a little copper and mundic.[130]

Late June 1884 saw ongoing progress at Watson's: Western Shaft was sinking below the 32 whilst Midway was sinking below the 20.[131] At the end of October

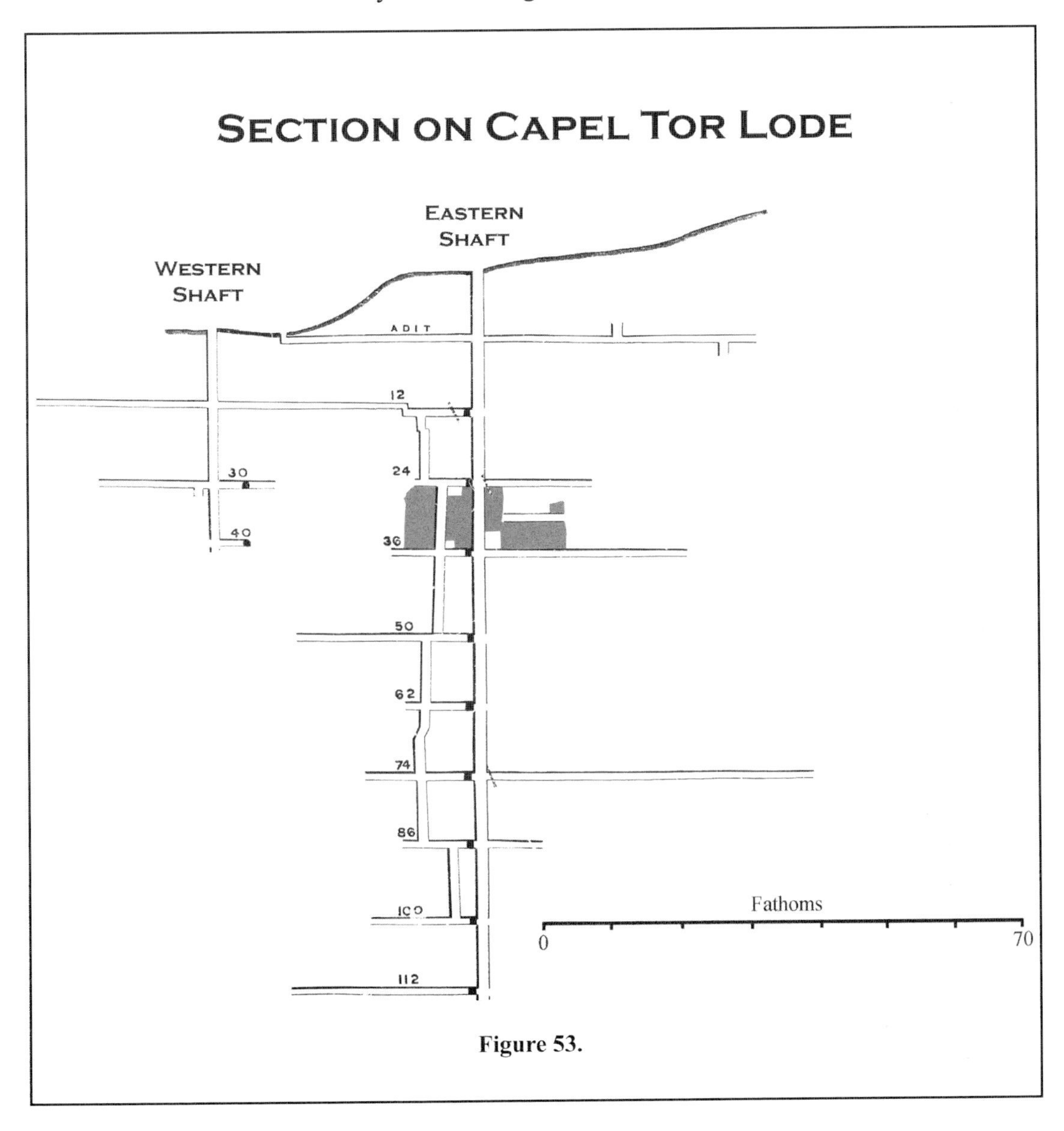

Figure 53.

it was reported that a plunger lift was being installed between the 88 and the 112 at Watson's Engine Shaft.[132] The work having been completed work commenced on sinking below the 112 by 18th December.[133]

Work on the eastern extension of Devon Great United's Capel Tor Lode at Wheal Maria was well underway by mid June 1884, by the 19th July the adit had been cleared as far as the shaft.[134] Work then focussed on opening up the shaft to a sufficient size for "future purposes".[135] By the beginning of September the shaft was being sunk below adit.[136] Interestingly by mid September the lode was being referred to as Capel Tor Lode.[137] Development continued throughout the Autumn of 1884, on 30th October the lode was described as being four to five feet wide and composed of capel and quartz with peach and some good quality mundic and copper ore.[138] Work ground to a halt early in December 1884 due to an influx of water and the decision was taken to install pumps before sinking could recommence however work on the flat rod run was seriously hampered by appalling weather.[139,140]

Due to the "depressed state of the Company's affairs" the use of boring machines was abandoned in August 1884.[141]

On October 27th Josiah Paull reported "with more regret than surprise" that about forty miners and labourers were discharged noting that:

> "They had nearly all been employed on ground which from its coarse quality and from the still declining value of copper it does not pay to work out".[142]

In addition to the October layoffs there was a general reduction in wages across the mine.[143]

The November 1884 meeting, held on Wednesday 26th, was quite a stormy affair. During the previous six months the average price per ton that the mine was receiving for its copper ore had slumped to £1 14s 4d, the lowest ever received. This was compounded by the fact that the trend was downwards, at the 23rd October ticketings the company had only received £1 8s 6d. At these prices the mine was actually making a loss on every ton of ore it raised. It was seriously mooted by a shareholder that they should cease producing copper and concentrate on arsenic. In response the meeting was informed that they had to raise the copper to get the arsenic. In effect a total reversal had taken place; copper had become a waste product whilst the mundic, formerly a waste product, was the prime economic mineral. In response Watson was pleased to announce that it had been possible to negotiate a £1,000 reduction in the Duke's dues during the following year. The mine's agents had responded to the situation by accepting a 10% reduction in their salaries. As was now the norm was shareholders were regaled with tales of future prosperity; tin would be the great saviour of the mine; Watson's would cut rich. Unfortunately the shareholders had repeatedly heard such promises with very little to show for them.[144,145] One shareholder, G. P. Shearwood, was moved to address the meeting; he thought that:

> the statements made from time to time by Capt. Richards in his reports were contradictory, He would not say that the reports were untrue, but it was "blarney" ; they were told of wonders which were going to happen , and they did not happen, and so it would go on.[146]

The Company ended the year in a very poor condition. The promised £1,000 rebate on 1885's dues, which, whilst welcome, was only a partial respite. Whilst the Estate recognised the problems the mine was facing, they were less than impressed by the fact that the Company had not made, nor intended to make, a call on shareholders. Paull noted in his annual report:

> By the end of 1885 the Company must come to one decision or the other – either to provide for further working and development by paying calls for a time or to wind up their affairs and dissolve, when the united ground would be set out for some other company who might feel inclined to go on with it as a new concern.[147]

1885

1885 continued in much the same vein as the previous year, efforts being focussed at Capel Tor, Watson's and Railway Shaft.

During January 1885 progress was made on the Capel Tor flat rod run, on the 15th Richards noted that the angle bob to connect the flat rods to the Gard's Shaft waterwheel had been erected and that shears and a capstan were in the course of construction on the shaft at Capel Tor.[148] In spite of rough weather towards the end of January work had been completed by mid February and on the 14th it was reported that the pumps were in operation and sinking had recommenced.[149,150] By February 19th Capel Tor Eastern Shaft was seven fathoms, three feet and five inches below adit.[151]

At Watson's work continued apace on all three shafts during the early part of 1885. On 19th February Richards wrote that driving of a 44 fathom level was about to commence at Western Shaft.[152] In early March, Midway Shaft communicated with the 32.[153] By April Watson's Engine Shaft had reached sufficient depth for a 124 fathom level.[154]

The deepest point reached by Railway Shaft was still the 220. In April 1885 the lode on the 220 was noted of being a good size and, encouragingly, was giving thirteen pounds of tin per ton.[155] During April work was underway to divide and case the shaft between the 205 and the 220 and install a skip road and a ladderway.[156] During 1885 the 220 was the only level driven in New South Lode in Wheal Emma.[157]

In February Moses Bawden received an anonymous letter, in which the sender threatened to shoot both Moses Bawden and Isaac Richards. Bawden read the letter, which was described as being of "a most infamous character", at the 14th February setting, he then read a further letter from Peter Watson condemning the anonymous letter. Addressing the men Bawden stated that if the man who had written the letter should "choose to attempt to carry out the threat contained therein, let him try it – there was a Higher power – for himself he was not at all afraid".[158]

At the May 1885 half yearly meeting there was mixed news. Copper prices remained to fall, averaging only £1 9s a ton. Also of note was the death of Lord Claud Hamilton, one of the directors. More positively the Duke's £1,000 rebate on his dues, announced at the November 1884 meeting, came into effect. Watson was also able to announce that they had managed to reduce costs on the mine during the preceding six months by £3,500. Railway Shaft was again touted as the future salvation of the mine. The meeting was informed that as the shaft went deeper it was getting poorer for copper which, rather than being as cause for alarm, was to be expected. As greater depths were attained the indications for tin were improving, Moses Bawden explaining that they were rapidly approaching the granite. Towards the close of the meeting the ever optimistic Isaac Richards stated that indications at Watson's, Wheal Maria and Wheal Emma were excellent. Unfortunately history does not record G. P. Shearwood's thoughts on the matter.[159,160]

Throughout the summer and autumn of 1885 work progressed in much the same way as before. In June 1885 work started on driving a 12 fathom level at Capel

Tor.[161] By 24th September Eastern and Western shafts had communicated on the 12 which greatly improved ventilation and sinking had commenced below the 12.[162] At Watson's the crosscut on the 124 had intersected the lode in early June and by October Watson's Engine Shaft was sinking below the 124 in search of the undoubted riches that lay at depth.[163,164]

There was some positive news from the 220 west at Railway Shaft. For example on 9th July 1885 Isaac Richards wrote:

> the lode proving 4 ft wide, of a strong masterly character, and yielding some good quality arsenical mundic, and some good quality tin ore, from which several hundredweights have been selected, making an assay of 1cwt of black tin to the ton of orestuff.[165]

The November 1885 meeting was held on Wednesday 25th; Peter Watson taking the chair as usual. Copper had continued its ever downward spiral which Watson attributed to a combination of high stocks already held by the smelters and foreign competition. During the previous six months the mine's copper ore had been averaging £1 6s 3d per ton although at the time of the meeting was only realising £1 0s 10d which, in real terms, meant that the mine's copper was less than worthless. In the directors' opinion the price would not pick up in the foreseeable future. To add to the tale of woes, receipts for arsenic were down in comparison with the same period in 1884, a reduction from £12,000 to £8,400. As usual the meeting was told that both Capel Tor and Watson's were looking promising, although by this time the platitudes, especially with regard to Watson's, must have been wearing thin. The elusive promise of tin was again dangled before the shareholders. The 220 was yielding better tin than the 205 and Watson believed that it was worth sinking Railway Shaft a further twenty or thirty fathoms to prospect the lode at depth.[166]

1886

Whilst Devon Great Consols had, arguably, made the transition from copper to arsenic it should not be forgotten that, in spite of the ruinous price, the mine was still a major copper producer. For example at the Truro ticketings on the 21st February 1886 the mine sold a not insignificant 700 tons of copper ore, realising £855.[167]

During 1886 exploratory work was pushed on apace, writing in January 1886 Isaac Richards noted that all driveages were "being continued by the side of the lode for more speedy progress".[168] As in the previous year attention was focussed on Capel Tor Lode, Watson's and Railway Shaft: in January 1886 a cross cut was being driven from Capel Tor Eastern Shaft to intersect the lode on the 24,[169] the lode, described as promising, having been cut by mid February.[170] By the beginning of April Richards was able to report that sinking had commenced below the 24.[171]

At Watson's, work focussed on sinking Engine Shaft. On 21st January Isaac Richards reported that the shaft was four fathoms, four feet and four inches below the 124 however work was severely hampered by water and, in consequence, "larger pump work" was being installed. It was also noted that drives on the 124 were yielding both copper and mundic.[172] By late February the larger pumps were in operation and sinking had recommenced below the 124.[173]

Work at Railway Shaft focussed on exploring the 220; the lode being noted as four to five feet wide, yielding small quantities of mundic and copper; hardly the tin bonanza the shareholders had been promised.[174]

The May meeting was held on Tuesday, 25th at the Company's offices at Austin Friars. Peter Watson, as usual in the chair, informed the meeting that the Company was only receiving an average of £1 3s 9d per ton for copper and that the ongoing trend was downward. There was also less than positive news regarding arsenic sales: in the half year ending 30th April 1886 arsenic sales had realised £11,200 which, as Watson told the meeting, was £1,000 less than for the previous corresponding half year. Watson did have some good news for the shareholders: Due to the depressed state of trade the Duchy of Cornwall had agreed to a 25% reduction in water dues. The Directors had also been trying to get the Estate to remit its dues from 1/18th with somewhat less success. In terms of mining the meeting was informed that Watson's mine had been vigorously developed and things were looking more "hopeful". However in terms of future prosperity the company was still pinning its hopes of finding tin at depth in Railway Shaft. Isaac Richards commented that indications on the 220 "have been far better than in the level above, and I am strongly of the opinion that on sinking the shaft from 40 fathoms to 50 fathoms deeper, tin in paying quantities will be found".[175]

The question on the Duke's dues was of considerable import, much of the correspondence between the directors and the Estate being reproduced in the *Mining Journal* of May 29th 1886. Whilst at the time of the May meeting nothing solid had been agreed between the company and the Duke, productive negotiations were evidently in progress. On Friday 4th June Peter Watson was able to write to Moses Bawden informing him that the Duke had agreed to forgo all royalties until such time as the mine became dividend paying. The news was announced to 300 miners who gathered at the account house the following day; the *Tavistock Gazette* noting that the news was greeted with "acclamation" and "enthusiasm". The meeting closed with "hearty cheers for his Grace the Duke of Bedford, Mr Peter Watson, Mr Moses Bawden, Mr I. Richards and the other agents of the mines".[176]

On Wednesday 23rd June 1886 a miner named John Gregory was killed in Richards' Shaft when a ladder guide collapsed. The *Tavistock Gazette* recorded that "he was precipitated a distance of about 80 fathoms, and had his brains knocked out".[177]

The intention, expressed at the May meeting, to sink deeper at Railway Shaft was rapidly put into effect. Isaac Richards reported on July 21st that the necessary preparations were underway to continue sinking for tin.[178] The work was overseen by Captain William Woolcock, the Wheal Emma Agent. Woolcock recorded the details of the contract in his notebook:

> August 24th: the Railway Shaft to sink below the 220 fm level to be carried 10 feet long and 5 feet wide by 16 men - viz 8 principal and 8 second rate. Stented 20fms certain at per fm £22.0.0; the takers to have £2 per fm extra for all ground sunk over 1 fm per week, or 4 fms per month for the 20 fms out. The takers to work (*unreadable*) men in each shift of six hours duration, commencing at midnight on Sundays, completing each week's work at 6pm on Saturdays. Taken by Thomas Bray, John Hill, John Hooper, Samuel Denner, John Cloak, William Rundle, Frederick Rice, John Dunn, Richard Cloak, Ernest Rundle, Joshua Denner, William Short, Edwin Hockaday, Thomas Voysey, Fred Newcombe, John Hockaday.[179]

Interestingly the decision was taken to sink by hand labour in order to avoid the expense of installing rock drills. By the time of the November 1886 meeting it was noted that, since work started on September 1st, fifteen fathoms had been sunk in hard rock. In typical style Peter Watson sang the praises of the men commenting:

> That, I think, is one of the best pieces of work ever done in Devon Consols or in any other mine by hand labour.[180]

The sinkers made rapid progress completing the contract in seventeen weeks; the 240 being reached by the last week of December 1886.[181,182] Woolcock recording the details in his notebook: First four weeks: five fathoms, two feet; second four weeks: four fathoms three feet; third four weeks: four fathoms, four feet; fourth four weeks: four fathoms three feet; in a final week one fathom was driven giving a total over the seventeen weeks of the contract of twenty fathoms.[183]

At the November 1886 meeting Watson had great news; the mine had made its first sale of tin. Two tons, two hundredweight had been sold realising £115 8s 9d. Watson, no doubt feeling vindicated, declaimed: "That will show you, at any rate, that there is such a thing as tin at Devon Great Consols". Not surprisingly this statement was met with cheers from the meeting.[184] Progress in sinking Railway Shaft has already been dealt with, although it is worth noting that Dolcoath's Captain Thomas backed Richards' judgement:

> I agree with Captain Richards that the lode at the 220 has a better appearance than that in the level above. It contains some very fine stones of tin such are frequently found in the vicinity of larger deposits. I hesitate to give a decided opinion as to whether tin in paying quantities is likely to be found "on sinking the shaft 40 to 50 fathoms deeper" but in my judgement the only hope of finding a profitable tin lode is by deeper sinking.[185]

Richards was in good form at the meeting stating that tin stuff was increasing with depth "from 25 up to 40lbs, 112lbs, 134lbs, and now to 210 lbs to the ton of stuff".[186] Could anyone doubt that Devon Great Consols was of the verge of becoming a great tin mine?

Whilst the prospects for tin were at the forefront of attention the meeting's attention was also drawn to the potential of both Watson's Mine and Capel Tor.

Josiah Paull, in his November report, was fairly cynical wondering how the Company's officials could be so confident given the falling reserves of both copper ore and mundic.[187]

In spite of all the positive news there was a down side, a call of ten shillings per share had been made of which between £115 to £120 was still outstanding.[188] The call was to allow deeper exploration to be pursued vigorously, although one could speculate that the call was a condition of the Dukes remission of dues.

1887

The speed with which Railway Shaft had been sunk from the 220 to the 240 was a major achievement. The management of the mine were evidently very satisfied with progress in Railway Shaft as the following extract from the *Tavistock Gazette* attests:

> Chipshop: On New Year's Day a dinner was provided at the Bedford Arms by Mr. and Mrs. Clemo, for the men employed at Devon Great Consols, who had completed their contract in sinking the Railway shaft at Wheal Emma by the time required. The men assembled at the Inn by one o'clock, and sat down to a bountiful spread, the chair being occupied by Captain Clemo. Dinner being over, the toast of "The Queen" was drunk, as was also the health of Mr. P. Watson, Mr. M. Bawden, and Captain Richards, the men expressing their gratitude to the donors for the good things provided. Songs were rendered by Messrs. Newcombe, Voyzey, Rundle, Cloke, and others, and a most convivial afternoon and evening were spent, the men returning to their homes highly pleased with the good treat given them by their employers.[189]

The intention to push exploration deeper expressed at the November 1886 meeting was being realised on the mine. On January 19th 1887 Richards reported that Capel Tor Eastern Shaft was three fathoms below the 36 and that a plunger lift had been installed.[190] Obviously no time was being wasted as on March 31st Richards was able to announce that Eastern Shaft had reached a sufficient depth for a 50 fathom level.[191]

On March 12th it was reported that Watson's Engine Shaft was sinking below the 136.[192]

At Railway Shaft work was underway to case and divide it and install a ladderway between the

Figure 54. A fine studio portrait of William Woolcock taken at the age of 86.

220 and 240 levels by the beginning of the new year.[193] On January 13th Isaac Richards reported that a 240 crosscut was being driven to intersect the lode.[194] by the end of February the lode had been cut. By March 12th the 240 crosscuts had been suspended, work being concentrated on driving west on lode.[195] On March 12th Woolcock recorded that he had:

> Set the Railway Shaft to sink, below the 240fm level to be carried 10ft long and 5ft wide stented 20fms certain by 16 men at per fm £22, - the contract to terminate if the lode come into the shaft. The takers to have £2 per fm extra for all ground sunk over 6ft per week and to work from 12 o'clock on Sunday night, to 6 o'clock on Saturday evenings, 4 men in each shift of 6 hours duration relieving in bottom of shaft.
>
> Taken by Thomas Bray, John Hill, Fred Reid, John Hooper, Samuel Denner, John Dunn, John Cloak, William Rundle, Richard Cloak, Ernest Rundle, William Short, Joshua Denner, Fred Newcombe, Thomas Voysey, John Cole, Thomas Palmer.[196]

On March 24th Isaac Richards, in his regular *Mining Journal* report, noted that sinking had commenced below the 240.[197]

Whilst Woolcock's time must have been largely taken up with his duties at Wheal Emma in general and supervising the sinking of Railway Shaft in particular he did not neglect his domestic responsibilities. On March 29th 1887 he recorded in his notebook that he had

> Put in potatoes by Thomas' Shaft, from hedge out to first strip Sunbeam Ashleaf; on to next strip George Kelly's early nuts; on to next strip Brownetts Beauty; on to next strip White Elephant; on to next strip Schoolmaster. Strips are paired unto front fence.[198]

No doubt an Agent of Woolcock's stature and length of service had earned certain perks!

At the May 25th Meeting Watson was in full flight regarding the sinking of Railway Shaft which was now ten fathoms below the 240. With regard to the contract to sink from the 220 to the 240 Watson stated that

> If ever I was proud of anything in my life I have been proud of the

Devonshire men who have accomplished that work in so short a time, in return for that we made them a present of half a sheep.[199]

Watson hoped that the shaft would reach the 260 within a couple of months. It was felt that New South Lode, which had split at the 115 would converge at the 260 and prove very rich. As usual Watson's and Capel Tor were full of promise, although, for once, Isaac Richards had his doubts commenting that "he was a little disappointed at Maria".[200] Whatever the situation no one could doubt that the Company was taking exploration seriously, £,4,500 having been expended on sinking shafts and driving levels during the past year.[201] As part of the drive to reduce costs a "rock breaker" had been purchased to speed up dressing.[202]

Long term shareholder H. Cattley Stewart, who would later prove to be something of a thorn in Watson's side, was elected to the board of directors at the meeting.[203]

The summer of 1887 was an unusually dry one; on June 30th Isaac Richards commented that the drought was hindering pumping, particularly at Capel Tor and Watsons'.[204] Whilst both Capel Tor and Watson's were flooded out work still continued at Railway Shaft. On July 7th Richards note that the lode had been encountered in the shaft fourteen fathoms, three feet and two inches below the 240.[205] Richards, writing on August 4th, reported the lode to be "of good size, nearly 5 feet wide, is composed of very strong capel, with quartz and small quantities of copper and mundic ores"[206] Unfortunately conspicuous in its absence was tin. The drought persisted throughout August, on the 11th, for example, Richards noted that water was still rising in the workings and that Watson's was flooded to the 124.[207] It was, therefore, with a great deal of relief that Richards was able to write at the beginning of September:

We are now......getting some fine showers of rain, and we hope that in a little while there will be a supply of water for working our pumping machinery, and a permanent draining of the workings may soon be effected.[208]

At the beginning of September a new shaft known as Victoria Shaft, presumably in recognition of the Queen's jubilee, was sunk sixty fathoms east of the Great Crosscourse on Watson's Lode.[209,210,211] On 20th October Isaac Richards noted that Victoria Shaft had cut a small lode six to nine inches wide, composed of capel, quartz, and a little gossan; Richards thought that it was probable that it was a portion of the Main Lode.[212] At the November meeting Charles Oxland, the largest shareholder, stated that he had found some tin at Victoria Shaft.[213]

Sinking Railway Shaft had continued throughout the long dry summer, the contract being completed on 7th September. On the following day the contract was set to cut the 260 shaft plat, case, divide, install ladders and skip road between the 240 and the 260. The contract was taken by Thomas Bray, John Hill, Fred Rice, John Hooper, John Cloak, William Rundle, Ernest Rundle, Richard Cloak, Frederick Newcombe, Thomas Voysey, John Cole, Thomas Palmer. The work was completed on 22nd October and included driving 3ft 8in in the 260 cross cut south. This was followed by a further contract dated the 22nd October let to "Thos. Bray & Son" for twelve men to drive the 260 crosscut north at £20 per fathom.[214]

The November 1887 meeting was held on Wednesday 16th. The shareholders were informed that output during the year had suffered due to drought and that it was not until October that things had got back to normal. Ironically at the time of the meeting the problem was not too little water but, rather, too much; Peter Watson commenting that "the wheels are clogged with water". Whilst the drought had affected output Watson was pleased to tell the meeting that the demand for arsenic was up. The meeting was also informed about progress at Railway Shaft. In spite of the sterling efforts of the miners in reaching and exploring the 260 no tin had been found and Isaac Richards had to admit that he was somewhat disappointed by the lack of results at Railway Shaft. In an attempt to locate tin the 240 was being driven to get under the 220 tin ground. Whilst this was a minor consolation it was not the great bonanza that the shareholders had been led to anticipate. The new Stannaries Act was also discussed, its most obvious impact being that miners would be paid fortnightly and labourers weekly.[215]

As 1887 drew to a close work continued much the same as always; in the vicinity of Railway Shaft crosscuts were being driven north and south on the 260, interestingly the north crosscut had cut a lode which contained some wolfram. On the 240 men were driving alongside the lode to get under the 220 tin ground.[216] Bray's winze on the 220 was producing some tin.[217]

At Watson's work was proceeding with the 148 crosscut.[218]

A "masterly lode" had been cut on by the 62 at Capel Tor Eastern Shaft, this "masterly lode" carried small quantities of copper and mundic.[219] On December 8th Richards noted that they were preparing to sink below the 62, the work having started by December 15th.[220,221]

1887 was a year which had started with high hopes, only to have them dashed by the failure of Railway Shaft to cut rich. As the year closed the shareholders were able to draw a small crumb of comfort from an unexpected direction. At the Redruth copper ticketing, held on December 22nd the mine sold 579 tons of copper ore for £1,605 16s. One parcel of Devon Great Consols ore realised £9 15s 6d per ton causing the *Tavistock Gazette* to comment that "this looks like old times".[222]

1888

As the new year started exploration continued on the 260 at Railway Shaft. Writing in his notebook on January 9th William Woolcock recorded that the "260 crosscut north of the Railway Shaft in a little over 7fms, gone through a small branch about 1 foot behind end underlying south. This end stopped, and men put to drive 260 west, from side of plat, £17 per fm".[223] On February 16th Isaac Richards observed that the lode on the 260 west was composed principally of capel with a little arsenical mundic.[224]

Capel Tor Eastern Shaft had reached the 74 by April 5th.[225]

On April 5th Isaac Richards wrote that sinking was due to commence below the 148 at Watson's Engine Shaft, work having started by April 19th.[226,227]

The May meeting was held on Wednesday 16th at the London office. Peter Watson was able to announce that copper prices were rising, the mine receiving an average of £1 19s 2d per ton of ore. During the half year ending April 1888 the mine had sold 2,105 tons of copper ore, realising £4,146 11s 6d. After protracted negotiations a good contract for the sale of the mine's arsenic had been entered into in April. Good news aside it was Watson's painful duty to inform the meeting that no tin had been discovered on the 260. There was some attempt to gloss over this by pointing to improving prospects at both Capel Tor and Watson's. As usual Isaac Richards was able to see a silver lining observing that whilst they had not made great discoveries they had opened up some valuable ground. However even Richards customary assurance that "valuable discoveries will at no distant period be made" could not hide the fact that the exploratory work at Railway Shaft had, to date, been an expensive failure.[228,229]

On July 1st Henry Youren, who had been the mine's accountant from 1845 died (see Note 3).[230]

Throughout the summer of 1888 work at Capel Tor, Watson's and Railway Shaft continued apace.

At the end of March the 74 at Capel Tor Eastern Shaft intersected the lode which yielded small quantities of copper and mundic.[231] On August 23rd Richards reported in the *Mining Journal* that the company intended to sink from the 74 down to the 84, work was well underway by October 11 when the shaft was reported as being four fathoms, two feet and nine inches below the 74.[232,233]

By June 7th a plunger lift had been installed to the 148 at Watson's Engine Shaft and sinking was continuing below the 148.[234] On September 6th Isaac Richards reported that Engine Shaft was 3 fathoms, 4 feet and 6 inches below the 148.[235]

At Railway Shaft the work on the 240 crosscut, and presumably the attempt to get under the tin ground on the 220 was suspended by September 6th.[236]

Thoughts at the November 1888 meeting were turned to the future exploration of Railway Shaft and whether to sink below the 260. Peter Watson estimated that it would cost somewhere in the region of £1,000 and seven or eight months to sink Railway Shaft to the 280. The directors felt that they had better consult Josiah Thomas before making a decision. Elsewhere on the mine it was reported that Capel Tor Eastern Shaft was looking very promising whilst Watson's was looking better than it ever had. The hope was expressed that Watson's would reach the 160 by the new year.[237]

Sinking in Watson's Engine Shaft must has progressed rapidly, *Mining Journal* reporting that work on cutting the 160 plat was underway by December 15th although a broken pump rod and heavy rain had slowed work.[238]

Progress was also being made at Capel Tor, Isaac Richards commenting that Eastern Shaft had reached the 86 by December 20th.[239]

The new stone breaker and engine had arrived by December 27th, Richards stating that it would be "got into position and set to work as quickly as possible".[240]

1889.

Isaac Richards' first *Mining Journal* report for 1888 contained the news that the ground on the 260 at Railway Shaft was poor.[241] In March the Company called in Captain Josiah Thomas to examine the deep workings in Railway Shaft.[242] At Capel Tor, Eastern Shaft was deep enough for an 86 fathom level whilst work in

Watson's was being hindered by water.[243]

As the year unfolded progress was reported from both Capel Tor and Watsons: at Capel Tor on March 28th Richards reported that work had been suspended on the 86 preparatory to sinking below the 86. Speaking of the Capel Tor Lode Richards observed that "it is confidently expected to be found profitably productive in the next level".[244]

Whilst the news from Capel Tor might be described as mixed, reports from Watson's were much more promising, in early May the lode on the 160 cut rich, yielding a reported eight tons of copper ore per fathom. The *Tavistock Gazette* commented thus:

"This is the best discovery made in the mines for many years, and being on a lode, which passes through the entire length of the property, further discoveries will, in all probability, be made as the workings progress on this promising range of mineral ground east and west of the present operations".[245]

The initial report in the *Tavistock Gazette* was confirmed by Isaac Richards writing on May 16th:

> At the 160 the lode continues to present the same very fine appearance, and for the short distance driven eat and west of the cross cut it is worth in each direction 8 tons of copper and mundic ores per fathom.[246]

The discovery at Watson's came just in time for the May 1889 meeting. Peter Watson was able to announce the discovery to the shareholders commenting that "This was the finest discovery they had made in the Devon Consols for the last 10 or 15 years". Watson also informed the meeting that the new stonebreaker would be able to deal with the 200,000 to 300,000 tons of halvans that had accumulated on the mine over the past 45 years. The good news was, however, balanced by the news that the decision had been taken to abandon work at Railway Shaft. This decision was based on a report by Josiah Thomas dated April 1st 1889 and a further report by the mine's agents dated 9th April. The meeting was told that "in the event of nothing being shortly discovered of importance, the pitwork will be taken out of the shaft..... and the miners employed at more favourable parts of the mines".[247,248]

On July 13th 1889 William Woolcock noted that work on the 260 crosscut south of Railway Shaft had been stopped. On July 15th work started taking out all materials from the bottom of shaft, Woolcock's orders being to clear the shaft to the 190 fathom level. The work was carried out by Thomas Bray, John Hill and C. Parsons, who had completed the work by autumn 1889.[249] Thus ended Devon Great Consols' attempt to emulate Dolcoath and re invent itself as a major tin producer.

In December 1902 Josiah Paull wrote:

> In 1885 and following years another trial was made for Tin by sinking one of the shafts on New South Lode to the 260 fathom level. It was done entirely on the initiative of the Directors and their Agents as we did not believe a weaker lode would change from Copper to Tin when the trunk lode of the mine and indeed of the Tavistock district had not been found to do so. The effort was quite unsuccessful, as depth increased the lode diminished in strength until at the bottom level reached it was so weak and divided that it was difficult to determine what part of it should be followed.[250]

Whilst the abandonment and stripping of Railway Shaft below the 190 must have been melancholy work for Woolcock he still able to record more pleasurable events in his notebook. On August 2nd 1889 Woolcock recorded a visit of "a company of Ladies and Gentlemen members of the Devonshire Association ...... for the purpose of viewing the different objects of interest". The guests were shown around the mine by Mr Thomas Youren, and Isaac Richards, William Clemo and William Woolcock (see Note 4).[251]

William Woolcock's duties as Captain at Wheal Emma were many and varied. The visit of the Devonshire Association must have been one of the more pleasant aspects of his job; others were less so: on October 2nd 1889 he wrote in his notebook that:

> Bright the Policeman took Frederick Garland in custody when he came up from underground at 2pm for criminally assaulting Reddicliffe's daughter near Lumburn on Sunday 29th September. The boy "Johns" also in custody for same offence.[252]

As a good chapel man William Woolcock must have found this incident particularly distressing.

The abandonment of exploratory work at Railway Shaft left only two points being actively explored, Capel Tor Eastern Shaft and Watson's Engine Shaft. On August 29th Isaac Richards noted that Capel Tor Eastern Shaft had reached the 100, whilst Watson's Engine Shaft was sinking below the 160. By the end of November 1889 it was noted that the 100 at Capel Tor had cut the lode which was three feet wide and "composed of capel with traces of copper and mundic ores".[253] Fortunately things were better at Watson's; the lode on the 160 west being described as promising, yielding eight tons of copper per fathom.[254]

At the November meeting held on Wednesday 18th there was little good news. The shareholders were informed that as a result of the copper crash of March and April 1889 when the copper standard fell from £78 to £80 to an all time low of £35 per ton the directors had taken the decision not to sample or sell any more copper ore until prices recovered. To offset the loss of income the directors took out a loan from the company's bankers of £3,500. Watson stated that if the company had not taken out the loan they would have been obliged to make a call of ten shillings per share. The company was banking on a revival in the price of copper in the New Year, the price having rallied to £47 per ton at the time of the meeting. The meeting was informed that the company had a stock of about 2,350 tons of copper ore which would be sold "when deemed desirable to do so". Peter Watson assured the meeting that the stock of ore in hand would "more than repay the £3,500 borrowed".[255,256] In terms of future development the meeting was informed that Watson's was the best prospect on the mine. The richness of the 160 indicated that the mine would prove rich at depth. Additionally the discovery of stones of tin on the 160 combined with the proximity to the granite was highly suggestive. Predictably Isaac Richards commented that "he thought it very likely, from the indications that as they went deeper they would come into a tin mine after the copper was done".[257] Elsewhere on the mine Capel Tor was not looking as promising as could be hoped, however Richards informed the meeting that "they were driving in the direction of where the great discoveries took place" (see Note 5).[258] At surface it was noted that the newly erected stonebreaker was working very well, particularly in dealing with arsenical ores.[259]

As the decade ended the mine was in a somewhat parlous condition. The search for tin, which was to have been the salvation of the mine had been an expensive failure. Capel Tor, in spite of Richards' assertions, was proving poor at depth. The only realistic hope of future prosperity was Watson's Mine which appeared to be improving with depth, and was even holding out hopes of tin. In financial terms the company was in a poor position having taken out a large loan against uncertain future copper sales. One can only presume that the shareholders would not have stood for a call as large as ten shillings and would have "knacked the bal". On the positive side of the equation were the considerable stocks of ore at surface and proven reserves underground, add to this the undying optimism and drive of Peter Watson and, arguably, the mine still had a future.

Chapter 8 Notes

Note 1

The 76 was the deepest point reached before work was suspended in the 1860s.[260]

Note 2

The Estate was strongly opposed to the proposed railway Paull noting that the railway, if constructed would:

> bisect this sett from north to south and sever the untried portion of it on the south side of the line from the pumping wheels on the eastern side, as well as cross the arsenic works and interrupt the delivery of ores to the works and the output of arsenic there from by existing railways.[261]

On the other hand Peter Watson was in favour of the railway arguing that improved transport links could only benefit the mine.

Note 3

Mr. Henry Youren, who for 43 years had been accountant at these mines, died on Sunday last (1st July 1888), after a painful illness, in his 65th year. Mr. Youren was one of those quiet pure-minded men who think no evil of others, and whose tender memories are enshrined in the hearts of those who are left behind. His nature was full of kindness for others, but being an extensive and thoughtful reader, he was of a retiring disposition, and unobtrusive to a degree. He was at the Devon Great Consols almost from its start, saw its rise to the proud position it once occupied of being the largest copper mine in the world, and had been there during all its varying fortunes since. What a book he might have written about the ups and downs of mining, its successes and vicissitudes? He was carried to his rest on Wednesday, by the agents of the mines, with whom he had worked with utmost cordiality. Among the exquisite wreaths sent, was one made of roses from the Devon Great Consols garden.[262]

Note 4

The 28th Annual General Meeting of the Association was held at Tavistock

Guildhall between Tuesday July 30th and Friday August 2nd 1889. Excursions included a carriage ride on Friday 30th July to Morwell Barton, Devon Great Consols and Endsleigh, the country house of the Duke of Bedford. The following is an extract from the Report of the Council of the Devonshire Association, as presented to the General Meeting at Barnstaple, July 29th 1890:

> About 60 members and associates proceeded in three large brakes to the Devon Great Consols and Arsenic Works, and thence to Endsleigh Cottage. After a very enjoyable drive the party alighted at the Devon Consols, which they were courteously shown over by those in authority, and the various interesting processes were fully explained. The working of the machinery of the mines, and the methods of ascent and descent of the miners were illustrated, and then the visitors were taken to the floor of the mine, where the various treatments to which the ore was subjected was witnessed. The most interesting parts of the works visited were, however, those in which arsenic was manufactured. The work was in full operation, and visitors were able to witness all the stages to which the mineral is subjected before it is made fit for commercial and chemical purposes. The furnaces for roasting the ore were viewed, and the long chambers in which the arsenic accumulated when given off in the form of a dense vapour from the ore were inspected. A number of fine arsenic crystals were shown, after which the final processes were inspected, and the workmen were seen engaged in packing the arsenic in barrels. The visit to these famous mining works was thoroughly instructive, and had the charm of novelty for the greater number of the visitors.[263]

Note 5

Richards' assertion at the November 1889 meeting that "they were driving in the direction of where the great discoveries took place" appears to be unmitigated nonsense. Capel Tor Lode and Main Lode (where the great discoveries took place) lie roughly parallel to each other and there was be no reason to drive a crosscut from Capel Tor to Main Lode, likewise there is not a shred of evidence to suggest that any attempt was made to drive from Capel Tor towards the Main Lode. One can only suggest that Isaac Richards got carried away with his own imagination.

Chapter 8 references

1. *Tavistock Gazette* 16 January 1880
2. Devon Record Office document 8187, Bedford Estate Mine Reports 1878-1886
3. *Tavistock Gazette* 30 January 1880

4. *Mining Journal* 29 May 1880
5. *Tavistock Gazette* 30 January 1880
6. *Mining Journal* 29 May 1880
7. Devon Record Office document 1258M/SS/MC
8. *Tavistock Gazette* 16 January 1880
9. *Mining Journal* 31 January 1880
10. *Mining Journal* 28 February 1880
11. *Mining Journal* 28 February 1880
12. *Mining Journal* 22 May 1880
13. *Mining Journal* 22 May 1880
14. Devon Record Office document 8187, Bedford Estate Mine Reports 1878-1886
15. Devon Record Office document 8187, Bedford Estate Mine Reports 1878-1886
16. Devon Record Office document 8187, Bedford Estate Mine Reports 1878-1886
17. *Mining Journal* 24 April 1880
18. *Mining Journal* 1 May 1880
19. *Mining Journal* 22 May 1880
20. *Mining Journal* 29 May 1880
21. *Tavistock Gazette* 28 May 1880
22. *Mining Journal* 23 October 1880
23. Devon Record Office document 8187, Bedford Estate Mine Reports 1878-1886
24. *Mining Journal* 4 December 1880
25. *Mining Journal* 19 June 1880
26. *Mining Journal* 3 July 1880
27. *Mining Journal* 3 April 1880
28. *Mining Journal* 19 June 1880
29. *Mining Journal* 29 May 1880
30. *Mining Journal* 12 June 1880
31. *Mining Journal* 3 July 1880
32. *Mining Journal* 11 September 1880
33. *Mining Journal* 18 September 1880
34. *Mining Journal* 20 November 1880
35. *Mining Journal* 11 December 1880
36. *Mining Journal* 8 May 1880
37. *Mining Journal* 22 May 1880
38. *Mining Journal* 29 May 1880
39. *Mining Journal* 26 June 1880
40. *Mining Journal* 3 July 1880
41. *Mining Journal* 4 September 1880

42. *Mining Journal* 18 September 1880
43. *Mining Journal* 20 November 1880
44. *Tavistock Gazette* 26 November 1880
45. *Mining Journal* 27 November 1880
46. Devon Record Office document 8187, Bedford Estate Mine Reports 1878-1886
47. Devon Record Office document 8187, Bedford Estate Mine Reports 1878-1886
48. Devon Record Office document 8187, Bedford Estate Mine Reports 1878-1886
49. Devon Record Office document 8187, Bedford Estate Mine Reports 1878-1886
50. *Tavistock Gazette* 4 March 1881
51. *Mining Journal* 1 January 1881
52. *Mining Journal* 22 January 1881
53. *Tavistock Gazette* 4 March 1881
54. *Mining Journal* 29 January 1881
55. *Mining Journal* 21 May 1881
56. *Mining Journal* 28 May 1881
57. *Tavistock Gazette* 4 March 1881
58. Devon Record Office document 8187, Bedford Estate Mine Reports 1878-1886
59. *Tavistock Gazette* 27 May 1881
60. *Mining Journal* 28 May 1881
61. *Mining Journal* 28 May 1881
62. *Mining Journal* 28 May 1881
63. *Mining Journal* 28 May 1881
64. *Mining Journal* 19 March 1881
65. *Mining Journal* 28 May 1881
66. *Mining Journal* 28 May 1881
67. *Mining Journal* 28 May 1881
68. *Mining Journal* 28 May 1881
69. *Mining Journal* 3 September 1881
70. *Mining Journal* 6 August 1881
71. *Mining Journal* 8 October 1881
72. *Mining Journal* 29 October 1881
73. *Mining Journal* 5 November 1881
74. *Mining Journal* 16 July 1881
75. *Mining Journal* 6 August 1881
76. *Mining Journal* 8 October 1881
77. *Mining Journal* 26 November 1881
78. *Tavistock Gazette* 2 December 1881
79. Devon Record Office document 8187, Bedford Estate Mine Reports 1878-1886

80. *Mining Journal* 4 February 1882
81. *Tavistock Gazette* 26 May 1882
82. Devon Record Office document 8187, Bedford Estate Mine Reports 1878-1886
83. *Mining Journal* 18 February 1882
84. *Mining Journal* 11 February 1882
85. *Mining Journal* 25 February 1882
86. *Mining Journal* 25 March 1882
87. *Mining Journal* 11 March 1881
88. *Mining Journal* 1 April 1882
89. *Mining Journal* 27 May 1882
90. *Mining Journal* 24 June 1882
91. *Mining Journal* 27 May 1882
92. *Tavistock Gazette* 26 May 1882
93. Devon Record Office document 8187, Bedford Estate Mine Reports 1878-1886
94. Devon Record Office document 8187, Bedford Estate Mine Reports 1878-1886
95. *Mining Journal* 25 November 1882
96. *Mining Journal* 25 November 1882
97. *Mining Journal* 25 November 1882
98. *Mining Journal* 25 November 1882
99. Devon Record Office document 8187, Bedford Estate Mine Reports 1878-1886
100. Devon Record Office document 8187, Bedford Estate Mine Reports 1878-1886
101. *Mining Journal* 2 June 1883
102. *Tavistock Gazette* November 30 1883
103. *Mining Journal* 6 October 1883
104. *Mining Journal* 1 December 1883
105. *Mining Journal* 22 December 1883
106. *Mining Journal* 29 December 1883
107. *Mining Journal* 11 August 1883
108. Devon Record Office document 8187, Bedford Estate Mine Reports 1878-1886
109. *Mining Journal* 22 September 1883
110. *Mining Journal* 14 July 1883
111. *Mining Journal* 11 August 1883
112. *Mining Journal* 13 October 1883
113. Devon Record Office document 8187, Bedford Estate Mine Reports 1878-1886
114. Devon Record Office document 8187, Bedford Estate Mine Reports 1878-1886
115. *Tavistock Gazette* 21 December 1883
116. *Mining Journal* 31 May 1884
117. *Mining Journal* 12 January 1884

118. *Mining Journal* 13 March 1884
119. *Mining Journal* 9 February 1884
120. *Mining Journal* 5 April 1884
121. *Mining Journal* 26 January 1884
122. *Mining Journal* 23 February 1884
123. *Mining Journal* 22 March 1884
124. *Mining Journal* 14 June 1884
125. *Tavistock Gazette* 30 May 1884
126. *Mining Journal* 31 May 1884
127. *Mining Journal* 28 June 1884
128. *Mining Journal* 16 August 1884
129. *Mining Journal* 13 December 1884
130. *Mining Journal* 20 December 1884
131. *Mining Journal* 28 June 1884
132. *Mining Journal* 1 November 1884
133. *Mining Journal* 20 December 1884
134. *Mining Journal* 21 June 1884
135. *Mining Journal* 23 August 1884
136. *Mining Journal* 6 September 1884
137. *Mining Journal* 13 September 1884
138. *Mining Journal* 1 November 1884
139. *Mining Journal* 6 December 1884
140. *Mining Journal* 13 December 1884
141. Devon Record Office document 8187, Bedford Estate Mine Reports 1878-1886
142. Devon Record Office document 8187, Bedford Estate Mine Reports 1878-1886
143. Devon Record Office document 8187, Bedford Estate Mine Reports 1878-1886
144. *Mining Journal* 29 November 1884
145. *Tavistock Gazette* 28 November 1884
146. *Mining Journal* 29 November 1884
147. Devon Record Office document 8187, Bedford Estate Mine Reports 1878-1886
148. *Mining Journal* 17 January 1885
149. *Mining Journal* 31 January 1885
150. *Mining Journal* 14 February 1885
151. *Mining Journal* 21 February 1885
152. *Mining Journal* 21 February 1885
153. *Mining Journal* 7 March 1885
154. *Mining Journal* 18 April 1885
155. Devon Record Office document 8187, Bedford Estate Mine Reports 1878-1886

156.*Mining Journal* 25 April 1885
157.Devon Record Office document 8187, Bedford Estate Mine Reports 1878-1886
158.*Tavistock Gazette* 20 February 1885
159.*Mining Journal* 23 May 1885
160.*Tavistock Gazette* 22 May 1885
161.*Mining Journal* 13 June 1885
162.*Mining Journal* 26 September 1885
163.*Mining Journal* 6 June 1885
164.*Mining Journal* 10 October 1885
165.*Mining Journal* 11 July 1885
166.*Mining Journal* 28 November 1885
167.*Tavistock Gazette* 29 January 1886
168.*Mining Journal* 23 January 1886
169.*Mining Journal* 23 January 1886
170.*Mining Journal* 20 February 1886
171.*Mining Journal* 3 April 1886
172.*Mining Journal* 23 January 1886
173.*Mining Journal* 27 February 1886
174.*Mining Journal* 23 January 1886
175.*Mining Journal* 29 May 1886
176.*Tavistock Gazette* 11 June 1886
177.*Tavistock Gazette* 25 June 1886
178.*Mining Journal* 24 July 1886
179.Woolcock MSS
180.*Mining Journal* 27 November 1886
181.Devon Record Office document 8187, Bedford Estate Mine Reports 1878-1886
182.Woolcock MSS
183.Woolcock MSS
184.*Mining Journal* 27 November 1886
185.*Mining Journal* 27 November 1886
186.*Tavistock Gazette* 26 November 1886
187.Devon Record Office document 8187, Bedford Estate Mine Reports 1878-1886
188.*Mining Journal* 27 November 1886
189.*Tavistock Gazette* 7 January 1887
190.*Mining Journal* 29 January 1887
191.*Mining Journal* 2 April 1887
192.*Mining Journal* 26 March 1887
193.*Mining Journal* 1 January 1887

194. *Mining Journal* 15 January 1887
195. *Mining Journal* 12 March 1887
196. Woolcock MSS
197. *Mining Journal* 26 March 1887
198. Woolcock MSS
199. *Mining Journal* 28 May 1887
200. *Mining Journal* 28 May 1887
201. *Tavistock Gazette* 27 May 1887
202. *Tavistock Gazette* 27 May 1887
203. *Mining Journal* 28 May 1887
204. *Mining Journal* 2 July 1887
205. *Mining Journal* 9 July 1887
206. *Mining Journal* 6 August 1887
207. *Mining Journal* 13 August 1887
208. *Mining Journal* 3 September 1887
209. *Mining Journal* 3 September 1887
210. *Mining Journal* 24 September 1887
211. *Mining Journal* 19 November 1887
212. *Mining Journal* 22 October 1887
213. *Mining Journal* 19 November 1887
214. Woolcock MSS
215. *Mining Journal* 19 November 1887
216. *Mining Journal* 26 November 1887
217. *Mining Journal* 17 December 1887
218. *Mining Journal* 26 November 1887
219. *Mining Journal* 3 December 1887
220. *Mining Journal* 10 December 1887
221. *Mining Journal* 17 December 1887
222. *Tavistock Gazette* 30 December 1887
223. Woolcock MSS
224. *Mining Journal* 18 February 1888
225. *Mining Journal* 7 April 1888
226. *Mining Journal* 7 April 1888
227. *Mining Journal* 21 April 1888
228. *Mining Journal* 19 May 1888
229. *Tavistock Gazette* 18 May 1888
230. *Tavistock Gazette* 6 July 1888
231. *Mining Journal* 2 June 1888

232. *Mining Journal* 25 August 1888
233. *Mining Journal* 13 October 1888
234. *Mining Journal* 9 June 1888
235. *Mining Journal* 21 July 1888
236. *Mining Journal* 8 September 1888
237. *Mining Journal* 24 November 1888
238. *Mining Journal* 15 December 1888
239. *Mining Journal* 22 December 1888
240. *Mining Journal* 29 December 1888
241. *Mining Journal* 5 January 1889
242. Devon Record Office document 8187, Bedford Estate Mining Reports 1887-1898
243. *Mining Journal* 5 January 1889
244. *Mining Journal* 30 March 1889
245. *Tavistock Gazette* 10 May 1889
246. *Mining Journal* 18 May 1889
247. *Mining Journal* 25 May 1889
248. *Tavistock Gazette* 17 May 1889
249. Woolcock MSS
250. Devon Record Office document 8187, Bedford Estate Mine Reports 1899-1906
251. Woolcock MSS
252. Woolcock MSS
253. *Mining Journal* 23 November 1889
254. *Mining Journal* 23 November 1889
255. *Mining Journal* 30 November 1889
256. *Tavistock Gazette* 20 November 1889
257. *Mining Journal* 30 November 1889
258. *Mining Journal* 30 November 1889
259. *Mining Journal* 30 November 1889
260. Devon Record Office document 8187, Bedford Estate Mine Reports 1878-1886
261. Devon Record Office document 8187, Bedford Estate Mine Reports 1878-1886
262. *Tavistock Gazette* 6 July 1888
263. Report of the Council. As presented to the General Meeting, Barnstaple, July 29th, 1890. *Report and Transactions of the Devonshire Association for the Advancement of Science, Literature and Art,* Vol. 22, p.18. Barnstaple, July 1890.

Chapter 9

1890 – 1899: Dissent, decline and the weather

"The production from the mines is good, but unfortunately the prices are poor"
– Peter Watson, May 1897.

As the new decade started it was fairly evident that Devon Great Consols was a mine in serious decline. The abandonment of deep exploration at Railway Shaft was an acknowledgement on the part of the Company that the mine would not reinvent itself as a tin producer. Given the collapse of copper prices the failure to find tin must have been a body blow. In addition Main Lode, South Lode and New South Lode had "all been worked much beyond the limits of ore producing ground both east and west".[1] The only thing keeping the mine afloat was the large sales of arsenic.

1890

The decision to abandon work at Railway Shaft left just two areas of the mine being explored: Capel Tor and Watson's. On January 2nd Isaac Richards reported that they were preparing to sink below the 100 at Capel Tor.[2] In his January 1890 report Josiah Paull observed that "the explorations on Capel Tor Lode are quite unproductive".[3]

In his January 2nd report Richards noted that Watson's Engine Shaft was six fathoms three feet below the 160.[4] In order to facilitate development at Watson's rock drills were being installed; their first use on the mine since August 1884. However Richards noted that this had not yet happened due to bad weather and the engineers being engaged elsewhere including repairs to the locomotive.[5] Some progress had been made by the end of January, Richards writing on January 30th that air pipes had been laid between the 160 and surface. However, due to bad weather, no progress had been made laying pipes between the compressor at Wheal Emma and Watson's.[6] Work on the installation of air pipes between Wheal Emma and Watson's continued during February; on the 27th Richards observed that the Wheal Emma air compressor had been repaired and the laying of pipes was being pushed on with good speed.[7] On March 12th Captain Woolcock recorded that they had finished laying air pipes between Wheal Emma and Watson's Engine Shaft "all ready for working rock drills".[8] By the end of March 1890, Richards noted that the installation was complete and that a drill was in use on the 160 East. However, Richards remarked, "at present we only have one shift of men at this point of operation. We hope, however, to be able to obtain a greater force soon".[9] By the end of May a crosscut was being driven north by drill on the 172 fathom level at Watson's.[10] In Josiah Paull's opinion the reintroduction of rock drills was not a success, writing in December 1890 he noted:

> ..... the speed obtained by it has not been much more than would be got by an equal number of men by hand labour and the cost of compressing air, machinery and miners has been twice as great.[11]

Development and exploration progressed uneventfully throughout the remainder of the year at both Capel Tor and Watson's:

At Capel Tor the 74 east had been driven throughout 1890, the lode proving poor and unproductive. Limited work was carried out on the 86; a winze being communicated between the 86 and the 100 in April. The 100 west was driven several fathoms, the lode yielding "a good deal of mundic and a little copper ore".[12] Capel Tor Eastern Shaft had reached the 112 by the end of May 1890 and a crosscut was being driven north to intersect the lode.[13] The 112 crosscut intersecting the lode in August.[14] By year's end Isaac Richards, writing on December 18th, was able to report that the 112 at Capel Tor Eastern Shaft was producing "saving work of copper and mundic ores".[15]

At Watson's the Engine Shaft had reached the 172 fathom level by the beginning of April and had been cased and divided between the 160 and 172 fathom levels.[16] Work cutting the plat on the 172 was interrupted on April 18th by the breakage of the angle bob connecting the Watson's Engine Shaft pit work with Richards' wheel, however this had been repaired by the April 23rd.[17] Minimal work was undertaken on the 148 and 160 fathom levels, the lode being poor. The 172 was little better: The lode on the 172 west was "of average size but it is quite unproductive whilst the lode on the 172 east carried "a little copper ore and mundic, but not enough of either to value".[18]

The May 1890 half yearly meeting was held on the Tuesday 20th at the mine account house; as usual Peter Watson was in the chair. The meeting was somewhat overshadowed by the previous year's "unprecedented and almost annihilating fall in the price of copper",[19] which occasioned the decision to cease trading copper ore on March 21st 1889. Due to the loss of income from copper the mine had borrowed and additional £1,000 on top of the £3,500 borrowed the previous year. Watson informed shareholders that the mine had sold a parcel of 1,400 tons of copper ore "early this year" which had realised £1,630. Slightly offsetting the depressed state of copper sales Watson was able to report that he had been able to secure a good annual contract for the mine's make of arsenic. Further to this returns from the reduction works were up due to the introduction of the stone breaker and engine. The meeting was also informed that £2,500 of the £4,500 loan had already been repaid. The depressed state of the mine was reflected in the number of men employed with averaged 108 men earning a monthly average of £3 2s 6d. The cost of running the mine was evidently to the fore in many shareholders minds. Mr Oxland, a major shareholder, raised the question as to whether a purser was necessary and whether the mine could do without one. In spite of general concerns about costs the mine continued to support the school which was reported to have 40 pupils.[20,21]

Charles Oxland's comments regarding the role of the purser were far from idle speculation; being a major shareholder his opinions carried a great deal of weight. On Saturday 9th August 1890 Moses Bawden informed the men in the "Public Survey" that:

> this would be the last time that he would pay them at Devon Great Consols, the Directors, for the sake of economy, had thought fit to dispense with his services and abolish the office of Purser.[22]

Writing in his notebook Woolcock commented that this had been brought about by Mr Charles Oxland in "pursuance of his advocacy at the meeting on the 20th May".[23] Moses Bawden was, by all accounts, a very competent man and one must seriously question the wisdom of his dismissal on practical grounds. Aside from the fact that Moses Bawden was a highly skilled man it should not be forgotten that he was also very much Peter Watson's man, in effect his eyes and ears on the ground. This must have been a severe blow to Watson's not inconsiderable pride. In the light of subsequent events one cannot help but speculate that the removal of Moses Bawden was merely the first step in a concerted campaign against Watson.

On September 6th William Woolcock recorded the visit of Major Cundell to inspect the explosives store at Wheal Emma. Woolcock noted that the store contained 200 detonators, seventy five pounds of dynamite and five pounds of powder.[24]

The plot against Watson, if plot there was, would further unfold at the November 1890 half yearly meeting. The Directors' Report, issued prior to the meeting, highlighted serious differences of opinion amongst the directors. Differences had arisen between Peter Watson on the one hand and H. C Stewart and Hugh Stanley Morris on the other. On several occasions Watson had voted counter to Stewart and Morris causing the latter two to offer their resignation to the shareholders. Stewart and Morris, in consultation with "several major shareholders" (see Note 1) asked Watson to resign "in the welfare of the company" which, not unsurprisingly, Watson refused to do. In consequence Stewart and Morris decided to put the matter before the meeting stating that:

> should it be decided to continue the services of the Chairman, Messrs. Morris and Stewart will then resign, as these gentlemen feel strongly that it would be impossible for them to continue to take part in the management unless perfect harmony existed.[25]

Thus the scene was set for an exciting Half Yearly Meeting; the *Mining Journal* of November 29th subtitled its report of the meeting "Stewart & Morris vs. Watson"; rather in the nature of a prize fight. The meeting was held at the London offices on Wednesday 26th November, 1890. H. C Stewart outlined his and Morris' grievances with Watson at one point stating that:

> One of the many reasons which rendered it impossible to work with Mr. Watson was the latter's "enormous and everlasting egotism.[26]

A number of issues were addressed including Watson taking credit for purchase of the stonebreaker, issues concerning Watson's salary (see Note 2), the treatment of burnt halvans and the role of Moses Bawden. In addressing the allegations Watson described himself as "an old and tired servant" who "only wanted fair play and what was right".[27] In the discussion that followed debate went to and fro between shareholders. The *Mining Journal*, reporting the rapier like cut and thrust of the shareholders debate, noted that:

> Mr Witt delivered a long and somewhat tedious speech in support of Mr Stewart and Mr Morris.[28]

T. N. Roberts, who reappears later in the story of the mine, proposed what was, in effect, a vote of confidence in Peter Watson which was carried by thirty votes to two; the two presumably being Stewart and Morris. His position assured, Watson announced that he had received a letter from Stewart and Morris stating that "they had decided for the present not to resign their seats on the board".[29] In response to this Watson was able to turn the tables on Stewart and Morris stating that since they had brought unsubstantiated allegations against him, could they continue to work along side him? After some discussion as to whether Watson could work alongside Stewart and Morris, Watson agreed to do so. The matter was decided when Watson pointed out that the company's articles of association would not allow them to pass a decision regarding the Directors at that meeting and, in the words of the *Tavistock Gazette*; "the question was allowed to stand over till May next, Messrs. Stewart and Morris to retain their position on the directorate in the meantime".[30,31] So ended the meeting; Watson's position at the helm of the company stronger than ever, leaving Stewart and Morris humiliated, their positions, in effect, untenable.

As if to underline the fact that Devon Great Consols was a mine in decline Josiah Paull's annual report contained a note that the Company was engaged in securing disused shafts (see Note 3).[32]

The only thing in favour of the mine's ongoing survival was arsenic. Fortunately the Main Lode in both Wheal Emma and Anna Maria was yielding "large quantities of good arsenical mundic". During 1890 both production and sales of arsenic had been large and the average price good. During 1890 the make of arsenic consisted of 2,441 ¾ tons. 2,615 tons were sold at an average price of £10 18s 1¾d per ton which realised £28,523.[33] As long as the mine had a market for its arsenic it had a chance of survival. In December 1890 Josiah Paull observed:

> The main stay of the mine is the arsenic works, it can be kept going for some time, I would venture to say four or five years, by the return of mundic from

underground and from the halvans heaps at surface, vigorous efforts to this end continue to be made by the Agents, they will leave nothing untried in the old workings and nothing will be left behind.[34]

1891

On January 14th 1891 Francis Charles Hastings Russell, the 9th Duke of Bedford shot himself, apparently as a result of insanity while suffering from "pneumonia". He was succeeded by George Russell, the 10th Duke.[35]

The start of 1891 was marked by very severe weather conditions, writing on January 1st 1891, Richards observed that the weather was "making sadly against all surface operations".[36] March 1891 was also notable for poor weather which affected operations on the mine. A storm on March 9th doing considerable damage.[37] On March 19th Isaac Richards wrote:

> I am very pleased to say that since the storm of last week has abated all surface operations are proceeding with regularity. The railway to Morwellham has been cleared of snow and running over the line was established on Friday last.[38]

In his six monthly report of May 1891 Richards reported to shareholders that the storm had, in addition to blocking the railway, damaged a number of buildings; the cost of repairs and clearing the snow amounting to £150.[39]

Richards' six monthly report contains further details of interest: An average of 91 men earning a monthly average of £3 3s were employed on the mine during the preceding six months. During that six month period eighty eight fathoms, four feet and six inches had been driven on the mine at an average cost of £9 0s 6d per fathom. As usual Richards felt that Capel Tor was looking promising as was Watson's. The 172 at Watson's yielding four tons of copper and mundic with a little tin per fathom. Richards also reported that two trials to the west of Watson's had produced good specimens of tin and he recommended that Watson's be driven west at depth towards the granite.[40]

The May meeting was held on Tuesday 25th at the London Office. Peter Watson, in the chair, was not a well man. Never one to do anything by halves, Watson informed the meeting that "I can only say it is something more than influenza....".[41] The report Watson made to the meeting was a catalogue of disappointments, for example the long and severe winter had militated against both surface and underground operations. More serious was the failure to secure a contract for the mine's arsenic. As a direct consequence of the failure to sell the arsenic the company was forced to take further loans during the preceding six months totalling £3,500; bringing the total owed to the bankers up to £5,500. Watson was quick to assure the meeting that the company would be able to meet their liabilities: The company was in credit with the bank to the sum of £1,199 18s 11d, it was due to receive £1,521 for a recent sale of copper and the stock of arsenic in store would cover the rest. The stand off between Watson, Stewart and Morris ended tamely at the meeting. Stewart had died since the November 1890 meeting whilst Watson and Morris, in accordance with the articles of association resigned. Watson was re-elected as a director whilst Morris did not stand for re-election. Morris and Stewart's places on the board of directors were taken by T. N. Roberts and Thomas Glen, both of whom appear to have been more amenable to following Peter Watson's "lead".[42,43]

Having consolidated his position Watson took the eminently sensible step of re-appointing Moses Bawden as the mine's purser; William Woolcock noted that Bawden was reappointed on June 16th, 1891 at £10 10s per month.[44]

The reappointment of Moses Bawden was merely a foretaste of further fundamental changes in the management of the mine. The following letter appeared in the *Mining Journal* of July 18th, 1891:

> To the Editor of the Mining Journal
> Dear Sir – I regret to have to inform you that my services as secretary to this company have been dispensed with by the board of directors who were so recently appointed as the 26th May last. I think it right to appraise you of this, as after 17½ years service in this company (and my uncle having previously served the company as secretary from its commencement in 1844 until his decease), this action of the directors seems unwarrantable and unjust, especially as no reason has been assigned – yours faithfully W. H. Allen.[45]

On July 2nd and 3rd 1891 the directors visited the mine on what might be termed a fact finding mission.[46] Captain Woolcock reported on their visit in an entry in his notebook dated July 2nd:

> Mr Peter Watson, Mr Glen, and Mr Roberts, Devon Consols Directors visited the mines at 10am, went to Wheal Anna Maria dressing floors (the

before named, Captain Isaac Richards, Mr Bawden and Captains Clemo and Woolcock) to the stone breaker and Arsenic Works. The Directors, and Mr Bawden went to Morwellham, returned to Count House at 2.30, held a meeting. Captain Isaac Richards called in first; Woolcock second; and Clemo third, to give their opinions concerning the prospects of the mine.[47]

The July 2nd meeting held between the directors and Isaac Richards was minuted by Peter Watson. The details of the meeting were outlined in a statement issued by T. N. Roberts to the shareholders in May 1892 and subsequently reproduced in the *Mining Journal*:

Mr W. To Capt. Richards: Can you reduce cost?
A: Not at all.

Q: Have you looked into this matter?
A: Yes, I have thoroughly.

Mr Roberts Q: Is there any machinery that can be dispensed with?
A: No I cannot see we can.

Mr. W, Q: If you had more men could you put them on?
A. Ask Captain Woolcock.

Mr Roberts, Q: How many more men could you profitably employ?
A. It would be difficult to say.

Q: What is to be done if we don't explore, cannot you try to meet expenses, because there is a loss monthly?
A. I have done my best to do so.

Mr Glen Q: Do you go down the mine often yourself?
A. No Sir, I don't.

Mr W. Q: How often?
A. About once a month, Captain Clemo and Captain Woolcock go underground.

Mr W. Q: Have you any suggestions to make as to increase of returns or decrease of cost?

A. No Sir I have not.

Mr Glen Q: Might a saving be affected by additional of improved machinery?

A. I cannot see that it can.

Mr R. Q: You have got one drill at work – why not put three drills to work at Watson's if our prospects are so good there, and is there any point elsewhere that could be developed?

A. There is no other place I could recommend at present.

Mr W. Q: Can any reduction at surface be made?

A. Not at a single place.[48]

On the following day the directors again met Captain Richards, the meeting again being minuted:

Mr W. Q: Can you suggest anything for the benefit of the company?

A. No I cannot.

Mr R. Q: Is there no way in which costs can be reduced and output increased?

A. I don't see that they can.

Mr R Q: Can the expenses at the dressing floors be reduced?

A. No, I cannot see they can.[49]

Richards' replies did not inspire much confidence and the directors "came to the conclusion that in the interest of the company the services of Captain Isaac Richards as an agent at the mines should be dispensed with". Richards' dismissal became effective from July 10th 1891 (see Note 4).[50] Richards was replaced as chief agent by William Clemo.

On July 13th a meeting was held at the mine's Count House attended by "Mr Bawden, Mr Mathews, Mr Thomas Youren, Captains Clemo and Woolcock". William Woolcock recorded the meeting for posterity in his notebook:

Mr Bawden read resolution passed by the Directors, hoping all would work together, and do their very best to lessen cost where possible and get more

> men if possible, at bigger wages, and raise more copper and mundic. Mr Bawden said Captain Clemo would take the lead, and he had no doubt Captain Woolcock would work harmoniously with him.[51]

Isaac Richards issued a circular dated July 22nd 1891, to shareholders outlining his achievements on the mine. These included laying out whole of surface arrangements including watercourses, dressing floors, the railway, the copper precipitation works and the arsenic works arsenic works. Richards also highlighted his general management of the mine and the analysis of copper and arsenic throughout mines. He also hinted that his dismissal was linked to the reinstatement of Moses Bawden the previous month (see Note 5).[52] On July 23rd William Woolcock noted Moses Bawden as saying that couldn't think why Captain Richards had made such pointed allusions to him (Bawden) as he hadn't tried to injure Richards in any way.[53]

In response to Richards' circular the directors responded with a circular of their own authorised on the July 22nd. The directors made it clear that what Richards did twenty, thirty or forty years previously was not relevant to the current situation. The circular made the point that:

> Captain Richards was afforded every opportunity of giving to the directors information and explaining anything he desired when the directors were at the mines on the 2nd and 3rd inst., the unsatisfactory answers to the questions they put were such that they had no other alternative but to dispense with his services at the following board meeting.[54]

The circular also highlighted the fact that whilst Richards was only entitled to a month's notice the company would pay him for the period of July to October; a sum of £61 4s. What the circular did not mention was that Richards had also been given notice to leave his house in Endsleigh Terrace, Tavistock, by August 7th.[55]

Whilst one can argue that the nature of Isaac Richards' dismissal did the directors little credit, it is hard to see what other option Richards had left them if the report of their meetings is accurate. Although Richards' dismissal was the main subject dealt with in the circular the shareholders were also informed of the re-appointment of Moses Bawden and also that the Company had managed to enter into an advantageous contract for their arsenic at the end of June.[56]

Captain Woolcock's entry in his notebook for Sunday September 27th 1891 struck a melancholy note:

> Captain Isaac Richards' last choir service at Ogbear Chapel (after 42 years regular attendance), in consequence of leaving the mines - (Devon Great Consols) to reside at Tavistock - left on Tuesday 29th inst. The once celebrated Ogbear Chapel Choir is now reduced to one only of its old members – viz. - W. Woolcock and is now assisted by the Misses Mitchell occasionally, and Mr Waterfield junior.[57]

William Clemo who had, for many years, been chief underground agent, quickly found his feet as chief agent. His first report in the pages of the *Mining Journal* appeared on July 18th 1891. This report is interesting as it takes a significantly different approach to the reports submitted by Isaac Richards. Whereas Richards' reports focussed on exploration and development, Clemo's focussed on production. This may have been a reflection of Clemo's career to date as underground agent where he would have been concerned with the day to day operation of the mine whilst Richards' position inclined him to a more strategic approach. Whatever the reasons may have been Richards' reports always seemed to promise the earth and deliver very little whilst Clemo's focussed on current production which was tangible and would have demonstrated to shareholders that, setbacks aside, Devon Great Consols was still a very productive mine.

Clemo's first report outlined current points of production in the mine: At Capel Tor Eastern Shaft a stope in the back of the 36 west was producing four tons of arsenical mundic with copper ore per fathom; a second stope in the back of the 36 east was producing similar amounts of ore. At Wheal Anna Maria a stope in the bottom of the 110 fathom level east of Engine Shaft was worth ten tons of mundic per fathom. There were a couple of reasonably productive stopes in Wheal Josiah: A stope on the bottom of the 130 east of Hitchins' Shaft on South Lode was producing three tons of copper and two tons of mundic per fathom whilst a stope in the back of the 130 east of Hitchins' Shaft was worth six tons of mundic per fathom. According to Captain Clemo Wheal Emma was by far the most active section of the mine. A stope in the bottom of the 110 fathom level east of Thomas' Shaft was producing an impressive twenty tons of mundic per fathom. At Inclined Shaft a stope in the back of the 150 east was yielding three tons of copper and two of mundic per fathom. Two stopes were being worked on the 112: The back stope on the 112 west was worth three tons of copper and one of mundic per fathom; the second stope in the back of the 112 east was less productive yielding one and a half tons of copper and two tons of mundic per fathom.

The stope in the back of the 100 west was producing eight tons of mundic and "saving work" of copper". A number of stopes were also active on New South Lode in the vicinity of New Shaft: The deepest stope on New South Lode, and indeed in the whole mine, was in the bottom of the 190 west and which was yielding five tons of mundic per fathom with "a little copper saving work". In the back of the 175 west of New Shaft a stope was producing two tons of copper and two tons of mundic per fathom. A stope in the back of the 145 east was recorded as producing one and a half tons of copper and one ton of mundic. On the 130 east a stope in the back of the level produced three tons of copper ore and two of mundic per fathom. Finally a stope in the back of the 110 east yielded six tons of mundic. Three stopes were active at Watson's Mine: The 172 west was producing "saving work of copper and mundic. A stope in the back of the 160 west of Watson's Engine Shaft was producing four tons of copper per fathom as was the 160 east.[58]

In mid October 1891 William Woolcock noted that all the night work at the halvan grinder and floors was stopped, and about thirty hands were discharged.[59]

The November meeting was held on Wednesday 25th, at the company's office at Winchester House, Old Broad Street. The *Tavistock Gazette* subtitled their report "Exciting Meeting". The excitement came largely from a shareholder called Edwin Sloper who had, prior to the meeting, produced a circular attacking the management of the mine. Sloper made a number of accusations relating both to allegations of financial irregularities and also the dismissal of Richards, all of which were refuted by the directors.[60,61] Josiah Paull considered the affair to be rather pointless; writing in December he commented that:

> I am sorry to say that in my belief the mine is so near its end as not to make it worth the while of the general body of the shareholders to quarrel over the management. It is strange to all of us who are familiar with the real condition and prospects of the mine that the shareholders do not see that year by year its returns are decreasing and that its reserves are so nearly worked out".[62]

Whilst his performance at the November meeting was Sloper's first real attempt to hijack a Devon Great Consols meeting it certainly was not his last. Throughout the 1890s he continued to interrupt meetings with a never ending stream of attacks against the mine's management. As the years passed Sloper increasingly became something of a standing joke whom no one (including the current author) could take seriously. For example in 1892 *Mining Journal* described Sloper as the "arch priest of discord and disturbance".[63] Due entirely to Sloper, Devon Great Consols was in danger of becoming a laughing stock; in June 1893 a *Mining Journal* editorial commented that:

> Of late the meetings of this company have latterly become famous for the lively and almost farcical nature of the proceedings; in fact it has even been said that many people bought shares in the company merely to have the privilege of attending the meetings. Mr Sloper was the sole cause of these protracted and stormy meetings.....[64]

William Clemo's final report of the year, dated December 31st is broadly similar to his first report of mid July. At Field Shaft a stope in the bottom of the 130 west has come into production, likewise Clemo reports on a stope the back of 115 west of Richards' Shaft not included in his July report. A new stope in the back of the 100 fathom level east of Thomas Shaft has come into use. At Inclined Shaft at Wheal Emma the stope in the back of the 112 west appears to have ceased production as has the stope in the back of the 100 east. On New South Lode at New Shaft the stopes of the 145, 130 and 110 appear to have ceased production whilst a stope on the 100 east has been added to Clemo's list. At Watson's Shaft, Clemo makes mention of production from the 148 fathom level west of Watson's Engine Shaft.[65]

Whilst William Clemo was keen to focus on production there was no getting away from the fact that the mine was living on borrowed time. The only exploratory work undertaken during 1891 was on Capel Tor Lode and at Watson's. Capel Tor Lode was, with the exception of some very rich ground in the back of the 36 west, unproductive from the 24 to the 112. Since June 1891 "no driveage or other exploration" had been carried out and Josiah Paull was "pretty certain that no further outlay will be made on it".[66]

Watson's was pushed with more energy during the year: the 148 east advanced thirteen fathoms, the 148 west an impressive sixty fathoms; the 160 east thirty eight fathoms, the 160 west twenty five fathoms, the 172 east sixteen fathoms and the 172 west forty four fathoms. A boring machine was in use on the 172 west. At the end of the year Paull was able to report that Watson's had been worked with “some spirit and with rather better results than I had expected”.[67]

During the year only around one thousand tons of mundic and copper had been discovered which was not enough to secure the mine’s future. What little future the mine had was dependent on stripping mundic from the Main Lode (see Note 6) and reprocessing the halvans dumps. In Paull’s opinion this would save the mine from any serious losses although, in his opinion, it could not hope to make a profit.[68]

1892

Industrial unrest surfaced on the mine in the spring of 1892. In a note dated “March 28th and after” Woolcock reports that “the Arsenic Men refused to work for the wages offered, and the hours imposed”.[69] The dispute between the mine and the arsenic workers appears, in part at least, to have been fuelled by outside influences. On April 4th William Woolcock observed that:

> The Gas Workers Union delegate – “Gardner”, again at the D.G.C. mines, said two of the men at Arsenic works had refused to join the union and unless those two joined he would call out all the others.[70]

The arsenic workers demanded that the two men who refused to join the union were expelled. In addition they also demanded an increase in pay “very much in advance of what has been paid”. Josiah Paull was of the opinion that whilst the work was “disagreeable and not healthy” a day’s pay of between 4s 3d to 4s 6d was “pretty good”.[71]

Relations between the Company and the arsenic men were tense, Paull noting that “the employees at the arsenic works (numbering some 80 altogether) have given a great deal of trouble to the agents”.[72] In spite of the tension an all-out strike of the arsenic men appears to have been avoided. On April 16th; Woolcock commented that the arsenic works was “not producing so much fine arsenic owing to the men (at day work), not calcining so much”. As a result of the reduction in production four coopers were discharged.[73] The laying off of the coopers must have set alarm bells ringing across the mine. Given that the mine was almost wholly dependent on arsenic a strike at the arsenic works would have had very serious consequences for the “300 other work people on the mine who had nothing whatever to do with the struggle between the arsenic employees and the Company”.[74]

The high water mark of the dispute came on Saturday April 23rd, the day the

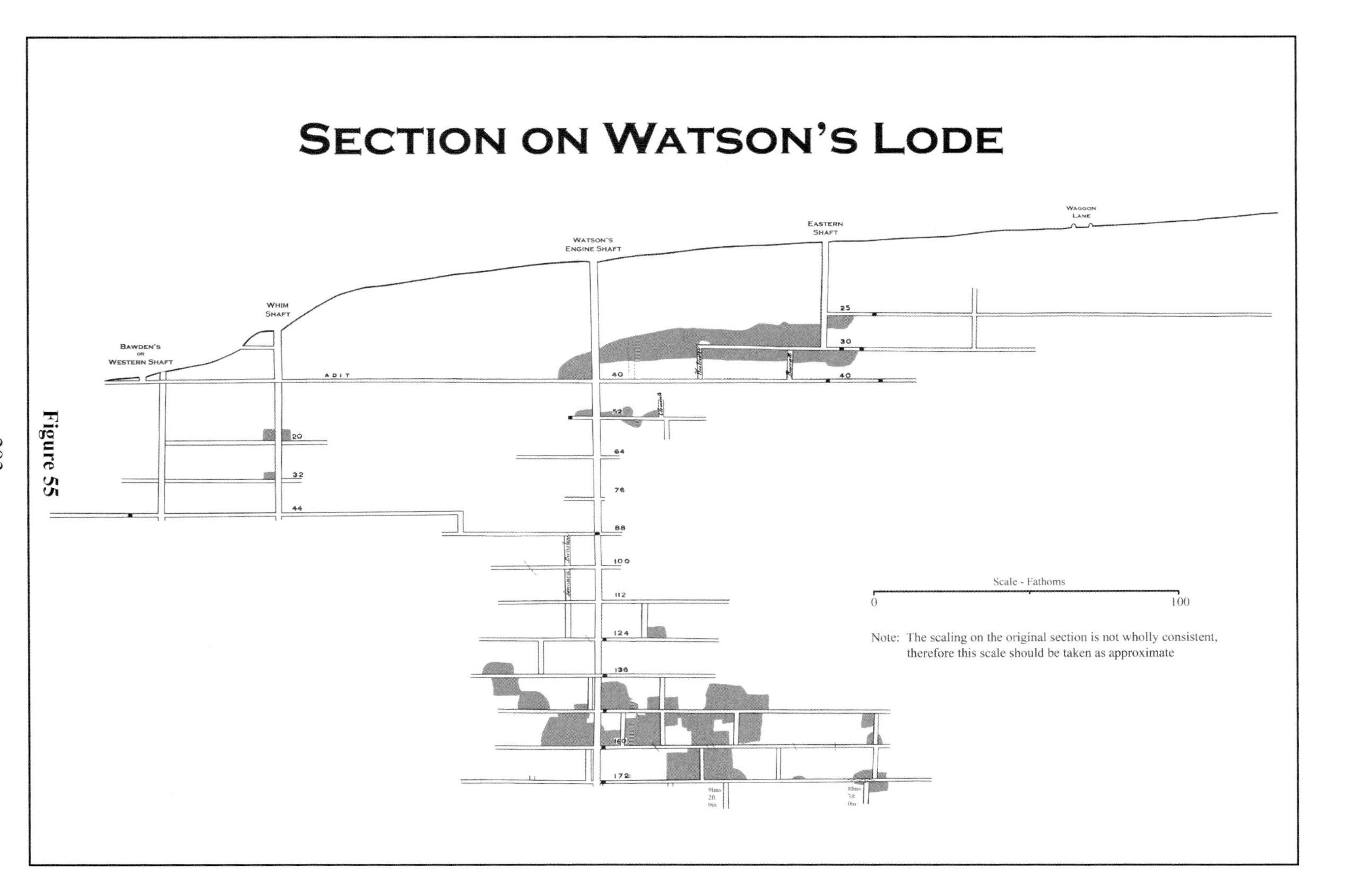
SECTION ON WATSON'S LODE
WAGGON LANE
EASTERN SHAFT
WATSON'S ENGINE SHAFT
WHIM SHAFT
BAWDEN'S OR WESTERN SHAFT
ADIT
25
30
40
40
52
64
76
88
100
112
124
136
160
172
20
32
44
Scale - Fathoms
0
100
Note: The scaling on the original section is not wholly consistent, therefore this scale should be taken as approximate

Figure 55

arsenic workers contracts were due for renewal. The arsenic men "seemed determined to have an advance in wages, payment for time not actually spent in the works and the expulsion of the non unionists".[75] However the "Manager and Purser", with the full authority of the Directors, stood their ground and refused to concede to the men's demands. The Company's firm stand served to highlight the differences amongst the men; Paull commenting that "their combination is weakened for a time at least, if not broken altogether".[76]

At the May 1892 meeting Moses Bawden informed shareholders of the situation regarding the dispute with the arsenic men. Bawden told the meeting that the arsenic workers had joined the Gas Workers Union:

> and thought they would be able to do what they liked with the company. The men were then preparing to demand extra pay or they would go on strike. To prevent that they took the whip hand, and gave notice to them that after a given time the company would do away with all night work on the dressing floors....[77]

Evidently the firm hand taken by the company worked; in July 1892 William Woolcock wrote:

> 4 of the ring leaders of the Unionists, Arsenic Men discharged, the others soon after took their contracts up to 3rd November - old prices and conditions as formerly.[78]

At the time the dispute with the arsenic workers was gathering pace Josiah Hugo Hitchins died aged 85; his death on April 4th passing unnoticed both on the mine and in the wider district. The *Mining Journal* commented that Hitchins lived "latterly much reduced in circumstances, being for many years previous to his death almost in poverty". For the last five years of his life he had been bed-ridden and had been nursed by his widowed daughter Mrs MacDonald. Their only source of income had been a quarterly allowance of £10 made by the Duke of Bedford (see Note 7). On Josiah's death the pension ceased and Mrs MacDonald was in real danger of becoming destitute. The *Mining Journal* appealed to its readers to "subscribe to the relief of Mrs MacDonald".[79] A pathetic and sad end to a sometime extraordinary life.

The May 1892 meeting was held on Thursday 19th with Peter Watson in the Chair. Watson highlighted that fact the changes that had been made in the local management of the mine had been highly beneficial, noting that "the monthly costs have been considerably reduced, whilst the returns of mineral have been increased".[80] During the preceding six months the mine had sold 1,285 tons of copper ore realising £1,616 at an average price of £1 5s 2d a ton whilst sales of arsenic had realised £11,361. The company had managed to reduce its debt to the bank to £1,000 from the £4,000 it had stood at the time of the November 1891 meeting.[81,82] This appears to have been achieved by a financial sleight of hand; the Company's credit balance had been reduced by "more than £2,000" since October 1891.[83] The rest of the meeting was "a very noisy one" taken up with petty and tedious squabbling largely driven by Sloper ably supported by Dr Oxland and Isaac Richards.[84,85,86] These squabbles spilled out into later editions of both the *Tavistock Gazette* and the *Mining Journal*.

The dissent at the May meeting was, in part, fuelled by the poor price the Company was receiving for its arsenic. It was noted that the mine's arsenic was realising between thirty and forty shillings less than it had done previously. This appears to have been a consequence of the dispute with the arsenic men. Josiah Paull forwarded the view that arsenic was of poor quality and that "this had been brought about....... by the unsatisfactory labour done by the employees at the works for some months past".[87]

The *Mining Journal* of June 11th 1892 commented that Devon Great Consols would be sampling 440 tons of copper ore, and increase of 120 tons compared with the mine's previous sale.[88]

On July 25th William Woolcock recorded an unusual occurrence:

> Robert Northey - in coming up from the 47 fm level at Wheal Emma, D.G.C. brought up a live Hare - half grown, which he found on the Man Engine Sollar between the 47, and 32 fm levels (see Note 8).[89]

The summer of 1892 was a notably dry one which had implications for operations on the mine. On September 8th William Clemo reported that:

> Owing to the very dry summer, the River Tamar has been so low, that the large pumping wheels have been working very slowly, and we were almost obliged to start the Anna Maria steam engine, but rain has recently fallen, and all the wheels are going again at full speed.[90]

On October 15th the *Mining Journal* carried news of the death Thomas Nicholls Roberts; his obituary described him as "a man of unswerving rectitude, a severe disciplinarian".[91] He was replaced as a director at the November meeting by F. G. Lane.[92]

Prior to the November 1892 meeting Captain Clemo submitted his report, dated November 14th, to the shareholders. On surface considerable repairs had been made to Latchley weir. Latchley weir, it will be recalled, was fundamental to the operation of the mine being the take off for the Great Leat which supplied the mine's large wheels.[93]

During the previous six months the mine had employed an average of 110 tutworkers who, on average, received £2 19s a month. Driveage on the mine totalled thirty four fathoms, three feet and eleven inches at an average cost of £10 11s 3d per fathom. Winzes had been sunk seven fathoms, one foot and ten inches during the preceding six months at an average cost of £11 14s per fathom. Such limited development work clearly indicates that the active exploration of previous years had been abandoned, emphasis being placed firmly on production with stopes active in Capel Tor, Anna Maria, Wheal Josiah, Wheal Emma and Watson's. At Inclined Shaft sixty fathoms of old pitwork had been replaced with new. Watson's Mine, as ever, was continuing to open up in a satisfactory manner. Clemo alluded to labour problems on the mine whilst reassuring shareholders that these had now been resolved.[94]

The November 1892 meeting saw a resumption of hostilities between the company on the one hand and the Sloper faction on the other. The flames of discord were fanned by two circulars issued by Sloper on November 14th and November 18th. Sloper's attacks on the financial management of the company were refuted by the directors in a circular dated November 17th and two further letters from Peter Watson dated November 20th and 24th.[95]

The meeting was held on Wednesday 29th November 1892. During the preceding six months up to October 31st, the mine had sold 1,189 tons of copper ore at an average price of £1 10s 6d per ton, realising £1,813 9s 9d. For the same period the mine sold £10,688 10s 9d-worth of arsenic. The loan account had been reduced to a manageable £500. Watson made the point that changes in local management had been vindicated "in the fact that whilst the expenditure has been largely reduced the reserves of mineral have been increased by about eighty per cent during the past twelve months". Peter Watson having presented his report, the

Figure 56. A mineral train, headed by the Estate's 20HP "Simplex", ready to leave Frementor during the 1925-1930 period.

Figure 57. A group of miners on the Frementor railroad in 1926.

Figure 58. A 1928 view from the New Dressing Plant across the Anna Maria dumps.

Figure 59. W. Honey on the Frementor railroad in 1926.

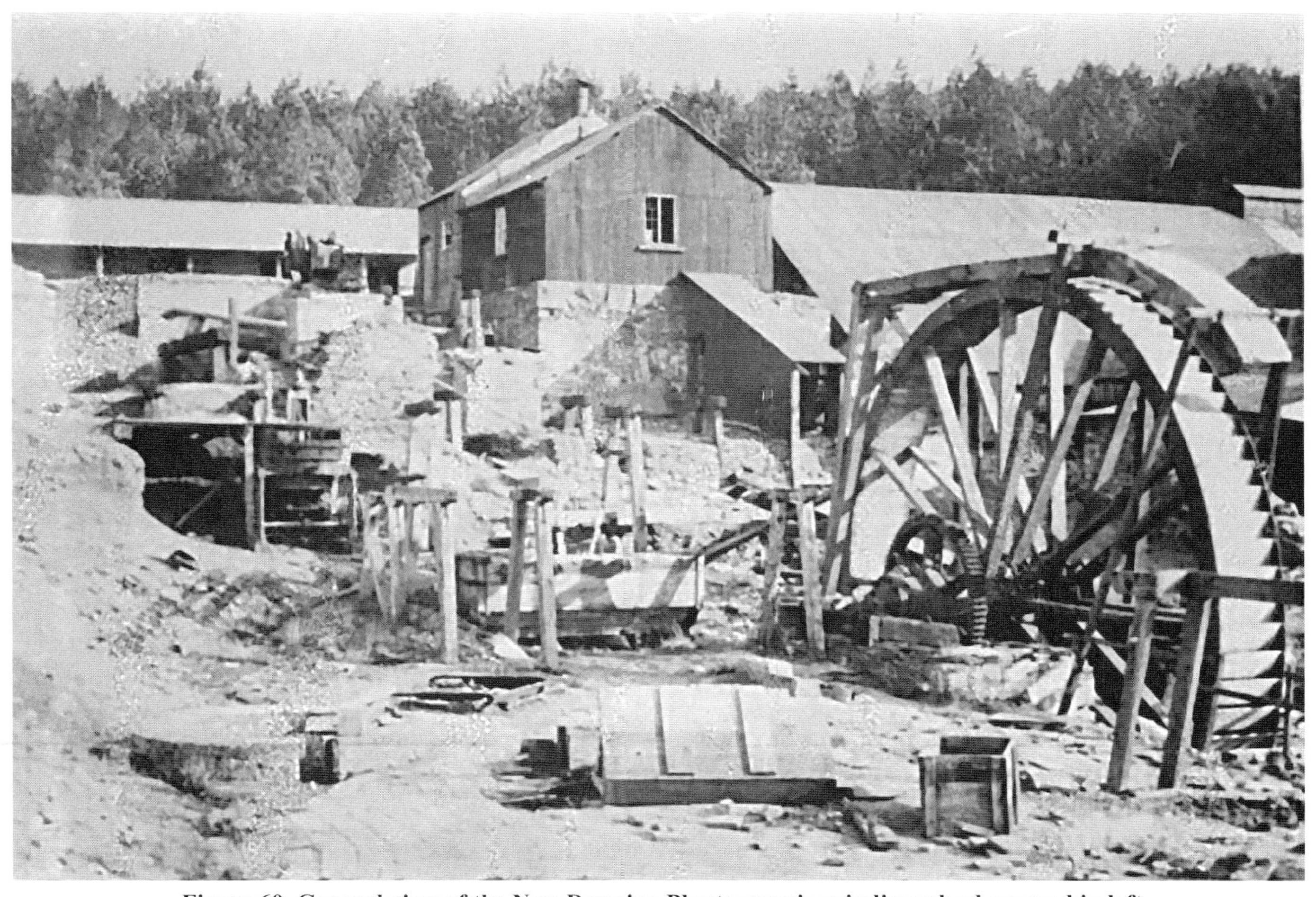

Figure 60. General view of the New Dressing Plant; arsenic grinding wheel exposed in left background, new mill buildings to right.

Figure 61. The Estate's 20HP "Simplex" complete with locally added canopy.

meeting, with Sloper acting as a catalyst, became, in the words of the *Tavistock Gazette*, "animated". Setting squabbles and argument aside Peter Watson touched on a matter of supreme importance, the renewal of the company's leases with the Duchy of Cornwall and the Bedford Estate which were due to expire in 1893. Watson intimated that they were negotiating for a seven year lease with the Duke of Bedford. The lease would be for a reduced sett, about 700 acres smaller than the current one. In terms of royalties Watson had negotiated the royalty down from one eighteenth to one thirtieth, to be paid from such a time as profits were made until such time as dividends of £15,000 had been paid to the shareholders, when it would increase to one twenty fourth.[96] Somewhat exasperated with the general mood of the meeting Watson addressed the gathered shareholders:

> Now gentlemen, do I deserve the treatment I have been subjected to after working for you and saving you thousands of pounds in this matter?[97]

One can certainly appreciate Watson's frustration; he had the difficult task of steering the mine through difficult and uncertain times, a task he was doing with no little success. It must have been galling for Watson to have to waste time on an idiot like Edwin Sloper when there were much more serious demands on his limited time.

1893

1893 started cold, William Clemo noting on January 5th that the frost during the preceding week had been severe and that the water wheels were "very much clogged". However by constant clearing they were kept at work without hindering operations on the mine.[98] On January 12th Clemo reported that a winze was being sunk below the 172 at Watson's.[99] At the end of January stopes were active at Capel Tor, Anna Maria, South Lode at Field's Shaft, Richards' Shaft, South Lode at Hitchins' Shaft, Thomas' Shaft, Inclined Shaft, New South Lode at New Shaft and Watson's.[100]

The 10th Duke of Bedford, George Russell, died of diabetes on 23rd March 1893 at the very young age of 40. His tenure as Duke was short, having inherited the title in 1891. He was succeeded by his younger brother Herbrand Russell who held the title from 1893 until 1940.[101]

Prior to the signing of the new leases an Extraordinary General Meeting was held on Wednesday April 12th, the purpose of which was to make "various alterations" to the company's articles of association. Since the November 1892 meeting the royalty had been fixed, according to the *Tavistock Gazette*, at one thirty eighth (the *Mining Journal* quoted one thirty sixth). The shareholders were also informed that the £500 debt had been paid off and that a twelve month contract "at a considerably increased price" had been entered into.[102,103]

The amended articles of association were confirmed by a further Extraordinary General Meeting held on April the 28th. Edwin Sloper, as usual, raised his objections.[104]

The May meeting of the company was held of Tuesday 30th at Winchester House. The directors' report noted that the mine had sold £1,818 12s 6d worth of copper ore and £13,457 16s 3d of arsenic during the six months to the 30th April. Mention was also made of the new arsenic contract and the clearing of the company's loan account. Leases had been agreed with the Duke of Cornwall and H.R.H. the Prince of Wales, presumably for water. The new leases from the Duke of Bedford had been prepared. Watson informed the meeting that some "little technical matters" needed to be rectified and he hoped to have the new leases in place "in a short time". The mine was in a sound financial position and the directors felt themselves in a position to authorise a three shilling dividend, a decision which was met with cheers from the meeting.[105] As usual Sloper had to have his say, however on this occasion he found himself short of willing allies, the *Mining Journal* noting;

> Mr Sloper was in evidence on Tuesday as usual but seemed so forsaken and alone, that those who remembered the support he was accustomed to receive from his former friends, felt inclined to sympathise with him.[106]

Like the preceding year the summer of 1893 was a dry one, this of course had implications for the mine. William Clemo noted on June 21st that:

> Owing to the prolonged drought and extreme heat the River Tamar has become so low that the speed of our water wheel is much reduced and we are obliged to run the steam engine at Wheal Anna Maria in order to keep the underground workings drained.[107]

The dry weather continued into July, William Clemo cataloguing matters on a weekly basis in his *Mining Journal* reports:

> June 29 – In consequence of the long drought the water has been so low

in the River Tamar that the wheels have failed to drain the bottom levels. Some heavy rain has, however, now fallen, and we hope to have all the bargains in full work in the course of a few days.[108]

July 6 – We have with the help of some heavy rain been draining the bottom levels at good speed, and we hope to have the water in fork in the course of a few days.[109]

July 13 – The water in the Tamar has scarcely been enough to drive the wheels with sufficient speed to fork the bottom levels, but the draining is now going on satisfactorily after a very long continuance of dry weather.[110]

July 20 – With the present favourable rain our wheels are working with tolerable speed and if the weather continues we hope to get the bottom levels in fork in the course of another week.[111]

July 27 – The mine will we hope be in fork to the bottom by the end of the week.[112]

Aug 3 – I am glad to inform you that the water is in fork throughout the mines, and all points of operation are again in full work.[113]

With the mine back in fork Captain Clemo may well have breathed a sigh of relief. However his battle with the weather and the mine's waterpower system was far from over. On September 7th Clemo reported the "breaking in of the leat over the lobby of one of our large wheels". This caused some ground to give way which, in turn stopped the wheel which was draining Wheal Josiah and Watson's. This stoppage, combined with further dry weather meant that Clemo had to resort to steam power to keep the workings at Anna Maria and Josiah drained.[114]

Apart from being temperamental, the mine's water power systems could also be dangerous. In November 1893 a Gunnislake man, Henry Burridge, was greasing the wheels which carried the flat rods. His jacket became entangled between the flat rod and the wheel he was greasing, fracturing his left arm close to the shoulder.[115]

The new leases were signed on 22nd November 1893 and were to run for seven years to the 25th March 1900. Of particular interest was the reduction in Lord's dues to 1/36th.[116]

A further dividend of three shillings was announced at the November 1893 meeting, held on Wednesday 29th. The *Mining Journal* noted that the dividend was the 156th in the mine's history. The shareholders were informed that, since the previous meeting, the Duke of Bedford had, with his London and local representatives, visited the mine. Watson was pleased to announce that the company had managed "execute" the new leases.[117,118]

1894

The year began, as was rapidly becoming traditional, with a William Clemo report decrying the prevailing weather conditions: On January 11th 1894 he wrote that the weather had been severe during the last week.[119] In the same *Mining Journal* report Clemo noted that Watson's was active on the 172, 160 and 148 fathom levels. He expanded his comments of Watson's on January 18th, commenting that "the stopes are all turning out large quantities of copper and mundic ores".[120] There was plenty of activity elsewhere on the mine in late January 1894. At Anna Maria Engine Shaft a stope was being worked in the back of the 110. On South Lode at Field Shaft at stope was active in the bottom of the 130. At Richards' Shaft a stope was active in the bottom of the 103 fathom level. Two stopes were being worked at Agnes Shaft; one in the back of the 103 west, the other in the back of the 90 west. As usual most activity was focussed on Wheal Emma. At Thomas' Shaft two stopes were noted in the bottom of the 100 east and one in the back of the same level. Four stopes were active at Inclined Shaft: two in the back of the 150 east, one in the back of the 112 and one in the back of the 110 fathom level west. At New Shaft on New South Lode two stopes were active on the 190 fathom level. At Watson's, in addition to the activity mentioned on the 172, 160 and 148 fathom levels Clemo also notes a stope on the 136 west.[121] It is interesting to note that no activity is reported at Capel Tor, presumably work had been suspended by this time.

On February 8th William Clemo reported that the mine was going to sample 340 tons of copper ore for sale at Redruth on the 22nd of the month.[122] The May 1894 meeting was held on Wednesday 30th at Winchester House. Peter Watson was able to reflect on a successful year during which the mine had sold £28,951 of copper and arsenic and had paid two three shilling dividends. Since November many men had been laid up with influenza, Captain Clemo noting that at one point nearly forty men, including some agents, had been absent, including both Moses Bawden and William Clemo. Peter Watson paid fulsome tribute to Clemo stating:

> there is no man in the two counties of Devon and Cornwall who has so vast an experience of mining as Captain Clemo, and none that stands in higher estimation of his fellow workmen.[123]

It was noted that during past six months the refiner furnaces had been rebuilt with several of the roasting furnaces, and the reduction works had been extensively repaired. The subject of coal was touched upon, the shareholders being informed that the mine was paying twelve shillings per ton compared with the market price of twenty to twenty one shillings. As usual Edwin Sloper was making a nuisance of himself.[124]

Unusually the reporting of the meeting included quite extensive reference to the engineers report:

> No. 2 boiler has just had some repairs. The 30-inch whim engine, No.1 boiler has been repaired but not put in work since, but the two boilers and engine are now in good condition, as well as the man engine.[125]

On South Lode:

> the boiler has been put into work, after a thorough repair, and the tube hooped to meet the Board of Trade requirements and is now, with the engine in good working order.
>
> In the arsenic works the mill engine is in good working order; the boiler, after examination, is to be removed; an 8-ton boiler and fittings have been purchased to replace it, and preparations are being made to do this without delaying the grinding.[126]

It was also reported that the calciners were working well and that two locomotives were in operation on the railway.[127]

On November 12th 1894 a disaster occurred; huge floods struck, the water in the River Tamar being thirty feet higher than usual. The flood carried away the miners' footbridge between Latchley and the mine, tore away fifty feet of the head weir and seriously damaged the Great Leat putting the mine's pumping wheels out of action.[128] Writing in May 1895 Moses Bawden noted that:

> although we worked our auxiliary steam engines to their utmost speed the

> mines were flooded to an alarming extent.[129]

By November 30th the deep levels in Watson's were already underwater, putting a halt to exploratory work. However the real concern was that the "main workings on the old lodes" would flood; these workings being the source of most of the mine's mundic. This would be a calamity: in the short term the affect on arsenic production would be disastrous. Assuming that the mine was able to weather the impact on arsenic production, which was far from certain, the flooding could have a catastrophic affect on the workings themselves.[130] Josiah Paull summarised the problem:

> It is to be feared that the cavities made by continuous working through nearly 50 years may collapse once the water got about them.[131]

It was imperative to get the weir repaired and the leat back in operation as quickly as possible. Captain Clemo and his men proved their mettle in no uncertain way. Forty men carried out temporary repairs on the leat during November. At the end of November an attempt was made to block the breach in the weir with huge baulks of timber. This was intended to be a temporary solution, until such time as the river dropped low enough to allow a proper repair. It was essential that the attempt succeeded as without a water supply the pumps could not operate and the mine would flood.[132]

Clemo's plan to block the breach in the weir with baulks of timber proved successful. At the November 1895 meeting Moses Bawden related the details

> .....Captain Clemo – who, with his son, was on the spot almost day and night – and in about five days we blocked this river, much to the astonishment of the people, because we had to block the whole of the river in one sweep, and men had to remain in November and December for some hours standing up to their waists in water.[133]

By December 1st the makeshift dam was in place and the pumping wheels were again turning.[134] In a *Mining Journal* report dated December 6th William Clemo wrote:

> I am glad to inform you that in repairing the head weir on the Tamar, damaged by the great flood, we have had good weather, and we have succeeded in sending on a full supply of water for the whole of the wheels

at Blanchdown. This was done, and all our underground operations will soon be in full operation.[135]

Further progress was reported in the *Mining Journal*:

> December 13 – A full supply of water is now conveyed from the head weir at Latchley to Blanchdown. The whole of the large wheels are at work, and the large accumulation of water in the underground workings – extending from Wheal Maria to Wheal Emma – is being drained with all speed.[136]

By the end of December Watson's had been drained to the 112 fathom level. Even better news was that both Main and South Lodes were drained and were "in full work again" "yielding large quantities of mundic".[137]

The November 1894 meeting was held on Wednesday 28th. Peter Watson, in the chair, was in good form. He informed the meeting that during the past six months the mine had sold copper and arsenic worth £13,752. In addition to the dividends of November 1893 and May 1894 Watson would announce a further dividend at the meeting. The meeting was informed that 416 people were employed on the mine.[138]

Watson reminded the meeting that this was the 50th anniversary of the mine and that in those fifty years the mine had paid £1,216,937 18s in dividends. In honour of the mine's jubilee the meeting authorised the directors to distribute an amount "not exceeding 200 guineas" amongst the officers and employees of the company.[139]

Never letting facts get in the way of a good story Peter Watson regaled the gathered shareholders with his view of the early days of the mine:

> certainly our mine is looking better today than it has done for a very long period, and shareholders should remember that it is not only the Jubilee of the mine, but also that of Captain Clemo. Having risked their lives many times going down into the mine, when there was not a speck of copper or mundic ore visible, Captain Clemo and his brother, one day came across a faint indication of copper, which grew larger and larger as they followed it until they last of all came to a lode worth £1,000 every six feet.[140]

Watson's revisionist view was echoed in a *Mining Journal* editorial:

> Fifty years of DGC. The inception of the undertaking was mean almost to meagerness. A few hundred pounds capital, and a couple of determined miners groping in the bosom of the earth..... For some time a severe test was put upon the racial determination of these two Cornish subterranean explorers. A weary period passed away before the discovery of the mineral; but at length the slenderest indications of copper and mundic, followed to their conclusion, led to the discovery of the vast stores of wealth.....[141]

Presumably neither commentator had access to the *Mining Journal* of October 26th, 1850! Josiah Hitchins must have been turning in his grave. The gift of two hundred guineas from the Company to its employees, supplemented by a further fifty guineas from the Duke of Bedford, was very well received:

> I have heard from many quarters that the kindness shown to them on this occasion was much appreciated by the employees, to many families having perhaps three or four of its members working under the Company the gift must have been very welcome indeed and the time it was given made it the more so being the pay day before Christmas – 22nd December.[142]

1895

Writing on January 10th 1895 Captain Clemo observed that the water in Watson's was "in fork 3 feet below the back of the 124 fathom level, leaving 48 fathoms to reach the bottom of the engine shaft". Readers of the *Mining Journal* were informed that the machinery was working well and that the draining of the workings was proceeding in a satisfactory manner.[143] Clemo's optimism suffered something of a reverse in mid January; a severe snowstorm struck, significantly slowing down work. On January 16th Clemo reported that "the forking of the water at Watson's has been slow owing to the partial blocking of the larger wheels at Blanchdown". In addition to exacerbating the problems at Watson's the snow had hindered operations throughout the mine and at Morwellham.[144] Towards the end of the month the snow melted and, combined with heavy rain, compounded Captain Clemo's problems. In a report dated January 24th Clemo noted that the wheels had been "considerably clogged". In spite of these problems Watson's was in fork to a point five fathoms below the 136 fathom level.[145] Further falls of snow at the end of the month caused "great inconvenience at surface", however by January 31st the water was half way down the 148.[146]

By February 8th Clemo was able to report that the water was fourteen feet below the 148 and men were at work driving eastwards. Elsewhere on the mine Clemo noted that "the stopes at Wheal Anna Maria, Josiah and Emma are in full work and continue to yield large quantities of mineral for the reduction works".[147] By the end of February the 160 fathom level had been reached.[148] On March 14th William Clemo reported that Watson's was in fork to the 172, a further report dated March 21st noting that driving on the 172 had recommenced.[149,150] This must have been a great relief to all concerned; one of the most productive areas on the mine having been inaccessible for nigh on four months.

The consequences of the November 1894 flood dominated the May 1895 meeting, held on Tuesday 28th. Watson outlined the problems faced by the mine during the preceding six months. The meeting was also informed that Moses Bawden had suffered a severe accident. On May 18th Bawden, whilst out riding, had been crushed by his horse and was still confined to bed. However Bawden was made of stern stuff and he did not let his injuries interfere with his duties, submitting his report to the meeting in a letter written from his sickbed. In his letter Bawden outlined the extensive damage caused both by the flood and one of the most severe winters on record. Bawden highlighted the damage the large waterwheels had sustained noting: "some of which we have been obliged to bolt together with massive iron plates, forming almost the complete circle each side of the wheel".[151] In spite of the difficult time the mine had been having, Watson was able to tell the meeting that the mine was in good health and, as evidence of this, a dividend of 1s 6d was announced.[152]

As if to highlight the problems of the mine's dependence on water power the summer of 1895 proved to be a dry one. An undoubtedly hard pressed William Clemo wrote on June 27th that:

> Owing to the long continuance of dry weather the River Tamar has become very low, and the water supplying our large wheels at Blanchdown is falling off. We are, however, preparing to start the steam engine at Wheal Anna Maria in case it should be required.[153]

On July 4th Clemo had some respite:

> Since the date of the last report we have had a heavy fall of rain, and our large waterwheels at Blanchdown are now all in full work.[154]

However the dry weather returned, Clemo writing on July 25th that:

During the past week the water in the Tamar became so low that we were obliged to start the Wheal Maria steam engine. We are now, however having continued showers of rain and our water wheels are again in full work.[155]

The mine was still counting the cost of the weather at the November 1895 meeting held on Thursday 28th. It was estimated that the November 1894 flood and the harsh winter and spring had cost the mine "over £2,000". In addition to the cost of repairing the weir there had also been the loss of production from Watson's. The meeting was informed that the weather had greatly reduced arsenic production. Captain Clemo reported that extensive damage had been done to the furnaces and flues commenting that "we can scarcely say we have that we have yet recovered.....".[156] The situation was compounded by problems with the mine's Oxland calciner which had been out of service for three weeks whilst repairs were made.[157,158]

1896

At the beginning of 1896 ore was being raised across the mine, William Clemo outlined points of production in a report dated January 23rd. At Wheal Anna Maria Engine Shaft two stopes were active: one in the bottom of the 124 east, the other in the back of the 110 fathom level east. Two stopes were also noted as active on the South Lode at Field's Shaft, both on the 130: one in the bottom of the 130 west, the other in the back of the 130 east. At Richards' Shaft Clemo noted a stope in the bottom of the 115 east. At Hitchins' Shaft a stope was active in the bottom of the 115 east. Moving eastwards to Agnes Shaft a stope was recorded in the bottom of the 103 east. As usual Wheal Emma was a hive of activity. A number of stopes were busy at Thomas' Shaft: one in the back of the 130 east, two in the bottom of 100 east and one in the back of the 100 east. Four stopes were in production at Inclined Shaft: Two in the back of the 162 eats and two in the back of the 100 west. On New South Lode two stopes were noted in the back of the 130 fathom level east of New Shaft. Watson's was, like Wheal Emma, also proving productive. Two stopes were active in the back on the 172 fathom level east of Watson's Engine Shaft. Clemo also noted two stopes in the back of the 160 east, one in the back of the 148 east and one in the bottom of the 136 west.[159]

The May 1896 meeting was held on the 19th at the company's offices which were now at 8 Finsbury Circus. The shareholders were left in no doubt that the preceding eighteen months had been extremely difficult through no fault of their own. Since the November 1895 meeting the mine had experienced ongoing problems at the reduction works. However the meeting was informed that the flues had largely been rebuilt and the furnaces repaired. In the light of the hard times the mine had been through recently, the company had been obliged to borrow £2,000 which had, by the time of the meeting, been paid back. The mine's financial situation had been somewhat mitigated by the Duke of Bedford who had agreed to give up half his dues to the end of February. There was some good news regarding the Latchley Bridge: Cornwall County Council had agreed to contribute to its rebuilding, Devon County Council had pledged £300 and the Duke of Bedford £200. As usual the meeting was "enlivened" by Edwin Sloper's contribution.[160,161]

The summer of 1896 was even more bereft of rain that 1895, being the worst drought in living memory; further demonstrating that the mine was being held hostage by the vagaries of the weather. On June 11th William Clemo wrote:

> After the longest continuance of dry weather known in Devonshire for a great number of years there has within the last week been a change, and heavy showers of rain have fallen throughout the neighbourhood.[162]

Unfortunately the dry weather returned, Clemo's report of July 23rd notes that:

> In consequence of the dry weather and the small quantity of water in the Tamar, two of our largest pumping wheels are idle, and at Watson's the workings are under water up to the 148 fathom level, and the stopes are stopped. At Wheal Anna Maria and Wheal Josiah, with the aid of the steam engine, all the points of operation are kept going. Wheal Emma is also in full work, the water here being drained by a steam engine. At the arsenic works we have only sufficient water for the working of Brunton's calciner, and the large tube is thrown out of work. We have increased the number of men on the hand furnace, and hope soon to place on some more. During the whole period of the working of the Devon Great Consols Mines the River Tamar has never been so low.[163]

By August 6th all the productive deep levels at Watson's were underwater, water having reached the 112. In order to maintain some production from Watson's a stope was being work in the back of the 50 fathom level east. Clemo noted that some rain had fallen but not enough. The Great Wheels were idle, however both Plunger Wheel and the Agnes Wheel were still working.[164] The dry weather continued into mid August. On August 13th Captain Clemo reported rising water levels throughout the mine. At Watson's water was up to the 70 fathom level, at Anna Maria it had reached the 124 whilst in Wheal Josiah it had reached the back of the 144. Clemo was able to report that the Plunger and Agnes wheels were still at work. The reduction works was also still in production, Clemo noting that the Oxland tube was working well.[165] Keeping the arsenic works in production was absolutely key to the mine's survival, so much so that operations were suspended elsewhere on the mine to ensure there was enough water for the works.[166]

One can feel the relief in Clemo's report of August 27th:

> We are glad to report that some heavy rain has fallen, that four wheels are in full work, and that the water is forking very satisfactorily throughout the mines.[167]

Unfortunately Clemo's next report contained mixed news:

> During the past week the Tamar has again fallen off, and we have only two large wheels at work. We are doing all we can to keep the various points of operation working in operation. Rain is again falling today, and we are anxiously hoping it may continue.[168]

The rain appeared to have continued for the next couple of weeks causing Clemo to observe on September 17th:

> We now have a strong supply of water from the Tamar, and all our wheels are in full work. The water is in fork below all our workings at Wheal Anna Maria and Wheal Josiah, and it is down to the 70 fathom level in Watson's. The steam engine at Wheal Anna Maria is no longer required.[169]

Much to William Clemo's relief, by the beginning of October he was able to report that all the pumping machinery on the mine was "in full work". By October 1st Watson's had been drained to the 112 and by the 15th was in fork to the 136.[170,171] On November 5th Clemo reported that Watson's had been drained to a depth of fifty three feet below the 148.[172] After a hard summer the water crisis was over.

In autumn 1896 some energy was being put into further developing South Lode. On November 5th William Clemo reported that a skip road had been installed in Field Shaft from the 130 to the 141 and that repairs had been made to the shaft, the intention being to "lay open more productive ground".[173] The reason for a renewed interest in this part of the mine was to increase arsenic reserves on the mine. For many years the mine had relied heavily on mundic recovered from the "burrows". The supply of dump material was, according to Clemo, coming to and end and the mine, if it was to continue arsenic production, would have to increasingly rely on its underground reserves.[174]

The November meeting was held on Wednesday 25th. Peter Watson took his customary position in the chair in spite of suffering from what he described as a severe cold. Not surprisingly he focussed on the drought which, according to himself, was "the longest drought ever known in England". In effect four months production had been lost. Watson estimated that they had to pump 10,000,000 gallons of water out of the mine and it was hoped to have the mine in fork by the end of the month. The loss of production occasioned by the drought meant that the mine had borrowed £1,000 to tide them over hard times.[175,176]

On December 2nd William Clemo was able to report that the 172 fathom level east of Watson's Engine Shaft was yielding three tons of copper ore and mundic per fathom; the mine was in fork.[177] Clemo's report of December 17th carried the ominous news that heavy rain was falling:

> wheels have been logged to some extent but so far our underground workings have been kept going".[178]

After the year that he had been through one can only sympathise with Captain Clemo when he submitted his final report for 1896, dated December 31st:

> I regret to say that at Wheal Anna Maria Engine Shaft the main bob gave way on Monday night and a number of pulley stands with the line of rods were thrown down for a considerable distance. We are making every effort to repair the breakage as quickly as possible, and we hope to get at work by the beginning of the next week. The days being short and the weather wet and stormy we are unable to proceed as fast as we could wish, but we have a full force of men, and our smiths and carpenters are busily engaged in preparing the iron and timber work for the new bob, pulley stands, and other necessary appliances.[179]

1897

William Clemo again demonstrated that he was an eminently capable and competent mining man, taking the collapse of the Anna Maria bob in his stride.

On January 7th 1897 he was able to report that

> The breakage at Wheal Anna Maria was so far repaired as to enable us to start the steam engine on Sunday, which will be kept at work until the line of rods is ready to be attached to the water wheel. This we hope to accomplish in the course of a few days, although the weather is very stormy, and has been very much against all our surface operations for some time past.[180]

By the time Clemo submitted his report on February 18th Anna Maria Engine Shaft was back in production, two stopes being noted. One was in the bottom of the 110 east, the other in the back of the same level. South Lode at Field Shaft was repaying expectations; Clemo describing the lode in the bottom of the 130 east as "yielding 12 tons of good mundic per fathom".[181] A stope in the back of the 130 east of Richards' Shaft was particularly productive, producing fourteen tons of mundic per fathom. A second stope at Richards' Shaft was noted in the bottom of the 115 east. A stope was also active on the 115 east of Hitchins' Shaft. At Agnes Shaft at stope was recorded in the bottom of the 70 west. Wheal Emma, as usual, was very active. At Thomas' Shaft Clemo notes five stopes on the 100 east; two in the bottom and three back stopes. Three stopes in operation at Inclined Shaft: two in the back of the 162 east and one in the back of the 100 west. At New Shaft two stopes were noted on New South Lode, both in the back of the 130 east. Work at Watson's was back in full swing, stopes being recorded in the back of the 172 east, the back of the 160 east, the bottom of the 148 west and the back of the 148 east. Clemo concluded his report observing that "the weather continues favourable, and all our works are in full operation throughout the mines".[182]

The Company had managed to secure a very favourable contract for its 1897 output of arsenic. They were able to negotiate a minimum price of £20 per ton, this was £6 10s higher than the previous year's contract.[183] By mid March the weather had deteriorated somewhat:

> During the present month we have had an almost continuous storm of wind and rain, which has caused some damage to our roofs, sheds &c., and has interfered with our surface workings. On the whole however, we have succeeded in getting on fairly well.[184]

The May meeting was held in the Company's London Office at 8 Finsbury Park on Wednesday 19th. Whilst the past years operations had been expensive due to the breaking of the weir, damage to the flues at the arsenic works and the previous year's drought, the mine had started on the road to financial recovery, the problems being offset by an "advantageous" arsenic contract. The company had paid off the £1,000 loan taken out during the drought. More significantly, for the shareholders at least, the directors were able to declare the mine's 160th dividend of four shillings per share, a decision that was met with cheers. In discussing the prospects of the mine Peter Watson lamented that "the production from the mines is good, but unfortunately the prices are poor". Moses Bawden noted that currently only three ends were being driven. The ever irritating Edwin Sloper advocated sinking for tin, demonstrating a woeful ignorance of recent history. He also continued his attacks on Watson. Watson replied, with some dignity, that "it was unnecessary for him to answer Mr Sloper, as he was continually asking questions which were both unkind and ungenerous". Watson went on to say that he had carefully guarded the mine's interest for twenty years and Mr Sloper's attacks against him were unwarranted.[185]

Sloper's attacks on the mine's management were on this occasion, as Watson suggests, totally unwarranted. Given the extremely trying times the mine had been through Watson and his colleagues had performed something of a minor miracle with the mine's finances. This was recognised by the *Mining Journal* of May 22nd which carried a fulsome editorial:

> In spite of the many serious difficulties which were encountered during the six months a good profit was earned and a large dividend declared. In fact the figures showed so wonderful an improvement that the corresponding period of last year will bear no comparison and sinks into utter insignificance.[186]

No doubt buoyed by their dividend the meeting agreed to provide a dinner for their employees in honour of the Queen's Jubilee.[187] This proposal was not well received:

> A dinner to the employees was voted on the occasion of H. M. Jubilee but this has been coldly received by the men so it is not to be carried out. They have plainly told Mr. Bawden that they require some solid addition to the very low wages that have been given now that dividends are being earned.[188]

Josiah Paull agreed with the men's grievance noting that the money spent on the dividend would be better spent raising the men's wages and repairing "the Company's premises and machinery".[189] Unlike the previous year the summer of 1897 proved to be an unremarkable one, which must have been a relief to William Clemo.

Held on Wednesday 17th at 8 Finsbury Circus, the November 1897 meeting proved to be as positive as the May meeting. The accounts were very satisfactory and Watson's Mine in particular, was repaying the confidence placed in it. Even the problems that mine had been having with its equipment was, according to Watson, a blessing in disguise as much of it had now been refurbished. The directors found themselves in the very pleasant position of being able to authorise the mine's 161st dividend; an impressive six shillings per share. Only Mr Sloper remained unimpressed and continued to advocate a return to deep exploration. F. G. Lane responded in terms that even Sloper could understand, explaining that "they had already experimented in sinking deeper, and had proved that it would be useless to attempt to find payable ore at greater depth".[190]

In his annual report Josiah Paull outlined the exploratory work being undertaken in Watson's: Limited work was undertaken on the 148 east, the 160 east and the 172 east. A rise was being put up between the 172 and the 160.[191]

The old workings in both Wheal Anna Maria and Wheal Josiah were providing good returns of mundic with some copper. The Wheal Emma workings were less productive however Paull commented that "even from these the returns have been considerable".[192]

Mundic recovery from the dumps was decreasing, although it was noted that the Wheal Fanny dumps were "pretty good". Josiah Paull noted that the high price of arsenic

> has stimulated the search for mundic in all parts of the mine heaps which had lain unnoticed for thirty years and more have been turned to in some instances with considerable success.[193]

Summing up the year 1897 Paull observed that:

> The year has been a prosperous one for the Company, the best they have had since 1880.[194]

1898

The Company was almost wholly dependent on securing a good contract for their make of arsenic to ensure the future of the mine. It was the "advantageous" arsenic contract that allowed the Company to pay ten shillings in dividends on each £1 share in 1897. It was, therefore a matter of fundamental importance that the mine secured a new contract on the ending of the 1897 contract which expired at the end of January 1898.[195]

Writing on March 30th 1898 Josiah Paull recorded a conversation he had with Moses Bawden. Bawden informed Paull that the Company had yet to sell the 1898 output of arsenic, Paull noting that "he speaks in the gloomiest terms of doing so". The problem appeared to lie with "one or two merchants" who usually bought the mine's output. These major players were determined to beat the price down and also to drive smaller dealers out of the market altogether.[196]

In mid April 1898 Peter Watson had a meeting in Tavistock with Messrs. Field and Lanyon, the largest arsenic dealers in Devon and Cornwall. Peter Watson offered to sell the mine's output at £17 10s per ton. Unfortunately the merchants refused to buy at that price or indeed at a lower price. Field and Lanyon asserted that the arsenic trade was in a very poor state. However Paull observed that the *Times* was quoting the price of "white arsenic" at £20 15s a ton. Josiah Paull was convinced that the real reason for the merchants' failure to buy Devon Great Consols arsenic was down to their plan to control the arsenic trade on their own terms.[197]

The Company had failed to sell its arsenic by the time of the May 1898 meeting, in consequence of which the Company was not in a position to pay a dividend.[198] At the end of June Moses Bawden commented to Josiah Paull that the state of the arsenic trade was becoming worse month to month, merchants not buying at all. Paull estimated that the Company had "quite 700 tons of unsold arsenic on their hands".[199]

By the end of September 1898, much to the relief of all concerned, the Company managed to sell its arsenic for the year ending January 31st 1899, albeit at the reduced price of £13 10s per ton.[200] Although not on such good terms as the 1897 contract the new contract saved the mine. By October 1898 the arsenic market had fully recovered, William Woolcock noting that the mine shipped more arsenic in the first week of October than had ever been done before.[201]

The uncertainty engendered by the lack of an arsenic contract was compounded by another drought which hindered both surface and underground works. William Clemo noted that there was not enough water from the Tamar for the wheels and that they had been obliged to start the Wheal Anna Maria Engine to keep the mine dry.[202]

To carry the mine through these hard times the mine was forced to borrow £8,000.[203] The November meeting was held on Wednesday 30th at the Company's office at 2 & 3 West Street Finsbury Circus. Peter Watson informed the meeting that "directors had experienced a good deal of anxiety in managing the company", given the uncertainty regarding the arsenic contract this might be taken as something of an understatement. The company had also spent a considerable sum repairing their buildings to the satisfaction of the Estate. Commenting on the drought Clemo observed that "at their mines they either had too little or too much water" adding ruefully that "at the present time they had rather more than they wished for". Fortunately since September the financial situation had significantly improved to such an extent that the company was able to repay its loans.[204,205]

1899

By 1899 the original locomotives which had worked the railway were worn out and the decision was taken to buy a replacement. Initially the intention had been to refurbish one of the old locomotives however all the builders they approached were too busy building new machines. Rather then spend £850 on a new locomotive it was decided to purchase a second-hand loco, seven or eight years old, for £400 and spend around £100 on repairs.[206] The new locomotive known as *Ada* was supplied by the firm of Thomas Spittle of the Cambrian Iron Foundry Newport. This may well have been the second locomotive by this builder to work on the mine, the first being *Hugo* of 1882.[207] The new loco was delivered on February 23rd 1899 (see Note 9).[208] On March 9th Captain Clemo noted that "the locomotive is being thoroughly overhauled, and some little alterations being needed to meet our requirements, it will take something like another week to complete this necessary work".[209]

During January, February and March of 1899 the mine was again plagued with appalling weather. William Clemo reported on March 9th that:

> owing to the long continued rain, and so many of the workings being underwater, we have been unable to keep up our usual returns and this hindrance is still being felt, especially at the reduction works. The water at

> Watson's cannot be forked for some weeks, and the principal stopes cannot be reached, and one of our best stopes east of Agnes Shaft will be idle for some weeks. While the water is being forked no returns can come from these parts of the mines. We are, however, doing everything possible in all the other points of operations. The main stopes at Wheal Anna Maria and Wheal Josiah having yielded great returns for so many years are becoming less productive. We are however, as we always have been, searching for further discoveries......[210]

The Devon Great Consols Band was still active under the leadership of one Mr Dingle. On Wednesday 31st May the band provided music for the Gulworthy Bazaar, held on the lawn at Gulworthy vicarage.[211]

Unusually the spring meeting was held on Thursday June 1st 1899 at the London office with Peter Watson in the chair. The fearful storms of the winter were alluded to, it being noted that the water had reached six fathoms above the 160 at Watson's. On a more positive note the company had made "large shipments" of arsenic since the November 1898 meeting. Demand had continued strong and the company had been able to secure a further contract in May 1899 at £17 per ton. The improved state of the arsenic trade allowed the directors to declare a dividend of 2s 6d; this would prove to be the last dividend paid by the Company. The question of the outstanding liability on shares was also discussed and the feeling of the meeting was that at £3 a share this was too high.[212,213,214]

In order to fully address the issue of the outstanding liability on the Company's shares an Extraordinary General Meeting was held on Wednesday 5th July. The consensus agreed that £3 per share was too high. The directors were of the opinion that the best course of action would be to reconstruct the company. The proposed new company would have a capital of £51,200 (the same as the current company). 40,960 shares of £1 would be issued to the existing shareholders, each share having been credited as having had fifteen shillings paid on them. This meant that each shareholder would receive four shares with fifteen shillings paid on each in return for one old share. Thus the liability would be reduced from £3 for one old share to £1 on four new shares of five shillings per new share. Peter Watson proposed that the old company be wound up voluntarily and replaced by a new company: Devon Great Consols (Limited).[215] Edwin Sloper was in combative mood; in the words of the *Mining Journal*:

> Mr Sloper rose, and pointing to the Chairman and addressing him by his

> Christian name, said "This will rest upon you and your death-bed", and repeating it several times, left the room.[216]

Drama over, Peter Watson put the proposals to the meeting, which duly carried them.[217]

The resolutions passed at the July 5th meeting were referred to a further Extraordinary General Meeting, the company being formally reconstructed on July 21st 1899.[218] The "old company" was put into voluntary liquidation under the provisions of the Companies Acts 1862 & 1890, George Hadlee acting as the liquidator. The assets of the old company were transferred to the reconstructed company. An indenture dated 29th November 1899 transferred the existing leases to the new company.[219]

During the summer of 1899 the district experienced a severe drought. William Woolcock refers to a "long, dry, hot summer" during which there was "not enough water for the wheels to keep the mine drained".[220] In August Josiah Paull observed that only one pumping wheel was at work. He also noted that there was not enough water to turn the wheel which drove the Oxland tube at the arsenic works. The drought continued into September during which month it was only possible to keep one pumping wheel turning. Watson's was flooded above the 136, Wheal Josiah was flooded, some of the mines most productive ore ground being underwater, Wheal Anna Maria and Wheal Emma were kept in fork by steam pumping engines. It was noted that by the end of October "the workings from which the supplies of mundic are got are again free of water". Unfortunately floods in November slowed down the draining of Watson's and it was felt that it would be the end of the year before this section of the mine was in fork.[221]

Woolcock's notebook provides some interesting detail on what must have been one of the last count house dinners held on the mine on October 3rd 1899:

> The Devon Great Consols Directors came today on their usual inspection, - before the November meeting and with them several shareholders. All dined at the Count House at 3PM, numbering 22 persons. A very nice dinner, - viz. Boiled leg of Mutton at the head of table, Ribs of Beef at bottom, two Beef Steak pies at middle, - with plum puddings after, and apples, pears, and grapes for desert [sic]. Waiters - Mrs Edgcumbe and three drivers. A liberal supply of Claret. Plenty of toasts, and speeches. A jovial party. All enjoyed themselves thoroughly. The party consisted of Messrs. - Watson,

> Glen, and Lane - (Directors), Hadlee, (secretary), Anderson - (clerk), Snell - (Solicitor) & Coppin and son, Chisholm, Clarke, Newberry, Way, Mathews and …….. Glascock.[222]

There does not appear to have been a formal Half Yearly Meeting held in November 1899. "A statutory meeting" was held on November 15th to confirm the resolutions passed for the reconstruction of the Company. Evidently the Directors felt that this was sufficient. Josiah Paull, at his cynical best, suggested it was the intention of the Directors "to stave off the production of accounts until May next".[223] This would appear to be a contravention of the "Memorandum of Association" of the new Company which required the Company to hold half yearly general meetings "in the months of May and November each year".[224] William Woolcock gives us a slight glimpse of the meeting noting that no dividend was paid in consequence of the winter storms and the summer drought.[225]

Chapter 9 Notes

Note 1

Whilst not made explicit one could speculate that one of the "major shareholders" consulted by Stewart and Morris was Charles Oxland.

Note 2

At the November meeting Watson reduced his salary from £450 to £250 until such time as the mine paid dividends. Josiah Paull, not one of Watson's warmest admirers, cynically observed:

> but that reduced rate of pay is to cease as soon as a dividend can be paid. This Mr Watson promises to do next year, and he can do so merely by a manipulation of the accounts. It only requires the sale of arsenic in advance of production, the starving of underground explorations, and letting the stocks of timber, coals and other materials to run low to give one dividend at least.[226]

Note 3

By the end of 1890 the following shafts had been "secured".[227]

In the Wheal Maria or Western Section.

Old Capel Tor Shaft	arched over
Kestel's (*Castle's Shaft? - author*)	arched over
Gard's	arched over

In the next section Eastward (Wheal Fanny)

Western Shaft		arched over
Engine Shaft	(*Eastern Shaft? – author*)	arched over
Eastern Shaft	(*Ventilating shaft? – author*)	arched over
Blakeways Shaft	(*Blackwell's Shaft? – author*)	arched over
Quarry Shaft		filled up

In Wheal Josiah

South Lode Shaft	filled up

In Wheal Emma

Ventilating Shaft – east of Waggon Lane	arched over

In Watson's

Eastern Shaft	arched over

Note 4

Woolcock states the 11th.[228]

Note 5

It would go beyond the realms of idle speculation to suggest that Peter Watson "sacrificed" Richards in order to secure Bawden's return. Richards continued to snipe at Bawden through the letters pages of the *Mining Journal*, the correspondence is somewhat tedious.

Note 6

In November 1891 Josiah Paull reported:

> On the 5th inst. I made an inspection of the workings in Wheal Anna and Wheal Josiah which have been going on for some time chiefly in the north part of the Lode in those sections of the mine. In the times of great prosperity this part of the Lode was allowed to stand unworked for a great length as it contained very little mineral other than arsenical mundic and there were no works for its reduction in those days, Now that copper ore has been so much less in quantity and the value of what is produced is so low as to make it hardly worth extracting the north part of the Lode which is large and rich in mundic has been bought into play.[229]

Note 7

History does not record what service Josiah Hitchins rendered to the Dukes of Bedford to warrant the payment of his pension. It does, however increase the mystery surrounding Hitchins' somewhat shadowy involvement in the early history of the mine.

Note 8

The quote from Woolcock's notebook confirms the somewhat elusive fact that the Wheal Emma man engine was located on Inclined Shaft – the only shaft which cuts both the 32 and 47 fathom levels.

Note 9

Remarkably two accounts of *Ada's* arrival on the mine survive: William Woolcock recorded the event in his notebook:

> February 23rd 1899 – The newly purchased Locomotive, (second hand) named "Ada", brought out from Tavistock G.W.R. Station, on the Tavy Foundry Waggon - by Mr Joel Harris, and T. Glanville, 15 Horses. When at lower corner of the gardens by Wheal Anna Maria Office, the under carriage broke and the bussel dropped down, fortunately very little damage. Next day repaired and taken down to station, by the loading place. Weight of Engine 12 tons, 4 cwt. Cylinders 10 inch.[230]

The second account was written in 1953 by S. Jackman, who described himself as the Chief Engine Driver at Devon Great Consols:

> Mr. Matthews, the Consultant Engineer and Mr. Moses Bawden, the Managing Director, went to Wales and bought the Engine Ada, and I came into Tavistock to supervise the transportation of "Ada" out to the mines. She was collected from the Tavistock (then "GWR") Railway Station, and unloaded from the railway truck onto a special waggon drawn by twenty horses, in pairs. The route I took was down through the town, along Plymouth Road and up over Callington Road.
>
> Now Sir, this is an incident I have never forgotten. On arriving at the Black Bridge (S.R.) which crosses over the road, the drivers stopped and told me the engine will have to be unloaded because they estimated she would not go under the bridge, and they refused to move. I made a bet with them that it would go under, and ordered them on. To prove my judgement I got up

on the boiler of the engine and sat on top – the engine cleared by 2 inches – and so I won my bet!

We proceeded without further incident on to the Mines, where we unloaded the Engine "Ada" onto the 4′ 8½″ track, and into the engine shed, where I personally overhauled her. She was then taken into use and performed the normal duties between the mines and Morwellham. Incidentally, she, too, was a four wheeler.

I drove the engine up to the time the Mines closed down and were taken over by the breaking up party, but by then I had secured a post as Engineer on the Gold Mines in India, in the year 1902, and from then on lost touch completely with the Devon Great Consols and have no knowledge as to what eventually happened to Ada.[231]

Chapter 9 references

1. Devon Record Office document 8187 Bedford Estate Mining Reports 1886-1898
2. *Mining Journal* 4 January 1890
3. Devon Record Office document 8187 Bedford Estate Mining Reports 1886-1898
4. *Mining Journal* 4 January 1890
5. *Mining Journal* 4 January 1890
6. *Mining Journal* 1 February 1890
7. *Mining Journal* 1 March 1890
8. Woolcock MSS
9. *Mining Journal* 29 March 1890
10. *Mining Journal* 31 May 1891
11. Devon Record Office document 8187 Bedford Estate Mining Reports 1886-1898
12. Devon Record Office document 8187 Bedford Estate Mining Reports 1886-1898
13. *Mining Journal* 31 May 1891
14. Devon Record Office document 8187 Bedford Estate Mining Reports 1887-1898
15. *Mining Journal* 20 December 1890
16. *Mining Journal* 5 April 1890
17. *Mining Journal* 26 April 1890
18. Devon Record Office document 8187 Bedford Estate Mining Reports 1886-1898
19. *Mining Journal* 24 May 1890
20. *Mining Journal* 24 May 1890
21. *Tavistock Gazette* 31 May 1890
22. Woolcock MSS
23. Woolcock MSS

24. Woolcock MSS
25. *Tavistock Gazette* 21 November 1890
26. *Mining Journal* 29 November 1890
27. *Mining Journal* 29 November 1890
28. *Mining Journal* 29 November 1890
29. *Mining Journal* 29 November 1890
30. *Mining Journal* 29 November 1890
31. *Tavistock Gazette* 28 November 1890
32. Devon Record Office document 8187 Bedford Estate Mining Reports 1886-1898
33. Devon Record Office document 8187 Bedford Estate Mining Reports 1886-1898
34. Devon Record Office document 8187 Bedford Estate Mining Reports 1886-1898
35. Francis Russell, 9th Duke of Bedford & George Russell, 10th Duke of Bedford.
36. *Mining Journal* 3 January 1891
37. Devon Record Office document 8187 Bedford Estate Mining Reports 1886- 1898
38. *Mining Journal* 21 March 1891
39. *Mining Journal* 23 May 1891
40. *Mining Journal* 23 May 1891
41. *Mining Journal* 30 May 1891
42. *Mining Journal* 30 May 1891
43. *Tavistock Gazette* 29 May 1891
44. Woolcock MSS
45. *Mining Journal* 18 July 1891
46. *Mining Journal* 25 July 1891
47. Woolcock MSS
48. *Mining Journal* 21 May 1892
49. *Mining Journal* 21 May 1892
50. *Mining Journal* 25 July 1891
51. Woolcock MSS
52. *Mining Journal* 25 July 1891
53. Woolcock MSS
54. *Mining Journal* 25 July 1891
55. *Mining Journal* 25 July 1891
56. *Mining Journal* 25 July 1891
57. Woolcock MSS
58. *Mining Journal* 18 July 1891
59. Woolcock MSS
60. *Mining Journal* 28 November 1891
61. *Tavistock Gazette* 27 November 1891

62. Devon Record Office document 8187 Bedford Estate Mining Reports 1886-1898
63. *Mining Journal* 3 December 1892
64. *Mining Journal* 3 June 1893
65. *Mining Journal* 2 January 1892
66. Devon Record Office document 8187 Bedford Estate Mining Reports 1886-1898
67. Devon Record Office document 8187 Bedford Estate Mining Reports 1886-1898
68. Devon Record Office document 8187 Bedford Estate Mining Reports 1886-1898
69. Woolcock MSS
70. Woolcock MSS
71. Devon Record Office document 8187 Bedford Estate Mining Reports 1886-1898
72. Devon Record Office document 8187 Bedford Estate Mining Reports 1886-1898
73. Woolcock MSS
74. Devon Record Office document 8187 Bedford Estate Mining Reports 1886-1898
75. Devon Record Office document 8187 Bedford Estate Mining Reports 1886-1898
76. Devon Record Office document 8187 Bedford Estate Mining Reports 1886-1898
77. *Mining Journal* 21 May 1892
78. Woolcock MSS
79. *Mining Journal* 1 October 1892
80. *Tavistock Gazette* 27 May 1892
81. *Mining Journal* 21 May 1891
82. *Tavistock Gazette* 27 May 1891
83. Devon Record Office document 8187 Bedford Estate Mining Reports 1887-1898
84. Devon Record Office document 8187 Bedford Estate Mining Reports 1886-1898
85. *Mining Journal* 21 May 1891
86. *Tavistock Gazette* 27 May 1891
87. Devon Record Office document 8187 Bedford Estate Mining Reports 1886-1898
88. *Mining Journal* 11 June 1892
89. Woolcock MSS
90. *Mining Journal* 10 September 1892
91. *Mining Journal* 15 October 1892
92. *Tavistock Gazette* 9 December 1892
93. *Mining Journal* 26 November 1892
94. *Mining Journal* 26 November 1892
95. *Mining Journal* 26 November 1892
96. *Tavistock Gazette* 9 December 1892
97. *Tavistock Gazette* 9 December 1892
98. *Mining Journal* 7 January 1893
99. *Mining Journal* 14 January 1893

100. *Mining Journal* 28 January 1893
101. George Russell, 10th Duke of Bedford & Herbrand Russell, 11th Duke of Bedford.
102. *Mining Journal* 15 April 1893
103. *Tavistock Gazette* 14 April 1893
104. *Mining Journal* 29 April 1893
105. *Tavistock Gazette* 2 June 1893
106. *Mining Journal* 3 June 1893
107. *Mining Journal* 24 June 1893
108. *Mining Journal* 1 July 1893
109. *Mining Journal* 8 July 1893
110. *Mining Journal* 15 July 1893
111. *Mining Journal* 22 July 1893
112. *Mining Journal* 29 July 1893
113. *Mining Journal* 5 August 1893
114. *Mining Journal* 9 September 1893
115. *Tavistock Gazette* 24 November 1893
116. Devon Record Office document L1258/SS/MC
117. *Mining Journal* 2 December 1893
118. *Tavistock Gazette* 1 December 1893
119. *Mining Journal* 13 January 1894
120. *Mining Journal* 20 January 1894
121. *Mining Journal* 27 January 1894
122. *Mining Journal* 10 February 1894
123. *Mining Journal* 2 June 1894
124. *Mining Journal* 2 June 1894
125. *Mining Journal* 2 June 1894
126. *Mining Journal* 2 June 1894
127. *Mining Journal* 2 June 1894
128. *Mining Journal* 1 June 1895
129. *Mining Journal* 1 June 1895
130. Devon Record Office document 8187 Bedford Estate Mining Reports 1886- 1898
131. Devon Record Office document 8187 Bedford Estate Mining Reports 1886-1898
132. Devon Record Office document 8187 Bedford Estate Mining Reports 1886-1898
133. *Mining Journal* 30 November 1895
134. Devon Record Office document 8187 Bedford Estate Mining Reports 1886-1898
135. *Mining Journal* 8 December 1894
136. *Mining Journal* 15 December 1894
137. Devon Record Office document 8187 Bedford Estate Mining Reports 1886-1898

138. *Mining Journal* 1 December 1894
139. *Mining Journal* 1 December 1894
140. *Mining Journal* 1 December 1894
141. *Mining Journal* 1 December 1894
142. Devon Record Office document 8187 Bedford Estate Mining Reports 1886-1898
143. *Mining Journal* 12 January 1895
144. *Mining Journal* 19 January 1895
145. *Mining Journal* 26 January 1895
146. *Mining Journal* 2 February 1895
147. *Mining Journal* 9 February 1895
148. *Mining Journal* 23 February 1895
149. *Mining Journal* 16 March 1895
150. *Mining Journal* 23 March 1895
151. *Mining Journal* 1 June 1895
152. *Mining Journal* 1 June 1895
153. *Mining Journal* 29 June 1895
154. *Mining Journal* 6 July 1895
155. *Mining Journal* 25 July 1895
156. *Tavistock Gazette* 6 December 1895
157. *Mining Journal* 30 November 1895
158. *Tavistock Gazette* 6 December 1895
159. *Mining Journal* 25 January 1896
160. *Mining Journal* 23 May 1896
161. *Tavistock Gazette* 22 May 1896
162. *Mining Journal* 13 June 1896
163. *Mining Journal* 25 July 1896
164. *Mining Journal* 8 August 1896
165. *Mining Journal* 15 August 1896
166. *Tavistock Gazette* 4 December 1896
167. *Mining Journal* 29 August 1896
168. *Mining Journal* 5 September 1896
169. *Mining Journal* 19 September 1896
170. *Mining Journal* 3 October 1896
171. *Mining Journal* 17 October 1896
172. *Mining Journal* 7 November 1896
173. *Mining Journal* 7 November 1896
174. *Tavistock Gazette* 4 December 1896
175. *Mining Journal* 28 November 1896

176. *Tavistock Gazette* 4 December 1896
177. *Mining Journal* 5 December 1896
178. *Mining Journal* 19 December 1896
179. *Mining Journal* 2 January 1897
180. *Mining Journal* 9 January 1897
181. *Mining Journal* 20 February 1897
182. *Mining Journal* 20 February 1897
183. Devon Record Office document 8187 Bedford Estate Mining Reports 1886-1898
184. *Mining Journal* 20 March 1897
185. *Mining Journal* 22 May 1897
186. *Mining Journal* 22 May 1897
187. *Mining Journal* 22 May 1897
188. Devon Record Office document 8187 Bedford Estate Mining Reports 1886-1898
189. Devon Record Office document 8187 Bedford Estate Mining Reports 1886-1898
190. *Mining Journal* 20 November 1897
191. Devon Record Office document 8187 Bedford Estate Mining Reports 1886-1898
192. Bedford Estate Mining Reports 1886-1898
193. Devon Record Office document 8187 Bedford Estate Mining Reports 1886-1898
194. Devon Record Office document 8187 Bedford Estate Mining Reports 1886-1898
195. Woolcock MSS
196. Devon Record Office document 8187 Bedford Estate Mining Reports 1886-1898
197. Devon Record Office document 8187 Bedford Estate Mining Reports 1886-1898
198. Devon Record Office document 8187 Bedford Estate Mining Reports 1886-1898
199. Devon Record Office document 8187 Bedford Estate Mining Reports 1886-1898
200. Devon Record Office document 8187 Bedford Estate Mining Reports 1886-1898
201. Woolcock MSS
202. *Mining Journal* 3 December 1898
203. *Mining Journal* 3 December 1898
204. *Mining Journal* 3 December 1898
205. *Tavistock Gazette* 9 December 1898
206. *Mining Journal* 3 June 1899
207. Hateley R., 1977. *Industrial locomotives of South West England.* Industrial Railway Society.
208. Woolcock MSS
209. *Mining Journal* 11 March 1899
210. *Mining Journal* 11 March 1899
211. *Tavistock Gazette* 2 June 1899
212. Devon Record Office document 8187 Bedford Estate Mining Reports 1899-1906
213. *Mining Journal* 3 June 1899

214. *Tavistock Gazette* 2 June 1899
215. *Mining Journal* 8 July 1899
216. *Mining Journal* 8 July 1899
217. *Mining Journal* 8 July 1899
218. London & West Country Chamber of Mines records 190-1903, Vol. 1, part 1 March 1901
219. Devon Record Office document L1258M/SS/MC
220. Woolcock MSS
221. Devon Record Office document 8187 Bedford Estate Mining Reports 1899-1906
222. Woolcock MSS
223. Devon Record Office document 8187 Bedford Estate Mining Reports 1899-1906
224. Devon Great Consols Memorandum of Association 1899
225. Woolcock MSS
226. Bedford Estate Mining Reports 1886-1898
227. Bedford Estate Mining Reports 1886-1898
228. Bedford Estate Mining Reports 1886-1898
229. Woolcock MSS
230. Woolcock MSS
231. Letter dated 23 April 1953 from S. Jackman to Frank Quant, Morwellham archive.

Chapter 10

1900 – 1905 Closure and scrapping

"This mine has not been stopped on account of the poverty of the mineral resources". Peter Watson, April 1903.

1900

In his first report of the new century Josiah Paull reported that although Watson's was "clear of water to the bottom of the mine" exploratory work had not been resumed. The lack of exploration was attributed to the "straitened condition of the Company's finances", all available men were employed extracting copper and mundic. At the end of January 1900 twenty eight men were extracting ore in Watson's. Elsewhere "in the old mines" there were significant problems with water which reduced output.[1]

On the 24th January a call of 1s per share was made on the 51,200 shares in the reconstructed company, the call amounting to £2,560. This call was significant in that it was the first call to be made on the shareholders for twelve years.[2] In February floods caused considerable problems, including some damage to the Great Leat. In Watson's the 172 was underwater, although men were able to work down to the 162. The older sections of the mine were still flooded, the Wheal Josiah working not coming back into production until April 1900. It would not be until May that all the workings were clear of water.[3]

There was some good news; on 1st February the company secured a good contract for the next twelve months' make of arsenic. The mine managed to agree a price of £18 5s per ton; 25s per ton more than the previous year.[4] The accounts presented to the May 1900 meeting for the period July 21st 1899 to April 31st 1900 showed that the mine had made a loss of £1,729 2s 9d, the loss, in part, being attributed to the rise in the price of coal. Watson was keen to point out that some of the costs were exceptional, relating to the reconstruction of the company the previous year.[5,6]

In July Josiah Paull noted that "a want of miners" meant that very little exploratory work was being carried out. Six men were intermittently employed driving the 172 in Watson's "when not employed on shaft work or in getting ore".[7] On 20th August the second call of the year, again of 1s per share was made. On August 2nd John Woodlock of Gunnislake was killed whilst oiling one of the crushers on the Wheal Anna Maria dressing floors.[8]

The following month the mine was shaken by another death, that of William Clemo (see Note 1). Captain Clemo died on Monday 3rd September at the age of 78 having been ill for a week following a seizure. The *Tavistock Gazette* described him as:

> A thoroughly practical and experienced miner, perfectly reliable, and a most valued servant of the company.[9]

William Clemo had been an employee of the company since its earliest days. How much truth there was in the *Mining Journal's* comment that he "was one of three who cut the first bunch of ore at the mine" is a moot point.[10] That aside Captain William Clemo served the company well during his fifty six years on the mine, filling both the role of underground captain and chief agent, posts which he discharged with distinction.

On October 2nd the directors Peter Watson, F. G. Lane and Mr Coppin (who had recently replaced Thomas Glen who had resigned due to ill health), held a board meeting on the mine, presumably to discuss the appointment of Captain Clemo's successor. The most senior agent on the mine and, arguably, the logical choice as Clemo's replacement was William Woolcock. Unfortunately for Woolcock he was passed over, although Peter Watson, to his credit, did his best to let him down lightly:

> The Chairman said you - Captain Woolcock - are the oldest Agent of the Company, and have been in that position for a great number of years, and your services are highly appreciated by the Company. We have lost poor Captain Clemo, - who was greatly beloved by the Company, - and especially so by the Directors. We have been talking the matter over of

appointing a successor. The Shareholders would like the name of Clemo to stand as before, as manager, and as you are pretty well advanced in years we thought that you would not object to his son, Captain W. H. Clemo taking his late Father's place as Manager, and that Joseph Clemo, 2nd, be appointed Agent in the Western Round to assist the newly appointed Manager, and that you - (W.W) go on at Wheal Emma as before - each one at same pay as before, and not any more.[11]

In October it was noted that no exploratory work was being undertaken at Watson's due to a "want of men". The lack of men was also hindering the extraction of ore at Watson's, Paull observing that there was work for thirty men but only sixteen were at work. By November a few more men had been put to work in Watson's in the stopes below the 172, which were said to be unusually rich in copper ore. Elsewhere sufficient men were stoping on both the Main Lode and South Lode.[12] With a degree of hindsight one could argue that the Company knew that the mine was in its last throes and were, therefore, focussing all their attention on extracting ore rather than on "dead work" such as exploration; in effect "picking the eyes" of the mine.

In spite of experiencing hard times the mine still continued to be a significant employer of labour, employing 364 persons on December 1900. This figure comprised 10 agents and sub agents, 101 miners, 32 labourers and boys underground, 141 tradesman, labourers and boys at surface, 13 women and girls and 67 employed on the arsenic works.[13]

1901

1901 started badly for the mine, the Company having problems securing labour. Mines on the Cornish bank of the river, such as Clitters, Drakewalls and Prince of Wales, were reopening and were paying higher wages than Devon Great Consols could afford. The "new" mines were employing some 250-300 miners "at most tempting wages". In addition working conditions in the newly reopened mines were better, not being as deep or as hot as Devon Great Consols. Josiah Paull estimated that the mine was 50% below its full complement. By February the problem was so bad that not only had exploration become a thing of the past but extraction was also suffering. This must have been galling for the Company as the stopes in the "old mine" were "yielding unusually well". This had serious implications: for whilst production, and hence income, was falling, fixed costs such as pumping still had to be met.[14]

At the end of January of 1901 the contract for the sale of arsenic had come to an end and the company were unable to renew it, being unable to agree satisfactory terms. The buyers were offering £5 a ton less than during the previous twelve months. If the Company sold their output of arsenic on these terms it would represent a reduction of £7,000 on a years output. The failure to make a contract at a reasonable price meant that the company had to borrow heavily against future sales. The situation was not helped by large increases in the price of materials such as coal, coke iron and steel. Thus by the year ending April 30th 1901 the Company was showing a loss of £1,258 3s 2d.[15,16]

By May labour shortages had started to have an impact on the arsenic works, Josiah Paull reporting that:

> I am sorry to say that the labour trouble has now reached the Arsenic Works. The refiners of arsenic who have the needful experience have nearly all left for the new workings in Gunnislake Clitters, Drakewalls and Prince of Wales where the wages are tempting and the work is more healthy than the Devon Great Consols Arsenic Works.[17]

On April 26th Peter Watson and Moses Bawden, in gloomy mood, met with the Estate's representatives to discuss the future of the mine. Various options were considered including suspending operations and putting the mine onto care and maintenance. Watson argued that the way forward was to increase arsenic production, an option which Josiah Paull thought unrealistic.[18]

At the 1901 Annual General Meeting held on June 4th, Peter Watson, Moses Bawden and William Clemo tried to sound positive, assuring the shareholders that new discoveries were imminent and that the mine was in better shape than ever, all of which smacked of desperation. Josiah Paull observed the company report depicted the "affairs of the company in much less gloomy colours than Mr Peter Watson thought fit to do when he was here on April 26th".[19] An interesting comment came from respected mining commentator J. H. Collins, who suggested that the mine might be run more efficiently if a modern system of electrical machinery was adopted throughout the mine.[20] If Collins' suggestion had been adopted it might have freed the mine from the dual problems of the mine's antiquated water power system and the random tyranny of the weather.

The stagnation of the arsenic trade deepened during the summer of 1901. In June it was noted that Mr Field, who effectively controlled the arsenic trade, did not intend to buy any more arsenic until such time as he had disposed of his large stocks of Devon Great Consols arsenic. These stocks comprised the make up to January 1901 amounting to 727 tons, for which the mine had already been paid. This left the mine with the make from February onwards which amounted, at the end of June, to 540 tons on unsold, and possibly unsaleable, arsenic in hand. In order to carry on the mine were borrowing heavily against the future sale of this arsenic. Josiah Paull estimated, in June 1901, that the Company's borrowing would, at current levels, reach the £10,000 limit allowed by its Articles of Association by the end of July.[21]

Things on the mine were looking quite promising. At Wheal Anna Maria the ground below the 124 fathom level was proving exceptionally rich in mundic, said to be worth twelve tons of good ore per fathom. This new discovery was felt to be exceptionally important given that both Peter Watson and Moses Bawden believed that the way out of current difficulties was to step up arsenic production. Such was the potential significance of this new find that Josiah Paull requested that this ground was trialled as soon as possible. In a report dated 19th October he estimated that the ground, as explored below the 124, would yield mundic to the value of £13 per ton. Unfortunately, in Paull's opinion the deeper part of the lode, in the vicinity of the 137 would not yield more than seven tons of mundic to the fathom, which, given current costs, would not pay working.[22] In an attempt to stave off further borrowing, a call of 1s was made in July.

To add to the Company's problems it had not proved possible to secure a contract to sell the Company's arsenic at a remunerative price. On August 27th it was noted that the stock of arsenic on the mine was 1,294 tons. Of this 519½ tons was Mr Field's which had been bought and paid for whilst 774½ tons was unsold. It was Josiah Paull's opinion that, if the mine wished to sell its arsenic it would have to drop its price considerably. Citing an article in the *Engineer*, Paull commented that America, which had been a major market for Devon Great Consols arsenic, was discovering and developing its own arsenic supplies. Additionally both Portugal and Spain were becoming significant producers exporting to America.[23]

On 18th October a further call of 1s was made.[24] In spite of the two calls on shareholders in July and October the Company had, by the end of October 1901, reached its £10,000 borrowing limit. Given the parlous state of the Company's finances a special meeting of shareholders was held on October 30th in London. After "much discussion of an unpleasant nature" the Directors were instructed "to carry on the mine for a time" by borrowing a further £10,000. Should the Company raise this sum they were empowered to spend it under the direction of an advisory committee of shareholders.[25]

After the October 30th meeting events unfolded rapidly; on November 2nd all work was suspended at Watson's with the exception of pumping.[26] A short paragraph in the *Tavistock Gazette* of the 15th November noted that underground operations had been suspended on the mine.[27] On the 16th all the men working in the old sections of the mine were given two weeks notice.[28]

At the Ordinary General Meeting held in London on Thursday November 28th 1901 the shareholders were informed that the company finances were in a parlous condition; the account up to the end of October 1901 showed a deficit of £1,590. The company's problems being compounded by its failure to secure a satisfactory contract for arsenic in September (see Note 2). In consequence the shareholders were informed that the decision had been taken to stop all surface and underground operations except for pumping and treatment of mineral already broken at surface. At the October 30th meeting the shareholders had given the Board the authority to seek additional finance, however the meeting was informed that the directors had been unable to borrow further sums to enable the Company to ride the depression in the arsenic trade. In the light of the situation the Directors informed that shareholders that unless they came forward with the money the only option left to them would be to wind up the Company and realise what assets they could. Having previously canvassed the shareholders the directors were able to tell the meeting that about 10,000 were in favour of winding the Company up, against 400 who wanted to continue.[29]

An Extraordinary General Meeting was convened and held on November 28th at 2 and 3 West Street, Finsbury Circus.[30] Mr Lane made the following proposal:

> That the company cannot, by reason of its liabilities, continue its business, and that it is desirable to wind up the same, that the company be wound up voluntarily, and that Mr. Peter Watson be appointed liquidator.[31]

The proposal was seconded by Mr Poupard and the motion was carried fourteen votes to three.[32] On the following Saturday, November 30th 1901, the decision to wind up the Company was formally announced to the employees on the mine.[33] It must have been a melancholy occasion, as the workers would have been well aware of the nature of the forthcoming announcement; the proceedings of the shareholders meeting having been fully reported in the previous day's *Tavistock Gazette*:

> As two o'clock approached in the Saturday afternoon the employees began to assemble in the yard beneath the balcony of the Account House, as some of them had been accustomed to do during the whole of the fifty five years that the mines had been in existence. All ages were represented in the crowd, from boys in their teens to the grand old miner of three score years and ten. The youths looked curious, the older men anxious, and the veterans sorrowful. The genial Vicar (Rev. E. E. B. Skinner) had previously been riding round on his bicycle, having a friendly chat with some of the men, and he expressed great concern at the turn matters had taken with respect to the mines.
>
> At the time appointed the following officials appeared on the balcony: Mr T. Youren, chief clerk, who represented the directors in the absence of Mr Moses Bawden, the much respected purser, through illness; Captain W. S. Clemo, resident manager, Captain W. Woolcock and Captain W. Clemo, Mr W. Matthews, engineer; Mr J. H. Carter and J. J. Harris, accountants:[34]

Mr Youren read the following letter penned by Moses Bawden the previous day:

> Dear Mr Youren - I most sincerely regret that I am unable to be with you tomorrow, and I would like you to convey to the men my sorrow at the turn things have taken at our mines and works. After being associated with them for over twenty years, it is a great blow to me to see these fine old mines and works shut down, when we have scarcely ever looked so well for the production of arsenic. I sincerely hope and trust that all those in the district who can employ the able bodied men will give preference to the heads of families. It distresses me more when I think of the coming winter with the large number of men suddenly thrown out of work.[35]

The miners were then informed that there was no money left in the sick club and, consequentially, those who had been dependent on it would no longer be able to claim relief from the fund.[36]

In the midst of such melancholy announcements there were some positive notes: Directions had been given to keep the "wheels and necessary machinery..... at work for draining the mine until further orders, should the price of arsenic appreciably improve before the mines are absolutely abandoned by being flooded".[37] Obviously maintaining the mine's pumps and treating the ore already at surface would only require a small number of men. There was some comfort for those for who had no future on the mine; in a notable act of philanthropy the Duke of Bedford undertook to employ many of the men. The *Tavistock Gazette* understood that they would be engaged at making a road between Endsleigh House and the Duke's yacht landing place on the Tamar.[38]

On November 30th, 330 people were still employed on the mine, comprising 8 agents and sub agents, 94 miners, 33 labourers and boys underground, 122 tradesman, labourers and boys at surface, 10 women and girls and 63 employed on the arsenic works.[39] During December "several men and lads" were employed underground sending copper ore and mundic, broken before the mine went into liquidation, to surface, likewise the arsenic works was kept in production processing the mundic.[40] Whilst a large number of hands had been discharged during December 148 remained employed on the mine. On December 31st this included 8 agents and sub agents, 94 miners, 33 labourers and boys underground, 68 tradesmen, labourers and boys at surface, 3 women and girls and 36 hands at the arsenic works.[41]

Peter Watson was taking his role as liquidator seriously, spending time in and around the mine during December 1901. Watson commissioned William Borlase of Penwith to examine the mine. Borlase had some experience of working electrical machinery on mines. At Devon Great Consols his remit was to report on ore reserves on the mine and to make recommendations as to the costs of replacing the mine's worn out pumping wheels and flat rods with water turbines and electric plant. Whilst Watson wanted to keep the mine in production during the liquidation period, he was unsure as to whether he was allowed, as liquidator, to authorise the ongoing working of the most productive parts of the lode (as opposed to extracting already broken ore). As a compromise Watson requested the agents to prepare estimates for the working of the rich ore ground in Anna Maria.[42]

Watson's major achievement during December 1901 was able to sell the 1901 and January 1902 stocks of arsenic. Just over 851 tons were sold at £13 10s 6d per ton realising £11,511 12s 4d, a further 296 tons were sold at £11 10s, giving a grand total of £1,4921 8s 9d. There was a distinct feeling of relief that the Company would be able to pay off the loans made by its bankers. Indeed it looked as if the Company would be able to address all of its liabilities without any further calls on the shareholders.[43]

Apart from meeting its liabilities there was a real prospect that the mine could work at a profit during 1902, Paull noting that:

> There is a prospect that the arsenic which may be made during January next will leave a profit of some hundreds of pounds to the company in as much as by the discharge of all the miners and a good many surface hands the working costs have been very much lessened, a sufficient number of men have been retained to keep the mine drained, and to carry on the Arsenic works so long as mundic can be got to feed the furnaces, and I think this can be done for some months in the new year.[44]

Whilst Watson had proved very successful in his role as Liquidator he had, thus far, failed to reconstruct the Company, which seems to have been his ultimate goal. Josiah Paull felt that this was partly due to the fact that there was uncertainty as to the term of years the Duke of Bedford would prepared to grant to a reconstructed company. A second problem was the current depressed state of the arsenic trade.[45]

The closure of the mine was seen as a body blow to the district; the mine spending £30,000 annually on "merchants bills, labour cost and dues".[46] The local bourgeoisie fulminated at length in various public meetings and in the pages of the *Tavistock Gazette*. Most vocal were the members of the Tavistock Mercantile Association who passed resolutions and formed committees (*e.g. Tavistock Gazette* December 27th 1901).

1902

During January 1902 men continued to work underground raising broken copper ore and mundic. By the end of the month work at Watson's had finished, although there was still some activity on both the Main and South Lodes. To process the mundic the arsenic works was kept running. In addition to raising ore good quality mundic was being recovered from dumps at Anna Maria and Wheal Emma. By the end of March the supply of mundic coming up from underground had been exhausted, however the dumps continued rich and it was estimated that the arsenic works could be kept running for a further three months.[47]

Apart from mundic some copper was also sold; for example on the 24th April 1902 at the Redruth copper ticketing Devon Great Consols sold 40 tons of copper ore at £1 15s per ton realising £70.[48] During the year the mine sold 267 tons, 19 cwt. of copper ore at an average price of £1 7s 7d, realising £370 4s 7d.[49]

In spite of being in liquidation the social life of the mine continued; on the 14th February 1902, for example, Moses Bawden chaired a meeting of the Tavistock Polling District Unionist Association in the Wheal Josiah Schoolroom when the meeting was addressed by Mr Arthur Taylor on "The British Empire".[50] Watson was still working hard to resurrect the mine. For example the *Tavistock Gazette* of April 25th contained the following advertisement:

> Devon Great Consols, Limited,
> In Liquidation.
> The Liquidator is prepared to receive offers for this property as a going concern, including plant and machinery. An extension of the Leases having recently been promised, will be granted to a responsible and acceptable Tenant.

During this time Peter Watson, in his role as liquidator, had not been idle. At a special meeting of shareholders held in London on May 28th 1902 Watson was able to tell the shareholders that the company's liabilities which were somewhere in between £4,000 and £5,000 had been discharged, leaving a deficit of only £360, apparently this had been achieved by a large and remunerative sale of arsenic.[51]

At the May 28th meeting Watson also presented details of a plan to reconstruct the company. The proposed company would have a capital of £30,720 comprising 122,880 shares of five shillings each. The meeting was not overly impressed by Watson's proposals. Of the three directors of the 1899 Company only Peter Watson was prepared to have anything to do with a new one. The meeting was adjourned for a period of two weeks to give the shareholders further time to consider the details of the reconstruction plan.[52,53]

The reconvened meeting was held at the Company's London Office on Wednesday 11th June 1902. An amended scheme involving the keeping of a reserve of unissued shares was put before the shareholders. The amended proposal met with a much more favourable response. The meeting, chaired by Watson, agreed the articles of association and adopted a scheme to reconstruct the Company with a capital of £50,000 in £1 shares.[54,55] The meeting passed the following resolutions:

> That pursuant to Section 161 of the Companies' Acts, 1862, the liquidator be authorised to sell and transfer all the mining rights, plant, and other assets of the company to a new company to be formed, for the purpose

of acquiring and working the same upon the terms set forth in the draft agreement approved by the meeting, and to enter into all necessary agreements for the purpose of carrying the said sale into effect, with such, if any modifications he may think fit.

That the liquidator be and he is hereby authorised to consent to the registration of a new company with the same or similar name to the existing company, with the memorandum and articles of association approved by this meeting.[56]

It was agreed that a further meeting should be held on Wednesday 2nd July. The shareholders who attended this meeting confirmed the resolutions made at the 11th June meeting. However the scheme fell through due to lack of support.[57]

A third scheme to resurrect the mine emerged in July 1902. Five "gentlemen" offered the Liquidator £7,000 for the concern. They asked for a months' option on the purchase for which they would pay £300 to keep the mine going during the month.[58]

In order to consider the proposal Peter Watson convened a shareholders meeting in July 1902. Unfortunately the meeting was "upset almost directly it assembled" by the receipt of a telegram from one of the five "gentlemen" in which he withdrew his support for the proposal. After some bickering the meeting broke up, the fate of the mine being left in the hands of the Liquidator "to do the best he could independent of any further meetings".[59]

By August 1902 the failure to resurrect the mine appears to have become obvious to all parties concerned and the decision was taken to dispose of the physical assets of the mine and surrender the leases to the Estate. Disposal of the plant would mean the cessation of pumping and consequent the flooding of the mine. Allowing the mine to flood was a final acknowledgment that the mine had failed. Cessation of pumping would also have implications for any future operations on the mine, the huge cost of bring the mine into fork seriously compromising the potential to work at depth.

During August negotiations had been ongoing between the Liquidator and Jabez Petherick (see Note 3) of the Tavy Foundry for the sale of the plant as scrap which Petherick would undertake to break up and sell by auction. The negotiations appear to have been fraught, Petherick being concerned about the "dangerous nature of stripping the shafts" and the long time that it would take to sell the material.[60]

An agreement dated August 29th 1902 was signed between a consortium of Tavistock businessmen headed by Jabez Petherick including William Holzapfel, Robert Stenner, and Devon Great Consols (Ltd) for the sale of the mine's physical assets. Apart from transferring some minor water leases the agreement also transferred all:

> plant machinery tools and utensils of the Company now on and about the Devon Great Consols Mines other than the following with are excepted (a) Furniture Books papers and other effects in the Counting House (b) Arsenic in the furnaces or in barrels or otherwise made or in the course of manufacture..... (c) The staves and headings for barrels.[61]

The sale agreement also included all the stores of coal, coke, tallow and candles. Petherick and his associates agreed to pay £7,200 for the assets of the Company. £1,000 was to be paid on the signing of the agreement and the balance on completion of the purchase which was to take place on or before September 29th 1902.[62]

Under the terms of the agreement the Company in Liquidation were permitted to continue producing arsenic until September 29th and reserved the right to remove arsenic for three months after. The Company was also required to use their "best endeavours to continue pumping the water from out of the Mine until twelve o'clock noon on the said twenty ninth September". The purchasers were granted free use of the railway and the locomotive.[63]

In addition to the sale agreement between Devon Great Consols and the Petherick consortium a tripartite agreement, dated September 29th 1902, was entered into with the Duke of Bedford.[64] The Company agreed to pay any royalties up to September 29th, and also any further royalties on arsenic manufactured after the 29th. It was also agreed that the company would transfer its Duchy of Cornwall water lease to the Duke.

The purchasers agreed to carry out remediation works on the mine. This included securing shafts and adits, filling flat rod trenches, wheel and bob pits, sloping off refuse heaps, throwing down all buildings which may be considered dangerous, filling in the tunnel at Morwellham, repairing bridges over the railway and pulling down and levelling the arsenic works. This work was to be completed within two years to satisfaction of the Government Inspector and Mr E. C. Rundle of the Tavistock Estate Office.[65]

Even before the appropriate paperwork was in place and without any ceremony Petherick and his associates began drawing the pumps in September 1902. Work was well underway by the time of Josiah Paull's report of September 27th in which he noted that:

> Messrs. Holzapfel and Petherick have begun to draw up the pumps and thus far there have been no breakage or hindrance in their doing so. For ten or twelve fathoms from the bottom of each shaft the columns of pumps have already been got to surface.[66]

The great mine was dying as, fathom by fathom, Devon Great Consols began to flood and the work of scrapping and disposal continued apace. As anticipated by Petherick, stripping the shafts was a dangerous business and there were a number of near misses caused by overwinds and breakages of chains and ropes. By the end of the year they had stripped nearly all of the shafts and underground workings of any material of value and some progress had been made in arching over the shafts.[67]

By the end of October "a considerable quantity of pumps, iron and wood rods" had been sold by private contract. A three day auction of materials was held, starting on Tuesday, 4th November. The November auction did not include the heavy plant such as pumps and steam engines. Paull commented that the results of the auction were satisfactory.[68]

As specified in the terms of the August 29th agreement the "Mining Company" continued to produce arsenic. In September the make amounted to 43½ tons, much of which came from reworking material recovered from the flues. In October 31¼ tons were produced and in November 31 tons. Due to arsenic sales the Liquidator was able to pay two dividends, on of 2s 6d per share, the other of 6d per share. December was the last full month in which arsenic was produced, 34 tons having been made. In addition to producing arsenic the Company also disposed of its existing stocks which, at the end of September amounted to 572½ tons.[69]

In his final report of 1902 Josiah Paull summarised the financial results of working the mine from 1844 until the end of 1902: The shareholders had answered calls amounting to £26,624, in return for which they received dividends of £1,232,105. The Duke of Bedford received £286,372 in dues, including the £20,000 fine for the granting of the 1857 leases. To his credit the Duke remitted £12,988 in dues.[70] Whilst these figures are impressive they pale into insignificance when one considers that the mine sold mineral to the value of £4,031,674. This comprised:

	Amount (tons)	Value (£)
Copper Ore	737,401	3,360,163
Refined Arsenic	71,607	662, 285
Mundic	15,307	8768
Ochre	81	172
Tin ore and tin stone	67	285
Lead ore	0.65	4

(*Source:* Bedford Estate Mining Reports 1899 – 1906, Devon Record Office)

1903

In the first few days of January 1903 14½ tons of arsenic soot were made "under the direction of the Liquidator". The completion of this work marked the end of Peter Watson's control of anything on the mine.[71] A week after the old company completed its final burn, Messrs. Petherick restarted work at the arsenic works. Their intention was to reprocess the bricks and other debris that made up the flues of the works as they demolished them. Up to the end of February 21 tons of arsenic had been produced. Production continued throughout the year, by years end Petherick's had sold 212 tons of arsenic, realising £2,403 10s 7d.

As Petherick's men scrapped the mine, material was dispatched either by the Great Western at Tavistock or via the Devon Great Consols railway and Morwellham. On the 9th March 1903 one of Petherick's men, Thomas Martin, who was tramming wagons on the quay at Morwellham, was killed when a section of the overhead railway collapsed and a wagon crushed him. The inquest, held in the Morwellham schoolroom on the 11th found:

> that the deceased died through injuries received through the negligence of Jabez Petherick, but that such negligence was not culpable.[72]

By early April 1903 Petherick & Co. decided to auction off the remaining plant on the mine:

> Parkhouse and Sons are instructed by Messrs. J. Petherick and Co. to sell by auction, on Wednesday and Thursday May 6th and 7th, the second portion of the plant, machinery, etc.: About 500 tons main line rails (bridge section) 30 to 40 lbs. per yard, with points and crossings; superior locomotive tank engine, by Spittle, 4ft. 8½ in. Gauge; 35 railway trucks, undertype engine by Fowler, 16 h.p.; horizontal and vertical winding engines, Cornish boilers, double gear screw cutting lathe, 16 in. centre; self – acting drilling machine, massive punching, shearing, and plate bending machines, several tons of heavy brass and gun metal, old lead and copper, about 100 tons wrought iron (B.B.H.) rods, 3 in. and 3¾ in. diameter, lot cast iron flange pipes, various sizes; good lot pine baulks and useful timber, large quantity of new stores, all the superior account house and other furniture, desks, iron safes, contents of laboratory etc.[73]

The auction, held on the 5th and 6th of May, was considered a success. Large quantities of "old iron and timber" and also "nearly all the railway iron" were sold. Paull noted that a large proportion of the material went to "the clay works in Cornwall". After the auction three steam engines remained unsold.[74]

During the last week of May 1903 the final meeting of the shareholders of Devon Great Consols (in liquidation) was chaired by Peter Watson at the Company's London Office. The final account was read and passed by the meeting, the shareholders receiving a final dividend of 5d. per share. Peter Watson was in a reflective mood as might be expected from a man had been involved in the mine for twenty seven years. In closing, Watson, in typical form, chided the shareholders for their failure to provide the required support:

> .....had the shareholders come forward and found the small amount of capital they were asked to provide, the company would even now with the price of arsenic be in a position to pay dividends. This mine has not been stopped on the account of the poverty of its mineral resources. It is very rich in arsenic, and I have not the slightest doubt that there is also a large quantity of copper ore in the mine. It has appeared strange to me that with a knowledge of the enormous wealth that has been obtained from it the shareholders should not have come forward and provided the machinery that was necessary to continue it.[75]

Josiah Paull's June 1903 report contained the sad news that Messrs. Petherick and Co. was "engaged in breaking up the large wheels at Blanchdown". This work continued until August when all four of the big wheels had been broken up. In June most of the railway from the mine to the head of the Morwellham incline had been lifted and the scrap sent away. The three steam engines noted as being unsold after the May auction included two at Wheal Emma and the engine at the Morwellham incline head. By the end of October two of the three remaining engines had been sold. At the end of the year one engine at Wheal Emma remained unsold.

In addition to scrapping and disposing of the material assets of the mine Petherick's men were also engaged in remediation works. These included filling in the railway tunnel at Morwellham which was carried out during November and December and sloping in wheel pits and flat rod cuttings. By the end of the year fifteen shafts had been arched over and three had been filled in.[76] By years' end Messrs. Petherick were employing forty people on the mine, seventeen of whom were working on the arsenic works.

1904

Scrapping and disposals continued into 1904. In January the castings which comprised the Anna Maria roller crusher were being broken up and sleepers were being removed from the railway. The surviving steam engine at Wheal Emma was broken up in February along with the remains of the Wheal Anna Maria crusher and the remaining castings which comprised the Blanchdown wheels.[77] During the spring the contractors were busy fencing the eastern side of the railway.

During the summer work focussed on demolishing the arsenic works. In July it was noted that the works had nearly been demolished with the exception of one flue which would be used for burning the residues from the demolished flues. On July 29th Josiah Paull noted that:

> burning of ores and residues will cease tomorrow, it will take another fortnight to do the refining. This will be the last make of arsenic we shall have to record.[78]

Refining continued during August. In September, with refining completed, the last flue was demolished. About forty tons of arsenic soot was recovered from the flue. The last vestige of the arsenic works to be demolished were the two chimney stacks which were levelled in December. During the year 107 tons 13 cwt of refined arsenic was produced realising £1,200 14s 7d.[79] On August 24th a third auction was held. The old iron work and timber offered at the auction found buyers, although at very low prices. In October thirty six granite markers were placed to mark the location of the capped shafts.

1905

Petherick's continued remediation works on site during the first months of 1905. In January seven men were employed on various jobs including sloping in wheel pits. In February the men were busy sloping waste dumps and flat rod cuttings.[80] In addition to recovering arsenic during their tenure Petherick's also managed to sell some seventy five tons of ochre recovered from the precipitation works. The "ochreous slime" was sold for five shillings a ton and realised £18 15s.

On 25th March 1904 the time period allowed to Messrs. Petherick for carrying out their contracted works expired. Outstanding work was commuted by the Estate in return for a payment made by Petherick.[81]

Chapter 10 Notes

Note 1

William Clemo 9th May 1822-3rd September 1900. Josiah Paull, who had at times been critical of William Clemo wrote a fulsome eulogy:

> On the 3rd inst the manager of the mine (Capn William Clemo) died after fifty six years service with the Company, he was the discoverer of the ore in 1844 when the mine was started, and from that time passed from working miner to timberman, pitman agent and Manager. Having been closely associated with him in the working of the mine since 1860 I have had full reason to appreciate his great practical knowledge of mining. He was a most agreeable man to work with and his removal is a loss to the safe and economical working of the mine of the mine which I very greatly regret.[82]

Note 2

A major contributing factor in the slump in the arsenic trade appears to have resulted from the death of 57 million sheep in Australia whose fleeces would have been cleaned with arsenic.[83]

Note 3

Jabez Petherick, at the inquest into the death of one of his workers at Morwellham, described himself as an ironmonger of Tavistock.[84] Petherick was something of a local worthy, being Vice-President of the Tavistock Mercantile Association and a local councillor.[85,86]

Chapter 10 references

1. Devon Record Office document 8187 Bedford Estate Mining Reports 1899-1906
2. Devon Record Office document 8187 Bedford Estate Mining Reports 1899-1906
3. Devon Record Office document 8187 Bedford Estate Mining Reports 1899-1906
4. Devon Record Office document 8187 Bedford Estate Mining Reports 1899-1906
5. Devon Record Office document 8187 Bedford Estate Mining Reports 1899-1906
6. *Tavistock Gazette* 1st June 1900
7. Devon Record Office document 8187 Bedford Estate Mining Reports 1899-1906
8. *Tavistock Gazette* 3rd August 1900
9. *Tavistock Gazette* 7th September 1900
10. *Mining Journal* 8th September 1900
11. Woolcock MSS
12. Devon Record Office document 8187 Bedford Estate Mining Reports 1899-1906
13. Devon Record Office document T158m/E44 a-b, 1901 Mining Report
14. Devon Record Office document 8187 Bedford Estate Mining Reports 1899-1906
15. Devon Record Office document 8187 Bedford Estate Mining Reports 1899-1906
16. *Tavistock Gazette* 7th June 1901
17. Devon Record Office document 8187 Bedford Estate Mining Reports 1899-1906
18. Devon Record Office document 8187 Bedford Estate Mining Reports 1899-1906
19. Devon Record Office document 8187 Bedford Estate Mining Reports 1899-1906
20. *Tavistock Gazette* 7th June 1901
21. Devon Record Office document 8187 Bedford Estate Mining Reports 1899-1906
22. Devon Record Office document 8187 Bedford Estate Mining Reports 1899-1906
23. Devon Record Office document 8187 Bedford Estate Mining Reports 1899-1906
24. Devon Record Office document 8187 Bedford Estate Mining Reports 1899-1906
25. Devon Record Office document 8187 Bedford Estate Mining Reports 1899-1906
26. Devon Record Office document 8187 Bedford Estate Mining Reports 1899-1906
27. *Tavistock Gazette* 15th November 1901
28. Devon Record Office document 8187 Bedford Estate Mining Reports 1899-1906
29. *Tavistock Gazette* 29th November 1901
30. National Archives document: Notice of Extraordinary Resolution November 28th, 1901,
31. *Tavistock Gazette* 29th November 1901
32. *Tavistock Gazette* 29th November 1901
33. *Tavistock Gazette* 6th December 1901
34. *Tavistock Gazette* 6th December 1901
35. *Tavistock Gazette* 6th December 1901
36. *Tavistock Gazette* 6th December 1901
37. *Tavistock Gazette* 6th December 1901

38. *Tavistock Gazette* 6th December 1901
39. Devon Record Office document T158m/E44 a-b, 1901 Mining Report
40. Devon Record Office document 8187 Bedford Estate Mining Reports 1899-1906
41. Devon Record Office document T158m/E44 a-b, 1901 Mining Report
42. Devon Record Office document 8187 Bedford Estate Mining Reports 1899-1906
43. Devon Record Office document 8187 Bedford Estate Mining Reports 1899-1906
44. Devon Record Office document T158m/E44 a-b, 1901 Mining Report
45. Devon Record Office document 8187 Bedford Estate Mining Reports 1899-1906
46. *Tavistock Gazette* 27th December 1901
47. Devon Record Office document 8187 Bedford Estate Mining Reports 1899-1906
48. *Tavistock Gazette* 2nd May 1902
49. Devon Record Office document 8187 Bedford Estate Mining Reports 1899-1906
50. *Tavistock Gazette* 21st February 1902
51. *Tavistock Gazette* 30th May 1902
52. Devon Record Office document 8187 Bedford Estate Mining Reports 1899-1906
53. *Tavistock Gazette* 30th May 1902
54. Devon Record Office document 8187 Bedford Estate Mining Reports 1899-1906
55. *Tavistock Gazette* 13th June 1902
56. *Tavistock Gazette* 4th July 1902
57. Devon Record Office document 8187 Bedford Estate Mining Reports 1899-1906
58. Devon Record Office document 8187 Bedford Estate Mining Reports 1899-1906
59. Devon Record Office document 8187 Bedford Estate Mining Reports 1899-1906
60. Devon Record Office document 8187 Bedford Estate Mining Reports 1899-1906
61. Devon Record Office document L1258M/SS/MC/1 Agreement for sale & purchase 29th August 1902
62. Devon Record Office document L1258M/SS/MC/1 Agreement for sale & purchase 29th August 1902 L1258M/SS/MC/1
63. Devon Record Office document L1258M/SS/MC/1 Agreement for sale & purchase 29th August 1902 L1258M/SS/MC/1
64. Devon Record Office document L1258M/SS/MC/1, Agreement 29th September 1902
65. Devon Record Office document L1258M/SS/MC/1, Agreement 29th September 1902
66. Devon Record Office document L1258M/SS/MC/1, Agreement 29th September 1902
67. Devon Record Office document 8187 Bedford Estate Mining Reports 1899-1906
68. Devon Record Office document 8187 Bedford Estate Mining Reports 1899-1906
69. Devon Record Office document 8187 Bedford Estate Mining Reports 1899-1906
70. Devon Record Office document 8187 Bedford Estate Mining Reports 1899-1906
71. Devon Record Office document 8187 Bedford Estate Mining Reports 1899-1906
72. *Tavistock Gazette* 13th March 1903
73. *Tavistock Gazette* 9th April 1903

74. Devon Record Office document 8187 Bedford Estate Mining Reports 1899-1906
75. *Tavistock Gazette* 1st May 1903
76. Devon Record Office document 8187 Bedford Estate Mining Reports 1899-1906
77. Devon Record Office document 8187 Bedford Estate Mining Reports 1899-1906
78. Devon Record Office document 8187 Bedford Estate Mining Reports 1899-1906
79. Devon Record Office document 8187 Bedford Estate Mining Reports 1899-1906
80. Devon Record Office document 8187 Bedford Estate Mining Reports 1899-1906
81. Devon Record Office document 8187 Bedford Estate Mining Reports 1899-1906
82. Devon Record Office document 8187 Bedford Estate Mining Reports 1899-1906
83. *Tavistock Gazette* 27th December 1901
84. *Tavistock Gazette* 13 March 1903
85. *Tavistock Gazette* 27th December 1901
86. *Tavistock Gazette* 20 March 1903

Chapter 11

The 20th century at Devon Great Consols:

The Bedford Estate started operations at Devon Great Consols in early 1906; work being carried out by a combination of labour employed directly by the Estate and tributers. In the early days work focussed on recovering mundic from the dumps and copper precipitation. Initially operations were fairly low key; the mundic for example was not processed into arsenic but simply sold on. However at the outbreak of the Great War arsenic prices rose to a point where the Estate felt it worth while to start processing the Devon Great Consols mundic at Coombe Works, Harrowbarrow. By mid 1914 the Estate had taken over the direct operation of Bedford United Mine which it worked in conjunction with Devon Great Consols. The Great War also saw the resumption underground work at Frementor and Wheal Fanny due to an increase in the demand for arsenic, tin and tungsten. The post war period saw further expansion including the construction of a new arsenic works at Devon Great Consols.[1,2]

1906

In a brief report dated 28th February 1906, T. S. Bliss, who appears to have replaced Josiah Paull as the Estate's mineral agent, noted that about forty tons of mundic had been picked out of the old dumps, broken and sold. This material contained 19½% arsenic. In mid March four men were employed picking over the Anna Maria dumps. These men, working under S. Remfree, appear to have been working on tribute. The mundic was sorted into three grades: 25% and over, 20% and over and 10% and over. On July 25th 1906 some 200 tons of mundic was sampled and tenders for its sale were invited. In August the first sale of mundic was made to Gunnislake Clitters Mine. In the same month Remfree settled his dues to the Estate paying 19s 11d. At the end of Remfree's tribute contract the Estate put their own men to work at Anna Maria, ten men working here in November 1906. In September Bliss noted that the burrows at Wheal Emma were being examined to see if they were worth working. Evidently they were, for in October Mr Remfree and his men started working the Wheal Emma dumps on tribute.[3]

In addition to reworking the dumps for mundic 1906 also saw the commencement of copper recovery. In October 1906 a copper precipitation works was set up and it was hoped that the works would be able to produce one to one and a half tons a year. During the first month some four to five hundredweight of precipitate was collected. The following month a second launder was set up and output increased to between ten and twelve hundredweight a month.[4]

1907

At the beginning of 1907 ten Estate men were still employed on dump recovery at Anna Maria, whilst Remfree and his tributers continued work at Wheal Emma, selling mundic worth £56 5s in January. In March a further five men were employed on the Watson's dumps. Like Remfree these men were working on tribute. In September 200 tons of mundic was sold by the Estate to Clitters mine.[5]

In March 1907 a second precipitation works came in to production, the two works being known as "Copper works No. 1" and "Copper works No. 2". The No.2 works produced its first precipitate in April. In May the new works was producing some fifteen hundredweight of precipitate a month.[6]

1908

At the beginning of 1908 10 men were busy picking mundic, presumably at Anna Maria. Six men including Messrs. Remfree, Sutton and Remfree were working the Wheal Emma dumps on tribute, whilst a further six men, also on tribute were working the Watson's dumps. In June the tributers at Wheal Emma stopped picking mundic and settled their dues: Messrs. Remfree & Sutton paying £8 12s 8d in dues whilst J. Jackman paid £1 4s 3d.[7]

By July 1908 mundic picking had been suspended across the mine. Clitters Mine had closed and, consequently, defaulted on a payment for 200 tons of mundic, leaving the Estate with, what in effect, was a bad debt of £402 16s 6d. All that the Estate was offered, should Clitters restart, was £402 in paid up shares. Given that the Estate had a stock of between 400 and 500 tons of mundic it made little point to continue working the dumps. In an attempt to sell its stock of mundic the

Estate had, in July, entered into negotiations with the British Mining and Metal Co. It must have been with some relief that Bliss was able to write in August that "the mundic on hand has been sold to the British Mining & Metal Co. and some 180 to 200 tons have already been carted away. Some half dozen men have begun picking mundic again". By September some 450 tons had been taken away by the B. M. & M. Co. In November five tributers were working the Wheal Emma dumps.[8]

In January 1908 both precipitation works were in full production. However during the summer the higher spring feeding No.1 Copper works dried up; fortunately the lower spring, feeding No.2 works, continued to run. By August "the springs had broken again" and both precipitation works were at full capacity, both works continuing in production throughout the remainder of the year. In November six tons, three hundredweight of copper precipitate was sold. During June and July a waterwheel and dipper wheel had been erected and was at work by August. The dipper wheel appears to have been used for recycling water that had already passed through the works. In December it was noted that "the dipper wheel has been at work all the month and some eight to ten hundredweight of precipitate collected from second treatment of the water".[9]

January 1908 saw the start of ochre recovery from "ochre pits". Whilst some ochre was worked by the Estate, some was let to "Messrs. Stiff & Co". In June the Estate sold twenty tons of ochre. In the same month Bliss commented that Messrs. Stiff & Co. "have so far done nothing with their ochre pits". Messrs. Stiff's lack of activity was a cause of some concern to the Estate which ended its contract in July and negotiations had started with another firm, a contract being agreed with the Golden Valley Ochre Co. In September and October attempts were made to air dry some ochre, presumably so they could sell it as a finished pigment rather than "raw" ochre. However the experiment was something of a failure, the weather being to damp and the process requiring too much space. During November 20 tons of ochre were "sent away".[10]

1909

In January of 1909 ten men were working the dumps, six at Anna Maria and four tributers at Wheal Emma. However due to the low price of mundic, £14 per ton, the Estate were not intending to sell until the price rose. In February three tributers started work on the Wheal Fanny dumps and continued until at least July. From October until the end of the year only three men were picking mundic.[11]

The precipitation works were in "full work" at the start of the year and continued with the odd hiatus throughout the year. In October the springs were noted as running very full due to rain and "a certain amount" of water was running to waste.[12]

The Golden Valley Ochre Co. sent away two truck loads of ochre in February. This consignment completed their contract with the Estate who received £14 7s 2d in settlement. After this date the Estate appear to have worked ochre directly.[13] During November and December four men were employed breaking stones for road metal.[14]

1910

At the start of 1910 only three men were picking mundic, numbers varying between three and four throughout the year typically producing four hundredweight of mundic a day. In August 16% mundic was worth fourteen shilling a ton. These men were earning an average daily wage of 2s 7d. In August 1910 T. S. Bliss wrote:

> The men employed are old miners two of whom are given this work rather than recommend them for the pension list. They work when they can and get more money in this way.[15]

At the precipitation works the springs were running "full" in January, however due to the frosty weather copper was only being deposited slowly. In March there was so much water coming out of the springs that the dipper wheels were stopped and the "surplus" water was directed in the launders usually used for secondary treatment. During the spring of 1910 the Estate undertook experiments with settling tanks in an attempt to reduce the loss of fine copper. In May it was noted that three men were engaged in collecting old iron for the copper works. It was estimated in August that the works were producing some £300 a month.[16]

During the year one man was usually employed breaking stone for road metal.[17] In November one man was employed recovering old bricks from the demolished arsenic flues as the mine had "an enquiry for a few thousand".[18]

1911

During the first few months of the year three to four men were engaged picking mundic. In April the four men who had been picking mundic were put to breaking road stone for Endsleigh, this work continued until July when it was noted that four men were working both mundic and road stone. In August the four men were back to breaking stone, this time for Latchley Concrete Works. This work evidently proved remunerative as in September eight men were engaged in the work, however by October this job had ended. During October, November and December four men were recorded as "breaking road stone, mundic and picking up scrap iron on various parts of the mine".[19]

In December the road between Wheal Maria and Wheal Fanny was being repaired.[20] Copper precipitation proceeded uneventfully through the year. Ochre production appears to have stopped in 1911. This may have been due to the problems of drying the ochre.

1912

During the early part of 1912 three men were busy picking mundic, recovering scrap iron and breaking road stone. As arsenic prices picked up during the year the men went back to picking mundic. In October, for example, it was noted that four men were picking mundic "for His Grace" whilst Remfree and Sutton were working on tribute. Remfree and Sutton had been working the dumps at both Wheal Fanny and Wheal Emma.[21]

In January and February two men were picking wolfram on tribute in Blanchdown Wood. Presumably this was Frementor ore. This represents the first instance of wolfram recovery on the mine.[22]

During spring 1912 many of the dumps on the mine, including the large halvans sand dump, were sampled by the West of England Tin Corporation. 137 samples were taken across the site. The sampling work involved driving tunnels, some one hundred feet long, into the dumps. To the displeasure of the Estate the tunnels were left in a dangerous state.[23]

The copper works continued to operate throughout the year.

1913

At the beginning of 1913 six men were picking mundic "on the Estate account", the mundic being sold to the Bristol Mining and Metal Company. From the late spring until the end of the year this had reduced to and average of two men picking mundic and breaking road stone. In addition to the Estate men two tributers were also at work recovering mundic.[24]

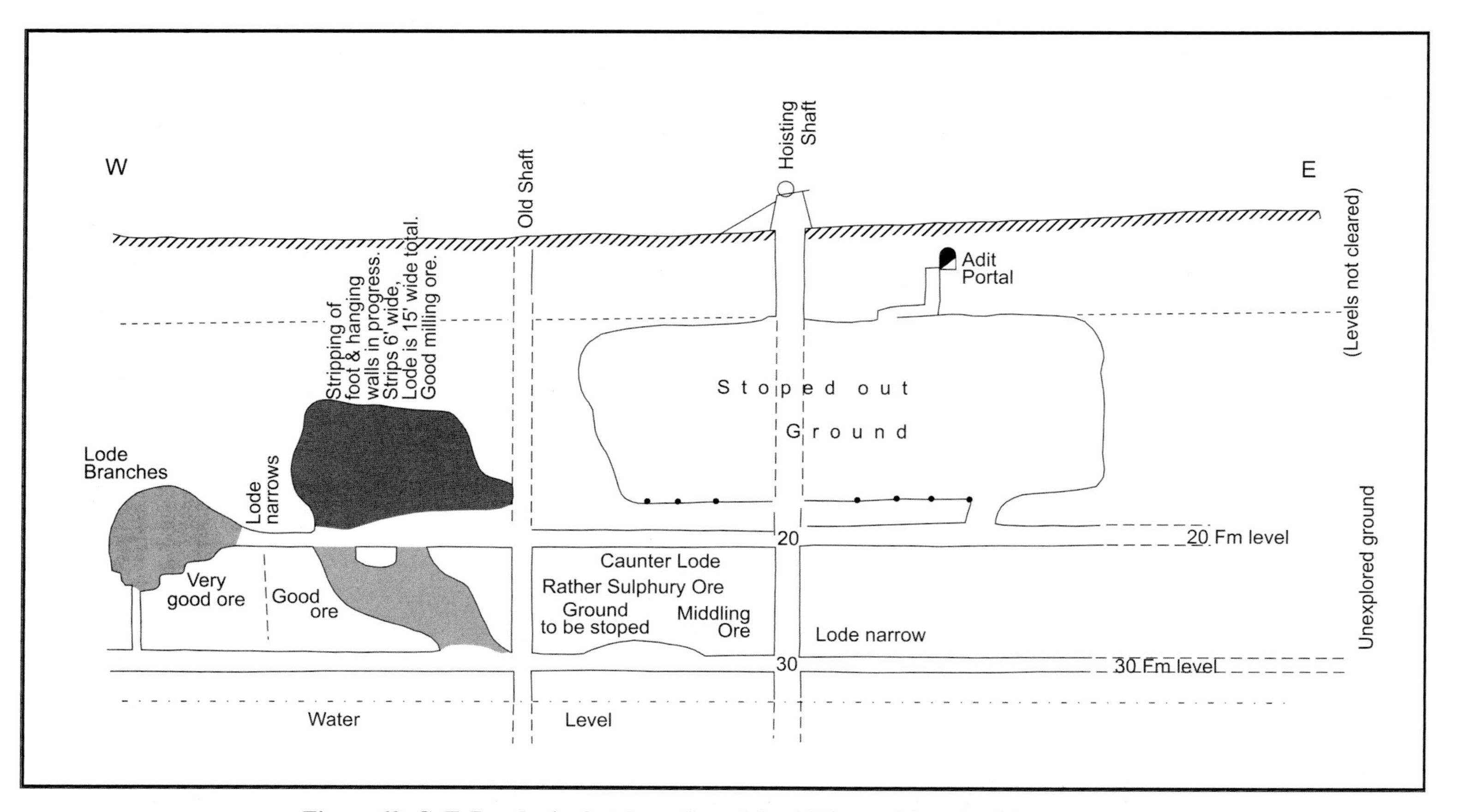

Figure 62. C. F. Barclay's sketch section of the 1920s workings in Wheal Fanny.

In February the lower or No.2 copper works was expanded with the addition of five hundred feet of new launders, the work being completed by March. In addition seven hundred and fifty feet of agricultural land drain was laid to feed the springs. It was hoped that this would give a more continuous supply and would also reduce the dilution of copper rich water by surface water. In September the leat from the Blanchdown Adit to the copper works was cleared, greatly increasing the amount of copper rich water passing through the works. An additional six hundred feet of launders was added in November. Also in November iron shavings were used in place of scrap iron in the launders. Iron shavings, having a much greater surface area than scrap metal, could recover more copper.[25]

1914

The year started with three men picking mundic and breaking road stone. In March and April it was noted that two men were picking mundic for the Estate with a further two men working on tribute.[26]

In August the price of arsenic rose to £18 per ton. In consequence the Estate decided to process its own mundic at Coombe Arsenic Works, Harrowbarrow in Cornwall. In September it was noted that the price of arsenic remained high and that "burning at Coombe would commence shortly"; the October report noting that "the burning of arsenic at Coombe has commenced". During November 190 tons of mundic were burnt producing forty tons, fourteen hundred weight of arsenic. In December fifteen men were employed at Coombe and 172 tons of Devon Great Consols mundic had been burnt during the month (see Note 1).[27]

In January 1914 the Estate was investigating two small springs which only ran during the winter months, one at Wheal Fanny, the other at Wheal Emma, with a view to setting up precipitation works. In February it was noted that the experiment at Wheal Emma was doing very well whilst that at Wheal Fanny was only partially successful. In March both Wheal Emma and Wheal Fanny were producing about five hundredweight of precipitate a month. At the original copper works a further four hundred feet of launders had been added and additional work carried out on the Blanchdown adit leat.[28]

1915

In August two men were employed at Frementor "picking out tin and wolfram ore". In September and October the number of men at Frementor had increased to four. This work had stopped by November.[29]

During the early months of 1915 five to seven men were working at Coombe arsenic works although refining seems to have been suspended at Coombe in April. Burning recommenced at Coombe in November when four men were noted as being at work here. From May it was noted that one man was precipitating copper at Coombe, presumably treating the burnt rinkle. This continued throughout the year.[30]

In May the report on Bedford United noted that two men were employed at surface "building a drying furnace".[31] This may well have been associated with tin recovery. The copper works continued to produce throughout the year.

1916

The tributers Sutton and Remfree were still busy working the dumps for mundic. Between 1st October 1915 and 10th February 1916 they sold two hundred and seventy six tons, seven hundredweight of mundic for £632 0s 8d, the Estate receiving £94 16s 1d in dues.[32]

In March and April 1916 two men were "prospecting" on the mine. This may well have been preliminary work prior with the intention of recommencing of underground working. In September two men were recorded as prospecting at surface for wolfram; presumably in the vicinity of Frementor. During October four men were recorded as prospecting underground at Frementor ("Frameter"), Wheal Fanny and Wheal Maria. In November two men were prospecting underground at Frementor and four at Wheal Fanny. Prospecting continued into December when two men were at Frementor and seven at Wheal Fanny.[33]

Coombe Arsenic Works was in production throughout the year as was the precipitation works.[34]

1917

Mining recommenced at Devon Great Consols in January 1917. In that month seven men were recorded as "Raising mundic" from Wheal Fanny. Typically nine or ten men were employed at Wheal Fanny although by December fourteen men were working there.

During January two men continued prospecting at Frementor. By February Frementor was also in production, six men are recorded as "Stoping". Work underground continued throughout the year. During the first few months of the year six men were working underground at Frementor, by July this had increased to nine and fourteen by October.[35]

During the nineteenth century the Frementor lodes had been largely ignored, work concentrating on driving the adit level to comply with lease conditions. Presumably the complex cassiterite-wolfram ores had defied satisfactory dressing by the traditional gravity separation techniques available to Isaac Richards (see Note 2). However by the turn of the century the invention of magnetic separators meant that it was possible to dress the Frementor ores.

Both the mundic from Wheal Fanny and the cassiterite-wolfram ore from Frementor were treated at the Bedford United mill. In 1920 Dewey noted that the arsenopyrite "is passed over a 2 inch grizzly, part broken in a Marsden jaw crusher, and sent to 16 head of Cornish stamps. There are three Buss tables, 1 Frue vanner, 1 Record vanner, 4 convex buddles and kieves and dumb pits. Wolfram and iron are separated from the other concentrates by a Wetherill magnetic separator" (see Note 3).[36] Copper precipitation continued during 1917.[37]

1918

Work underground continued at both Wheal Fanny and Frementor during 1918, likewise arsenic refining at Coombe and copper precipitation continued apace.[38]

December 1918 was the last month in which Frementor was worked (see Note 4), eight men being employed A number of factors contributed to closure including the post war slump in the demand for tungsten, the erratic nature of the ore body and difficulty of transporting the ore from the mine.

During 1917 and 1918 the Frementor "main lode" (see Appendix 2) was extensively developed down to the 12 Fathom Level. A report dated 28th April 1919 outlined the work undertaken at Frementor:

> Mining operations to date consist of a shaft sunk on the lode to a depth of 72 feet, and a level driven East and West at this depth for a total distance of 102 feet, and also a little stoping above the level. About 1,000 tons of ore were raised from the shaft, level and stope, and the yield in mineral from these 1,000 tons was 56 lbs of black tin and wolfram per ton of ore in the proportion of 2/3rd tin and 1/3rd wolfram. This is a high average, much above that of the Cornish Mines now working.[39]

C. F. Barclay, who visited the mine on April 25th 1918, observed that the copper precipitation works, below the halvans dump comprised five launders.[40]

1919

Work underground at Wheal Fanny continued on a small scale throughout 1919.[41] As in previous years copper continued to be recovered at the precipitation works.[42]

With the end of the war the Estate began to give consideration to future operations across its property including Devon Great Consols. To this end a report, probably written by Josiah Paull, was commissioned by the Estate. Frementor ("Fermator") was considered to be a "very excellent prospect". Paull advocated connecting the Frementor workings to the Frementor – South Fanny Adit which came in seventy five feet below the current bottom of the Frementor workings. This would require a drive of some five hundred and fifty yards and a connecting rise or winze between the two sets of workings. The report also recommended that the adit should be cleared as far as the South Fanny workings and the old workings tried for "arsenical ore". The report suggested that Frementor and Bedford United ore should be treated "at one mill and concentrating plant". It was recommended that "a light tramway" should be laid down to transport ore to the Bedford United mill for treatment.[43]

1920

Wheal Fanny continued to be worked throughout 1920. During the latter part of the year the workforce gradually increased, peaking at 19 men in November and December.[44] An inventory dated December 1920 noted that a portable boiler valued at £35 and a steam hoist and house valued at £45 were located on the shaft at Wheal Fanny.[45]

In August 1920 development work was underway at Wheal Maria; a new crosscut adit was in the process of being driven to dewater the mine ten fathoms below the then current water level (see Note 5).[46]

Although work had stopped at Frementor at the end of 1918 the mine was receiving limited attention. When C. F. Barclay visited Frementor in August 1920 the mine appears to have been kept on care and maintenance:

> the mine is kept on fork by means of a baling bucket working up and down in....the shaft.... When the market recovers and the costs of working are lower a small five head battery and dressing plant will probably be erected

to treat the ore on the spot.[47]

The December 1920 inventory recorded that the an "Engine Drum, kibbles and house" valued at £65 and a "Headgear and sundry ladders" valued at £35 were located at "Frameter Shaft".[48]

Recovery in the market appears to have been slow in coming, in 1923 Frementor was noted as being idle and full of water.[49] It would not be until 1925 that Frementor reopened.

By 1920 arsenic production was being stepped up, output being around 360 tons per annum.[50] This appears to have been a result of an increased demand for arsenic as a pesticide in the United States. By 1918 B. R. Coad working on behalf of the United States Department of Agriculture had established that calcium arsenate was effective against the Boll Weevil which periodically ravaged the cotton crop. The mild winter of 1919 meant that there was a particularly bad infestation in 1920 and consequently a greatly increased demand for calcium arsenate pesticides.[51] Devon Great Consols was well placed to meet this demand: The potential for arsenic production at the mine was truly colossal, and in 1920 the dumps were estimated to contain 750,000 tons of waste, much of which contained arsenic, additionally there were large reserves in the stopes.[52]

To facilitate increased production extensive railroads (see Note 6) and a new arsenic plant were being developed. The existing arrangement of transporting the ore to Coombe had proved to be too cumbersome and the Estate decided to erect its own arsenic works at Anna Maria. By August 1920 work was well underway, the stack already having been completed.[53] As built the arsenic works appears to have comprised a primary crusher, a single Brunton calciner, condensing chambers, refining chambers, flues, a grinding mill and an associated coopers' shop.

Visiting in August 1920 Barclay reported that "the old railway is being re-laid from Wheals Fanny and Josiah into Bedford United where the dressing plant is situated. The haulage will be done by a petrol-driven locomotive".[54] This involved relaying the Devon Great Consols railway to a nominal two foot gauge using war surplus equipment including a Motor Rail and Tramcar Co. Ltd 20HP Simplex (see Note 7). The purpose of this railroad was to transport the poorer "milling grade" ore (see Note 8) for initial treatment at the Bedford United mill. Best quality ore not requiring initial treatment was transported directly to the arsenic works.[55] The railroad did not extend as far as Wheal Fanny, terminating in the vicinity of the new arsenic works. Ore was transported from the mines to the railhead by two Estate owned lorries: a Garrett steam lorry and a Clayton, both of which, if contemporary reports are to believed, were in need of constant repair, usually by sundry members of the Higman Clan.[56]

1921

Little if any concrete information has survived for 1921, however quite a lot can be inferred.

Underground work continued at Wheal Fanny and, possibly Wheal Maria.

1921 would also have seen the arsenic works coming into production. Typical jobs, as recorded in later workbooks include: feeding mill, working calciner, cleaning out arsenic, cleaning out arsenic soot, building refiner chamber, putting stuff to battery, papering barrels, bringing mundic to grinder and making casks.[57]

The copper precipitation works would have continued to operate.

1922

By mid 1922 work was underway to expand the arsenic works effectively doubling its output. This expansion appears to be related to the planned reopening of Wheal Maria and South Wheal Fanny and the increased amount of material being recovered from the dumps. A second Brunton, known as No.2 calciner was under construction coming into service in autumn 1922 (see Note 9). An important part of the expansion of the arsenic works was the erection of a "sprinkler" or waterfall chamber on the arsenic flue.[58]

The unmetaled lorry roads between Wheal Fanny and the arsenic works required regular maintenance, C. Friend and J. Gregory spent a considerable amount of 1922 repairing lorry roads at Fanny.[59] The wear and tear on both the lorries and the lorry roads must have been considered excessive as the decision was taken to build a railroad from the arsenic works to Wheal Fanny, work starting in the latter part of 1922. Unlike the 1920 DGC-Bedford United railroad which was able to utilise an existing trackbed the 1922 Wheal Fanny railroad required a new formation. To add to the problem the railroad would have to cross an intervening ridge necessitating the construction of two rope hauled inclines: one from the arsenic works to the top of the ridge, the other from Wheal Fanny to the top of the ridge. The winder and drum gear was located on the top of the ridge roughly halfway between Wheal Fanny and the arsenic works. Work on the railroad continued over several months during the summer and autumn of 1922.[60]

1923

At the time of Barclay's April (?) 1923 visit Wheal Fanny was being worked down to the 30 fathom level. Three stopes were being worked. The first stope was working the back of the 20, west of the shaft, the work consisting of stripping the walls of an old copper stope for "milling" grade ore. Two pares were employed on this work. The second stope was an underhand stope in the bottom end of the 20 working high grade ore. One pare was working in this stope. Ore from this stope was dropped down via a winze to the 30 fathom level which was the main tramming level. The third stope was located in the back of the 30 and was worked by two pares of miners. A steam winch hoisted ore out of the mine in a kibble. The best quality ore was trammed directly to the new arsenic works at Anna Maria whilst milling grade ore was trammed to the mill at Bedford United for initial treatment.

In April 1923 work at Wheal Maria was less advanced than at Fanny: the new adit had recently been driven on North Lode; a crosscut being driven from the adit drive to intersect a 19th century level on the main Lode. Unfortunately owing to a surveying error the crosscut hit the back of the level which meant that water had to be drained via a siphon! A pre-existing air shaft had been reopened for hoisting and a new raise put up as a footway. Barclay noted that there was a plan to reopen and refurbish Gard's Shaft for hoisting and pumping. Whilst Fanny was only worked down to the 30 the intention appears to have been to work Wheal Maria down to a depth of fifty or sixty fathoms. To this end a small-petrol driven pump, located in a large flooded stope on the Main Lode, was at work. Stopes were being worked on both North Lode (Adit level) and South Lode.[61]

In addition to work at Wheal Fanny and Wheal Maria there was also some activity at South Wheal Fanny in the summer of 1923.[62,63] The workbook for July to September 1923 records that N. Willcock and G. Willcock were employed on South Fanny Lode during July. Work continued apace during August and into September 1923; J. Symons snr. and jnr. were busy timbering at the mine, an old building was being converted into a dry and the shaft was being reconditioned and deepened.[64] In September 1923 a couple of men were busy "cleaning up Framator adit" (*i.e.* the South Fanny deep adit.[65,66] All these activities are indicative that there was an intention to seriously work the mine, however and for whatever reason, all activity at South Fanny appears to have ceased.[67]

After Barclay's visit in April 1923 the historical record is uneven; underground activity at Wheal Maria was recorded in the series of workbooks which have survived, unfortunately underground activity at Wheal Fanny appears to have been recorded elsewhere; these records have not survived.

Activity at Wheal Maria increased throughout 1923, the workbooks recording the transition from development work to production. For example the workbook entries during July 1923 included items such as testing the lode, timbering, removing dirt and raising air shaft. By the end of August there are references to landing kibbles, sollaring Wheal Maria Shaft, spalling mundic and grinding mundic. The first reference to stoping appears in early September 1923 when five miners were thus engaged. This seems to have been a common number throughout the rest of 1923 and into 1924. The mine was evidently working seriously: on 16th November 1923 for example 90 kibbles were hoisted, this equates to 45 tons. On the 19th November 50 full kibbles were hoisted comprising 10 tons of "No.1" ore and 15 tons of "seconds". Hoisting was done using a lorry, possibly the Garrett, at this stage there does not appear to have been a proper head frame, reference being made to "gin legs". This arrangement may not have been ideal as by the end of the year a drum was recovered from Frementor for use at Wheal Maria. Throughout this phase of working the mine, as Barclay noted, was being pumped and there are ongoing references in the workbooks to attending to the pumps. In addition to keeping incoming water in check the pump was also capable of lowering water levels, the pump itself being lowered in December 1923, by all accounts quite an involved operation.[68]

As part of the expansion of the arsenic works a new refiner chamber was started in 1922 and completed in the late spring or early summer of 1923. Presumably associated with the expansion of the refining capacity, a new furnace was constructed (see Note 10) this appears to have been completed by the end of September 1923, a note in the work books referring to collecting coke from the "gas house" for refining.[69]

Whilst both Wheal Fanny and Wheal Maria proved to be important sources of ore the massive potential of the dumps was not ignored. As has been noted, dump recovery had been an important factor since the pre war period however there appears to have been an increase in activity from early 1923. In mid February 1923 a "State" crusher (see Note 11) was being set up on a dump. The Garrett lorry was also moved to the dump at this time, presumably to drive the crusher. From that date on the workbooks contain ongoing references to recovering mundic from "crusher dump". The workbooks also contain references to recovering "burnt rinkle", that is to say arsenic ore which had already been processed; this material would have come from the extensive dumps associated with the 1860s arsenic works.[70]

Whilst No. 1 grade ore only required limited dressing, probably only crushing and jigging prior to calcining, the poorer No. 2 grade ores, dump material and burnt rinkle would have required more extensive treatment before it could be calcined. Originally this treatment would have taken place at the Bedford United mill. However as production stepped up during 1923 moves were taken to establish a more sophisticated dressing plant, often referred to as the "New Dressing Plant" or simply the "N.D.P.".[71]

The Estate continued to operate copper precipitation works; in 1923 Dewey noted that about 45 tons of copper was being recovered annually.[72]

1924

Work on the New Dressing Plant continued into 1924 and the workbooks contain numerous references to erecting a waterwheel, pulveriser, jigger, "buttles", strip, tables, and a bottle furnace all of which would facilitate the dressing of lower grade ores.[73] Underground work continued at Wheal Maria and Wheal Fanny throughout 1924.

Unfortunately no information regarding of work at Wheal Fanny during 1924 appears to have survived, however the workbooks give considerable detail concerning Wheal Maria. Some idea of the scale of workings in this section of the mine can be gained from the size of timbers being used: In June 1924, 41′ by 2′ 1″ timbers were brought over from Rubbytown for use at Wheal Maria. In addition to stoping some development work was being undertaken for example during May and June 1924 are a number of references to driving and crosscutting. Work continued apace throughout the summer of 1924 with both development work and extraction taking place. Wheal Maria was evidently productive; for example on the 5th September 1924 fifty eight kibbles were raised, this was surpassed on the 8th September when 65 kibbles were raised. Extrapolating, this would suggest a daily output in the region of thirty tons of ore.[74]

Mundic continued to be raised from both Wheal Fanny and Wheal Maria until "December 1924 or January 1925" when underground working was suspended due to a severe slump in the price of arsenic.[75] During the "arsenic boom" of the early 1920s the price of arsenic reached £70 per ton, in 1925 the price slumped to £10-£15 per ton and did not pick up throughout the inter-war period.[76] After the suspension of underground mining at Maria and Fanny it would seem that arsenic production continued on a much smaller scale, and comprised of reworking dump material.[77]

In June 1924 "oker" was recovered from Wheal Josiah, possibly from the Blanchdown adit.[78] This does not seem to have been a successful exercise as it does not appear to have been repeated until the 1940s.

Frementor 1924 – 1930

The collapse in the market for arsenic caused the Estate to re start tin ore and wolfram recovery from Frementor ("Framator"), development work starting in September 1924. The workbooks record details such as opening up the adit (see Note 12), putting in and commissioning pumps, driving, timbering, putting in ladders.[79,80]

Frementor was back in production by February 1925; however the end of large scale arsenic production must have been a severe blow which the reopening of Frementor could hardly have compensated for. The bleak mood of the time is captured in a rather pathetic letter dated 11th February 1925 from Captain Cloke to Captain Josiah "Paul" (Paull) at South Crofty offering to sell two to three tons of tin and wolfram concentrate:

> I am writing to ask if you are able to buy this material to help us in these hard times and keep the wolf from the door.[81]

During the post 1924 reworking a crosscut adit was driven from the western end of the 12 fathom level. The shaft was deepened to twenty fathoms and a new level, the 20 fathom level, driven at this depth. Much of the ground above the 20 was stoped out, leaving a very impressive void. By February 1929 eighty fathoms had been driven on the 12 fathom level and thirty four fathoms on the 20 fathom level. In addition to the shaft two winzes had been sunk between the 12 and the 20. A further winze had was sunk to a depth of seven fathoms below the 20. Rather than putting in a long drive to connect with the Frementor-South Fanny adit, as recommended in the April 1919 report, a crosscut adit thirty six fathoms long was put in on the 20, coming to grass just above the Great Leat (see Note 13).[82,83,84,85]

Figure 63. This wheel was erected in the 1920s to provide power to the New Dressing Plant.

An attempt was made to put in a crosscut on the 20 fathom level from the Frementor "Main Lode" to intersect North Lode at depth. This, however, was abandoned after only five fathoms, fifteen fathoms short of North Lode.[86]

In line with the April 1919 report into the future working of the mine the ore was trammed to Bedford United. By May 1925 the Anna Maria-Bedford United railroad had been extended to Frementor. From the arsenic works an incline dropped down to the foot of the 19th century arsenic dumps and then followed the course of an old leat, which skirted South Fanny Engine Shaft and ran across the top of Frementor Granite Quarry. Facilities at the Frementor railhead were limited to an ore bin and the last mortal remains of the long suffering Garrett Lorry which was being used for hoisting ore and baling water.[87,88]

During 1925 and 1926 10,830 tons of ore were mined at Frementor yielding 65 tons 10 hundredweights of black tin and 18 tons 5 hundredweight of wolfram; this represented a recovery rate of 14.5 pounds of tin and 4.1 pounds of wolfram per ton.[89]

Although Frementor produced some very rich ore the lode was bunchy and erratic, this, in conjunction with a fall in tin prices in 1929-1930 led to the final abandonment of Wheal Frementor by November 1930.[90] The February 1929 report was less than sanguine regarding future prospects at Frementor:

> As regards the Fermator ore reserves above the 20 Fathom or Adit Level, I should not consider these exceeded 1,000 tons of ore similar to that mined in 1926-27, which at the present price of Tin and Wolfram represents a value of £1 per ton. To open up any further considerable tonnage it would be necessary to sink and develop below the Adit or water level and the average grade of Ore that has so far been found does not warrant this at present prices of Tin and Wolfram.[91]

1930s – 1940s

After the suspension of work at the mine by the Estate, a well known local mining engineer R. W. Toll (see Note 14), who had been managing Bedford United, acquired a lease from the Bedford Estate to work the dumps below the 1860s arsenic works for tin during 1930. The "rinkle" from these dumps assayed 10-15 lbs of cassiterite by vanning assay. This material was treated at the Bedford United mill. However a further drop in the price of tin towards the end of the year led to the abandonment of these operations.[92]

Harry Higman, who had worked at the mine during the 1920s reworking, operated a copper precipitation works during the 1930s.[93]

During the winter of 1940-1941 the consultants Terrell, Davis and Toll prepared a scheme to rework dumps across the mine. A series of tunnels were driven into the dumps to prospect them at depth. Plans were also drawn up for a mill to process the material recovered from the dumps. The intention was to locate the mill at the base of the Anna Maria arsenic and sand dumps, on the site of the precipitation works. (During the 1970s reworking the mill was established on this site). A plan dated 20th March 1941 suggests that the trackbed of the railway could be used as road access to the mine via Bedford United.[94]

For whatever reasons the plan to rework the dumps on a large scale did not proceed; although Toll was reworking dumps on the mine in the autumn of 1944.[95] Richardson notes that material from the Devon Great Consols dumps was treated at Prince of Wales Mine at Harrowbarrow.[96]

Correspondence from H. G. Dines of the Geological Survey to Toll dating from 1944 is of particular interest. In a letter a letter dated 7th September 1944 Dines is in speculative mood:

> .....it makes one wonder to what extent the wall rock of the Devon Great Consols Lode contains tin, this is a very wide lode and it was unnecessary to break the walls when only copper and arsenic were being worked, in fact the sides of the lode were left in place.[97]

Booker notes that ochre was recovered from the Blanchdown Adit during the war, presumably for use as a paint pigment.[98]

In August 1947 the *Mining Magazine* published an interesting, if brief, article, authored by R. W. Toll, on precipitation methods employed on the mine by the Bedford Estate. At the time Toll was writing copper was being recovered from both the dumps of calcined material below the demolished arsenic works and the large halvans dump below the Anna Maria floors. Of the two the calcined material proved the richest, averaging 0.59% copper. Water was channelled over the top of the dumps, the course of which was changed from time to time. The water, having percolated through the dump, passed into launders containing "iron turnings and small iron scrap". At regular intervals the iron was raked over. When sufficient copper had been deposited on the iron the precipitate was collected and passed through a 24 mesh sieve. Any undissolved iron was returned to the launder. At Devon Great Consols a mechanical jig had been devised to sieve the precipitate:

> The machine is carried on a hutch, which is fitted with rollers and runs over the launders. One man shovels the precipitate into the machine, while another operates the jig by means of a crank and gearing. By this process the precipitate is passed through the jig screen into the hutch and the oversized iron turnings pass automatically through a chute back into the launder, the machine being moved along as the clean up proceeds.[99]

Toll noted that the precipitate produced in this way averaged 70% copper. Over time the acidity of the water coming off the dumps declined to a point where it was almost neutral. This had a significant impact on copper recovery rates. The reduction in acidity allowed a "gelatinous precipitate of alumina" to coat the iron impeding the deposition of copper. The alumina contamination reduced the grade of the precipitate to 45% copper. In order to remedy this, the acidity of the water was artificially raised by the addition of sulphuric acid in the form of "nitre cake". The cake was placed in the water issuing from the dump allowing the acidity of to be controlled. By raising the acidity it was found possible to keep the alumina in solution, allowing the continued recovery of copper precipitate.[100]

In 1949 further material was recovered and put through the mill at New Great Consols, Luckett.[101]

Change of ownership

In October 1953 Hastings Russell the 12th Duke of Bedford died as result of gunshot wounds, possibly self inflicted. The consequent financial problems including death duties placed John, the 13th Duke in a difficult financial position which necessitated the sale of the Bedford Estate's property in the Tamar Valley including the former Devon Great Consols sett. The land was purchased by Gerald, the Sixth Earl of Bradford, in 1959. The Earl of Bradford's acquisition stemmed from his keen interest in forestry rather than the mining potential of the estate.[102,103]

Thus ended the Bedford Estate's long association with the Tamar Valley, dating back to Henry VIII's land grant to the Russell family at the time of the dissolution of the monasteries four hundred years earlier.

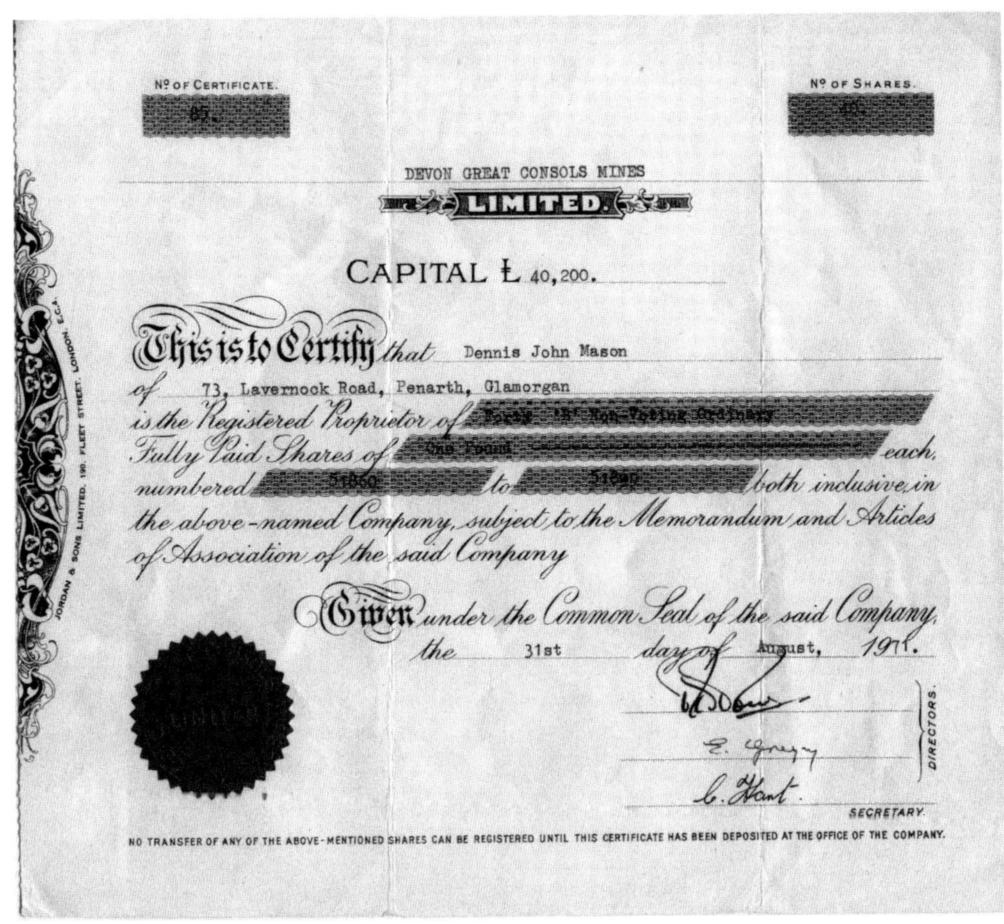

Nº OF CERTIFICATE. 85.

Nº OF SHARES. 40.

DEVON GREAT CONSOLS MINES

LIMITED.

CAPITAL £ 40,200.

This is to Certify that Dennis John Mason

of 73, Lavernook Road, Penarth, Glamorgan

is the Registered Proprietor of Forty 'B' Non-Voting Ordinary

Fully Paid Shares of One Pound each,

numbered 51860 to 51899 both inclusive in the above-named Company, subject to the Memorandum and Articles of Association of the said Company

Given under the Common Seal of the said Company, the 31st day of August, 1971.

DIRECTORS.

E. Gregory

C. Hart

SECRETARY.

NO TRANSFER OF ANY OF THE ABOVE-MENTIONED SHARES CAN BE REGISTERED UNTIL THIS CERTIFICATE HAS BEEN DEPOSITED AT THE OFFICE OF THE COMPANY.

JORDAN & SONS LIMITED, 190, FLEET STREET, LONDON, E.C.4.

Figure 64. 1971 Devon Great Consols share certificate.

1960s / 1970s Dump Reworking

During the late 1960s and 1970s the dumps at the mine were extensively reworked for tin ore and copper. This activity marks the final phase of ore recovery at the mine. The seeds of this operation date from the late 1940s when a colourful character called Ernest Gregory (see Note 15) was prospecting the dumps at the mine. Gregory went to work in Africa, taking a quantity of Devon Great Consols samples with him. After working on mines in the Gold Coast and Nigeria, Gregory returned to the U.K. and by the mid to late 1960s had established a presence at Devon Great Consols. During the latter part of the 1960s Gregory was operating a copper precipitation works and also a tin recovery operation taking material from the mine's dumps (see Note 16) During this early phase of tin recovery Gregory erected a wooden shed in which was housed a shaking table.

In 1969 Gregory's "mill" was remodelled and expanded by Roger Harrisson.

The wooden shed was replaced by a pair of Nissen huts containing four shaking tables (possibly Holmans). At the top of the dump above the mill, an ore bin and a jig was erected, material being transported in suspension from the jig to the mill via a four-inch pipe. During this second phase of operation consideration was given to producing fluorspar, however, due to its high arsenic content, the Devon Great Consols fluorspar was unsuitable for use as a flux in the steel making process and thus was never produced.[104] For some or all of this early period of working the company appears to have been operating under the name Devon Great Consols Mines Ltd.

By August 1969 material was being extracted from the large halvans dump and by Summer 1970 significant amounts of material were being removed for reprocessing from the Anna Maria dumps.[105]

By 1971 a new company Redcaves Ltd was formed by Devon Great Consols Ltd and "a Canadian group" probably with links to Rio Tinto.[106,107] The formation of Redcaves saw a large influx of Canadian money which led to improvements in plant. This included the removal of the Nissen huts and the erection of a much larger purpose built, steel framed building during late 1971 and early 1972.[108] The new mill was in operation by August 1972. The mill contained a ball mill, a cyclone, spiral concentrators, four shaking tables (this would be increased to eight and, latterly, fifteen) and a magnetic separator.[109,110] To power the plant a three phase supply had been installed, the transformer being located on poles adjacent to the mill.[111] During 1974 the mill was in continuous production, working a three shift system, and by the end of 1974 18 men were employed and 30.9 tonnes of tin concentrate was produced.[112] During the latter part of 1974 and during 1975 further improvements were made to the mill.[113] During 1975 it was noted that material for reprocessing was being extracted from the "arch dump" which straddled the trackbed of the railway.[114]

In March 1976 the company came under new ownership and changed its name to Nottsvale Ltd. During 1976 14 men were employed and 40 tonnes of tin concentrate were produced. In addition to the Devon Great Consols operation the company also held a number of mineral leases around Callington and Gunnislake. Towards the end of 1976 negotiations were in progress to separate the Devon Great Consols part of the operation from the mineral holding side of the business.[115] This was completed early in 1977, the Devon Great Consols side of the business being run by yet another new company, Tamar Valley Metals Ltd. During 1977 production dropped to 27 tonnes due to a change in the type of feed material. This led to a considerable amount testing and research. It was realised that due to the change in feed characteristics the mill would have to be extended and the classification and slimes recovery processes modified. To do this would require a period of shutdown in early 1978, the intention being to resume three shift working on completion. However before this work was finished a catastrophic fall in the price of tin resulted in the closure of the operation with little or no fanfare.[116]

On closure the Devon Great Consols tin plant was purchased and removed from site by Tolgus Tin. The plant was eventually acquired by Cornish Tin and Engineering Ltd which installed it at its Tolgarrick Tin Streams.[117]

Deep drilling 1979

The last chapter in the story of the search for tin at depth took place in 1979. Cominco (UK) Ltd. undertook a deep drilling programme on the central section of the mine targeting tin with copper as a secondary objective.

In addition to deep drilling there was also a programme of exploration of the accessible sections from Wheal Fanny (Western Shaft) to Anna Maria (Engine Shaft). The purpose of this work was to allow the geologists to examine the lode structures and to undertake a limited sampling programme.[118]

Cominco's contractor Drill Sure Ltd drilled three holes on the mine: D1, D2 and D3 (see Note 17) Hole D1 was sunk between August 8th 1979 and August 14th 1979, with the intention of intersecting the Main Lode at depth. The hole was initially sunk at angle of 57 degrees, however as the hole went deeper it steepened to 63 degrees and in consequence was abandoned at a depth of 204.5 feet. A second borehole D2 was sunk ten metres to the east of D1. D2 was sunk in September 1979 and reached a depth of 660.75 feet before being abandoned, again due to an "excessive steepening". The final hole, D3 was sunk during September – October 1979 at a point north east of the previous two holes to "test the possible down dip extension of the DGC lode structure for high grade Cu Sn W mineralization". D3, sunk on a dip of 50 degrees on a bearing of 358 degrees, cut the granite at a depth of 1825.5 feet. Whilst D3 was a technical success the results it revealed were disappointing. At 1624 feet to 1632 feet the hole cut a quartz vein carrying "minor ZnS". In the vicinity of the granite killas contact between 1814.6 feet and 1825 feet a quartz-fluorite vein was encountered with traces of copper, iron and selenium. In the granite itself the hole passed through quartz carrying traces of copper, lead and zinc sulphides. The bottom of D3 lay at 2197.25 feet; D3 had not cut tin ground.[119]

Having satisfied itself that no payable tin reserves existed at depth Cominco took no further interest in the mine.

Cornish Tin and Engineering Ltd

At the time Tolgus acquired the Devon Great Consols tin plant in 1978 Richard Williams, then Assistant General Manager at Tolgus, vanned the 1970s tailings and was surprised to find that they contained significant amounts of "free tin", suggesting that "there had been a lack of control during treatment of the waste dumps".

By the 1980s Richard Williams had moved on and had set up Cornish Tin and Engineering Ltd. By the mid 1980s C. T. & E. Ltd had established a successful tin recovery operation at Tolgarrick and was in a position to expand. In 1984 the Company gained permission from the Bradford Estate to sample the tailings dam. A special sampling auger was constructed by Bob Orchard, a former Holmans apprentice and C. T. & E. Ltd's engineering expert. The tailings dam was divided up into a grid and samples taken at ten foot intervals. The samples were analysed by C. T. & E.'s assayer, Ross Perry, who established that the tin was recoverable using gravity methods as opposed to more expensive flotation. Whilst not a "gold mine" the 1970s tailings were certainly a paying proposition and the decision was taken to go ahead. The intention was to set up a gravity plant operated by some of the skilled men from Tolgarrick and implement close monitoring and process controls. In addition to sampling the tailings C. T. & E. also considered the possibility of copper precipitation and a small scale trial was undertaken "purely for interest". Unfortunately before work could progress any further the project fell victim to the 1985 tin crisis.[120,121]

The author is particularly grateful to Alasdair Neill for his assistance with this chapter.

Chapter 11 Notes

Note 1

Due to the wholesale clearance of the arsenic works undertaken by Jabez Petherick it was necessary to use the facilities at Coombe.

C. F. Barclay notes that, in addition to the facilities at Coombe, a calciner was erected at Bedford United in 1915.[122] The advantage of this would be that it

Figure 65. 1970s view of the jigger shed.

Figure 66. A rather confused view taken inside the 1970s mill showing some of the shaking tables and associated pipe work.

Figure 67. An internal view of the 1970s tin recovery plant showing a set of spiral concentrators.

Figure 68. A 1962 view of the copper precipitation launders at Wheal Fanny.

Figure 69. Copper precipitation launders at Wheal Anna Maria.

Figure 70. A general view of the upper section of the 1970s mill with the spiral concentrators on the right and classifiers on the left.

Figure 71. 1970s mill being erected.

Figure 72. Tunnel for the DGC railway running under the Wheal Josiah dumps.

allowed the first phase of the arsenic refining process, the production of arsenic soot, to be carried out at Bedford United, greatly reducing the amount of material being transported to Coombe. This method of working appears to have been used throughout the war period; for example in August 1918:

> the rough ore was carried by cart to the stamps at Bedford United Mine about 1 mile to the east and there treated. The crude arsenic or soot is afterwards refined at the Coombe Refineries, which lie between Calstock and Callington, where it is raised to 99% white arsenic.[123]

The current author is not wholly convinced that arsenic soot was produced at Bedford United during the Great War. The Bedford Estate Mining Reports for May 1915 do note that two men were employed at surface "building a drying furnace",[124] tying in with Barclay's observation. Likewise an inventory dated 3rd December 1920 notes that a "drying and calcining furnace" valued at £30 had been added to the plant after 1914.

However the Estate Mining Reports contain no details either of the operation of an arsenic calciner at Bedford United. The reports do record that copper was being precipitated by the Estate at Coombe throughout the war. For this to happen the mundic ore must have found its way to Coombe, which would not be the case if the mundic was had been processed at Bedford United. Finally an inventory of Coombe arsenic works dated 3rd December 1920 records the presence of both an Oxland tube and a Brunton calciner. Given that this plant was at Coombe it would seem odd to duplicate the facility at Bedford United.

Remains including structures which might be interpreted as furnaces, a crude flue and the vestiges of a stack at Bedford United may well be the remains of Barclay's 1915 "calciner". It is probable that this was associated with tin rather than arsenic recovery.

Note 2

Given the Oxlands association as major shareholders at Devon Great Consols it is surprising that the Oxland process for treating tin-wolfram ores was not employed on the mine in regard to the Frementor ores during the nineteenth century. The process had been in use at Drakewalls from the late 1840s.

Note 3

The Bedford United magnetic separator was acquired in autumn 1908, possibly from Clitters, the intention being to treat burnt material from Coombe "as the cost of power at Coombe is so great".[125]

Note 4
Both C. F. Barclay and Dewey & Dines assert that work at Frementor stopped in 1919.[126,127]

Note 5
In the 1920 memoir of the Geological Survey relating to arsenic, Henry Dewey notes that both Main Lode and New South Lode were being worked at Anna Maria, a statement which he echoes in the 1923 memoir on copper ores. Given that New South Lode does not extend as far west as Anna Maria one can only assume that Dewey has somehow got his details confused. Barclay, who visited the mine regularly makes no mention of work at Anna Maria, neither do the surviving work books or Estate mining records

Note 6
The term railroad is used in preference to the more common term tramway. Railroad is used throughout the DGC workbooks 1922-25.

Note 7
In addition to the 20HP Simplex there is a possibility that a Fordson tractor based loco, such as a Muir Hill or Hudson "Go-go", was also in use post 1922.

Note 8
Barclay's reference to "milling grade ore" seems to be his own usage, the workbooks refer to this as No.2 ore, No.1 being the best quality ore.

Note 9
The workbooks included an interesting reference to "erecting wind screens" at the calciner. Possibly the wind was affecting draughting.

Note 10
Presumably this is the flat bed refiner, the remains of which are still extant and have recently been inexpertly "conserved". The use of coke avoided the soot contamination during this latter phase of arsenic oxide production.

Note 11
Richardson renders this as "Estate crusher",[128] however this is a misreading of the

entries in the workbooks.

Note 12

The relevant workbook notes that on 10th September 1924 men were engaged in opening up "Framator" adit. Whether this referred to re commissioning the Frementor South Fanny adit, or driving a new adit, possibly on the Frementor 12 Fathom Level is open to speculation. If it refers to the Frementor-South Fanny Adit this might imply that the intention was to connect it to the main Frementor workings as recommended in the April 1919 report.

Note 13

Dines[129] notes that "the 20-fm level connects eastwards with the adit of South Fanny Mine". On balance it would appear that Dines has misinterpreted the situation.

Note 14

R. W. Toll is very probably the most important figure in 20th century mining in the Tamar Valley. Reginald Warmington Toll was born at Devonport in early 1895. He was educated at Devonport Technical School. Toll was involved with re workings of Tavy Consols, Little Duke, South Crebor, Bedford United, Devon Great Consols and New Great Consols. During the 1930s Toll was involved in the re commissioning of the Tavistock Canal and the associated hydro-electric power scheme. In addition to being an eminently practical mining engineer Toll also found time to produce a number of articles on mining topics, mainly, although not exclusively, published in the *Mining Magazine*. Sadly Reg Toll died in 1962, apparently committing suicide by taking a dose of arsenic.

Note 15

Ernest Gregory was a somewhat eccentric character with a penchant for hard drinking, flatulence, powerful motor cars and cross dressing. Rumours that he had a sex change in later life are unconfirmed.

Note 16

In addition to tin and copper recovery, there is a suggestion that the good Mr. Gregory planned to recover germanium using crab shells;[130] there is a distinct possibility that this was an elaborate joke. It has been suggested to the author that "not all of Ernie's ideas were based on respectable science".[131]

Note 17

Borehole locations: D1: SX 42697320, D2 SX 42707320, D3 SX42757320

Chapter 11 references

1. Devon Record Office document 8187, Bedford Estate Mining Reports 1899-1906
2. Devon Record Office document 8187, Bedford Estate Mining Reports 1908-1928
3. Devon Record Office document 8187, Bedford Estate Mining Reports 1899-1906
4. Devon Record Office document 8187, Bedford Estate Mining Reports 1899-1906
5. Devon Record Office document 8187, Bedford Estate Mining Reports 1899-1906
6. Devon Record Office document 8187, Bedford Estate Mining Reports 1899-1906
7. Devon Record Office document 8187, Bedford Estate Mining Reports 1908-1928
8. Devon Record Office document 8187, Bedford Estate Mining Reports 1908-1928
9. Devon Record Office document 8187, Bedford Estate Mining Reports 1908-1928
10. Devon Record Office document 8187, Bedford Estate Mining Reports 1908-1928
11. Devon Record Office document 8187, Bedford Estate Mining Reports 1908-1928
12. Devon Record Office document 8187, Bedford Estate Mining Reports 1908-1928
13. Devon Record Office document 8187, Bedford Estate Mining Reports 1908-1928
14. Devon Record Office document 8187, Bedford Estate Mining Reports 1908-1928
15. Devon Record Office document 8187, Bedford Estate Mining Reports 1908-1928
16. Devon Record Office document 8187, Bedford Estate Mining Reports 1908-1928
17. Devon Record Office document 8187, Bedford Estate Mining Reports 1908-1928
18. Devon Record Office document 8187, Bedford Estate Mining Reports 1908-1928
19. Devon Record Office document 8187, Bedford Estate Mining Reports 1908-1928
20. Devon Record Office document 8187, Bedford Estate Mining Reports 1908-1928
21. Devon Record Office document 8187, Bedford Estate Mining Reports 1908-1928
22. Devon Record Office document 8187, Bedford Estate Mining Reports 1908-1928
23. Devon Record Office document 8187, Bedford Estate Mining Reports 1908-1928
24. Devon Record Office document 8187, Bedford Estate Mining Reports 1908-1928
25. Devon Record Office document 8187, Bedford Estate Mining Reports 1908-1928
26. Devon Record Office document 8187, Bedford Estate Mining Reports 1908-1928
27. Devon Record Office document 8187, Bedford Estate Mining Reports 1908-1928
28. Devon Record Office document 8187, Bedford Estate Mining Reports 1908-1928
29. Devon Record Office document 8187, Bedford Estate Mining Reports 1908-1928
30. Devon Record Office document 8187, Bedford Estate Mining Reports 1908-1928
31. Devon Record Office document 8187, Bedford Estate Mining Reports 1908-1928
32. Devon Record Office document 8187, Bedford Estate Mining Reports 1908-1928
33. Devon Record Office document 8187, Bedford Estate Mining Reports 1908-1928
34. Devon Record Office document 8187, Bedford Estate Mining Reports 1908-1928

35. Devon Record Office document 8187, Bedford Estate Mining Reports 1908-1928
36. Dewey H., 1920, *Arsenic and Antimony Ores,* Special Reports on the Mineral Resources of Great Britain, Vol. XV, H.M.S.O.
37. Devon Record Office document 8187, Bedford Estate Mining Reports 1908-1928
38. Devon Record Office document 8187, Bedford Estate Mining Reports 1908-1928
39. Devon Record Office document 8187, Bedford Estate Mining Reports 1908-1928
40. Barclay C. F., 2004, *Mines of the Tamar & Tavy* (ed. Stewart R. J.), Tamar Mining Press.
41. Devon Record Office document 8187, Bedford Estate Mining Reports 1908-1928
42. Devon Record Office document 8187, Bedford Estate Mining Reports 1908-1928
43. Report to the Trustees of the Bedford Estate, 28th April 1919
44. Devon Record Office document 8187, Bedford Estate Mining Reports 1908-1928
45. Estate Mine Inventory, 3rd December 1920
46. Barclay C. F., 2004, *op. cit.*
47. Barclay C. F., 2004, *op. cit.*
48. Estate Mine Inventory, 3rd December 1920
49. Barclay C. F., 2004, *op. cit.*
50. Dewey H., 1920, *Arsenic and Antimony Ores,* Special Reports on the Mineral Resources of Great Britain, Vol. XV, H.M.S.O.
51. Burt R., 1988, Arsenic - Its significance for the survival of South Western metal mining in the late nineteenth and early twentieth centuries, *Journal of the Trevithick Society*, No. 15, pp. 5 – 26.
52. Dewey, H., 1920, *op. cit.*
53. Barclay C. F., 2004, *op. cit.*
54. Barclay C. F., 2004, *op. cit.*
55. Barclay C. F., 2004, *op. cit.*
56. Devon Record Office document 6728, DGC workbooks 1922-1925
57. Devon Record Office document 6728, DGC workbooks 1922-1925
58. Devon Record Office document 6728, DGC workbooks 1922-1925
59. Devon Record Office document 6728, DGC workbooks 1922-1925
60. Devon Record Office document 6728, DGC workbooks 1922-1925
61. Barclay C. F., 2004, *op. cit.*
62. Barclay C. F., 2004, *op. cit.*
63. Devon Record Office document 6728, DGC workbooks 1922-1925
64. Devon Record Office document 6728, DGC workbooks 1922-1925
65. Devon Record Office document 6728, DGC workbooks 1922-1925
66. Cornwall Record Office document MRO 4406, South Fanny mine plan
67. Devon Record Office document 6728, DGC workbooks 1922-1925
68. Devon Record Office document 6728, DGC workbooks 1922-1925
69. Devon Record Office document 6728, DGC workbooks 1922-1925

70. Devon Record Office document 6728, DGC workbooks 1922-1925
71. Devon Record Office document 6728, DGC workbooks 1922-1925
72. Dewey H., 1923, *op. cit.*
73. Devon Record Office document 6728, DGC workbooks 1922-1925
74. Devon Record Office document 6728, DGC workbooks 1922-1925
75. Barclay C. F., 2004, *Mines of the Tamar & Tavy* (ed. Stewart R. J.), Tamar Mining Press.
76. Toll R. W. 1938, The arsenic industry in the Tavistock district of Devon, *Sands Clays & Minerals*, April 1938, pp 224-227.
77. Devon Record Office document 6728, DGC workbooks 1922-1925
78. Devon Record Office document 6728, DGC workbooks 1922-1925
79. Barclay 2004, *op. cit.*
80. Devon Record Office document 6728, DGC workbooks 1922-1925
81. Cloke to Paull 11 February 1925 www.aditnow.co.uk
82. Barclay 2004, *op. cit.*
83. Dines H. G, 1956, *The Metalliferous Mining Region of South West England*, H.M.S.O.
84. Personal communication, A. Neill,
85. Cornwall Record Office document CMDA 1/6/1, Report to the Trustees of the Bedford Estate, 7th February 1929
86. Cornwall Record Office document CMDA 1/6/1, Report to the Trustees of the Bedford Estate, 7th February 1929
87. Barclay 2004, *op. cit.*
88. Devon Record Office document 6728, DGC workbooks 1922-1925
89. Cornwall Record Office document CMDA 1/6/1, Report to the Trustees of the Bedford Estate, 7th February 1929
90. Barclay 2004, *op. cit.*
91. Cornwall Record Office document CMDA 1/6/1, Report to the Trustees of the Bedford Estate, 7th February 1929
92. Barclay 2004, *op. cit.*
93. Richardson P. H. G., 1995, *Mines of Dartmoor & the Tamar Valley after 1913*, (2nd edition) Devon Books.
94. Devon Record Office document 672A/HS/15314
95. Cornwall Record Office document AD159, Toll papers
96. Richardson P. H. G., 1995, *op. cit.*
97. Cornwall Record Office document AD159, Toll papers
98. Booker F., 1967, *Industrial Archaeology of the Tamar Valley*, David & Charles.
99. Toll, R. W, Copper recovery from low-grade waters, *Mining Magazine* August 1947
100. Toll, R. W, Copper recovery from low-grade waters, *Mining Magazine* August 1947
101. Richardson P. H. G., 1995, *op. cit.*
102. Booker F., 1967, *op. cit.*

103. Hastings Russell, 11th Duke of Bedford & John Russell, 13th Duke of Bedford
104. Personal communication, Roger Harrisson
105. Beament R., 2008, Devon Great Consols in the 1970s, Unpublished manuscript.
106. Beament R., 2008, *op. cit.*
107. Brooke J. Index to Devon & Cornwall mines held in the Westcountry Studies Library
108. Beament R., 2008, *op. cit.*
109. Personal communication, Roger Harrisson & Garry Metters
110. Buck C., 2002, *Devon Great Consols, archaeological assessment,* Cornwall Archaeological Unit.
111. Beament R., 2008, *op. cit.*
112. Cornish Mining Development Association Annual Report 1974
113. Cornish Mining Development Association Annual Report 1975
114. Beament R., 2008, *op. cit.*
115. Cornish Mining Development Association Annual Report 1976
116. Cornish Mining Development Association Annual Report 1977 &1978
117. Williams R. 2011, The Tolgarrick tin streams of Cornish Tin and Engineering Ltd., *Journal of the Trevithick Society,* No. 38, pp. 40 – 74.
118. Personal communication, Barry Gamble
119. British Geological Society, Borehole record viewer
120. Personal communication, Richard Williams
121. Williams R. 2011
122. Barclay 2004, *op. cit.*
123. Dewey, H., 1920, *op. cit.*
124. Bedford Estate Mining reports 1908 - 1928
125. Bedford Estate Mining Reports 1908 - 1928
126. Barclay 2004, *op. cit.*
127. Dewey H, and Dines H. G., 1923, *Tungsten and Manganese Ores*, Special Reports on the Mineral Resources of Great Britain, Vol. I, H.M.S.O.
128. Richardson P. H. G., 1995, op. cit.
129. Dines H. G., 1956, *op. cit.*
130. www.aditnow.co.uk
131. Personal communication, Charlie Daniel

Appendix 1

Management, workers, welfare and housing

"we do not let very little boys go underground." – Thomas Morris, 1863

The Directors

At the very top of the tree were the directors who were appointed on an annual basis at the Annual General Meeting. The Directors were responsible to the shareholders who had increased to about 80 by 1850.[1] Typically the Board of Directors comprised of three individuals, one of whom would fill the role of Managing Director. During the life of the mine there were only two Managing Directors; William Alexander Thomas who managed the mine from the beginning until he was replaced by Peter Watson who became Managing Director in 1879, having been Chairman since 1877, a post he held until the final closure.

The Chief Agent

The Directors appointed an agent to oversee the mine. From the commencement of operations in 1844 until 1848 Josiah Hitchins filled the role. In 1848 the role was split: Thomas Morris taking over the day to day running of the financial side of the business as Resident Director, whilst Hitchins became "Principal Superintendant of the Mining Department" (4th Annual report, 1848). After Hitchins' resignation in 1850 Captain James Richards took the job of Principal Managing Agent,[2,3] effectively Mine Manager in modern terms. James Richards held the post until 1878 when he was replaced by his brother Isaac. Isaac Richards was dismissed in 1891 and was replaced by William Clemo who held the position until his death in 1900. The final Chief Agent was William Clemo's son, W. H. Clemo, who held the post until closure.

The Purser

The role of the Purser became increasingly important after Peter Watson took over the management of the mine and appointed Moses Bawden Purser in 1879.[4] In practical terms Moses Bawden took over many of the duties formerly undertaken by Thomas Morris, the Resident Director who had retired in 1879, at which time the role of Resident Director was abolished.[5,6]

The Agents

The day to day running of the mine was the responsibility of the "Agents" or "Captains" of whom there were seven in 1850, five underground at two at surface.[7] As the mines developed so did the number of Agents, by 1860 for example the number had increased to nine.[8]

Agents' duties were fully outlined in a memorandum issued in May 1871:

> The agents are required to be on the mines on all occasions throughout the year, commencing at 5 o'clock on the Monday mornings, for the delivery of materials, candles, powder and other requisites. To attend daily at 6 in the morning, and again at 10 o'clock at night, to see the men down and up and from their different places of work, in addition to their daily duties underground. Two agents to watch from Saturday evenings until Monday mornings, and in case of accident (which must occur occasionally in the extensive range of machinery throughout the mines), the other agents are called upon to assist if required. To attend at the monthly sampling, which sometimes happens on the measuring day at the mines, on which occasion the measurements underground have to be made before leaving the mines. Two locomotives to be kept in repair. All the machinery throughout this whole range of great mines, both at surface and underground, to be kept in thorough repair. Railway wagons to make and to be kept in repair. Manufacture of coke for foundry, locomotives and arsenic works. From the great widths of the lodes, and the dangerous nature of some portions of the ground, great care and attention are at all times required in connection with timbering of the underground workings, in order that the mines be kept well open for advantageous working, and ensure safety to the men. At the reduction works, from the nature of the different processes, the furnaces require to be kept in work both night and day, and strict supervision is at all times necessary, in order that no lag (?) may accrue (especially in the production of pure arsenic) from inattention or neglect of the men engaged in the works.

MEMORANDUM OF DUTIES OF AGENTS

Underground Agents – Daily visits to, and monthly measurements of, 60 bargains, including shafts, winzes, rises and stopes, extending from Wheal Emma to Wheal Maria – a distance underground of nearly two miles, and the superintendence of pit and timbermen, and 72 trammers, fillers, and landers, besides the care of machinery underground and at surface throughout the mines during the year, including Saturdays. On duty throughout the year from 6 0'clock in the morning until 10 o'clock at night.

Surface Agents – Superintendence of dressing departments, the dispatch of all the ores from the mines to the quay, and the receipt of coals, timber, and all other materials there from, the superintendence of smiths, carpenters, sawyers, masons, enginemen, labourers, and the foundry and railway. Dividing and weighing the ores at quays, and constant attention to the precipitate and arsenic works. Attendance in case of breakages by night or day, the receipt and delivery of stores on Monday Mornings at 5 o'clock, and on all other mornings at half past 6.

Underground and Surface Agents – Attendance on the mines from Saturday evenings to Monday mornings, attending to leats, watercourses, and machinery, including steam – engines, water – wheels, and lines of rods, with various other incidental duties throughout the year.[9]

Underground workers

Various classes of workers were employed underground at Devon Great Consols, including: Tutworkers, tributers and various types of labourer. Both tutwork and tribute were, in effect, piece work systems allocated to groups or "pares" of miners under a competitive tendering system known as setting.

By far the most numerous class of employee at the mine was the "Tutworkman". At Devon Great Consols in the 1840s development work such as sinking shafts and winzes and driving levels were carried out by tutworkers. During the 1840s tutwork was "set" for a period of one month at a fixed price per fathom with the exception of shaft sinking where a specified depth, for example ten fathoms, was contracted for. In addition to payment per fathom tutworkers at the mine were also paid one shilling in the pound for any ore they might raise.[10]

An example of a tutwork setting relating to November 1849 was fortunately recorded by J. H. Murchison:

Tutwork. The 80 fathom level, to drive west of Richards' Engine Shaft. To be carried out seven feet high and four feet wide, with dead levels, by six men for the month out. At £7 per fathom, and one shilling in the pound on the ores. Taken by Richard Osborne, James Sandercock, John Lawrey, R. Hooper, J. Sandercock junior, William Duntsone.[11]

Once the work had been completed the Company would settle its accounts with the men:

Devonshire Great Consolidated Copper Mines.
Pay November 1849 – Paid January 5th 1850.

Richard Osborne and partners, driving the 80 fathom level, west of Richards' Engine Shaft:

3 fathoms, 3 feet, four inches	£24	17s	9d
Putting in rails	£0	2s	0d
	£25	0s	3d[12]

The £25 0s 3d was not outright profit for Osborne and his partners. From that sum deductions were made by the Company for the supply of materials such and candles and powder and services such as that of the mine blacksmith:

Deduct			
48 lbs. candles at 8d per pound	£1	12s	0d
50 lbs. powder at 8d per pound	£1	13s	4d
Safety fuse	£0	6s	0d
Hilts	£0	0s	3d
Cans	£0	2s	6d
Smith cost	£2	6s	10d
Drawing at 10s	£1	15s	6d
Doctor and sick	£0	9s	0d
	£8	5s	5d
Leaves	£16	14s	10d[13]

This represented a wage for each member of the partnership of £2 15s 9d or £33 9s per annum. Whilst "tutworkmen" were largely engaged on what might be termed development work, i.e. driving levels and sinking shafts, tributers were concerned with the extraction of ore. Tributers would be paid at a rate of so much in the pound based on the value of ore raised.

During the early years of the mine tribute was typically set for a month, the tributers contracting for a certain quantity of work known as a "stent", the agreement being the "bargain". If the work was not completed within the month the bargain was considered to be "out" and a new setting took place. Likewise if the work is completed within the month, again a new bargain was set.[14] Setting a tribute bargain required a great deal of skill on both the parts of the mine agents and the tributers, both parties having to assess to a nicety the value and potential of the lode.

In addition to the costs met by tutworkers, tributers had to meet the costs of dressing the ore they had raised. To ensure this was done fairly each company of tributers would appoint one of its number to oversee the dressing.[15]

Details of a tribute bargain in Wheal Fanny were recorded by the ever informative Mr. Murchison:

Tribute: Robert Trethewey's pitch, in the back of the 35 fathom level, east of the Eastern Engine Shaft, so high as the bottom of the 25 fathom level, to extend from Matthews winze 10 fathoms east; the takers being bound to work and secure their ground as directed by the agents, and be subject to the general rules and conditions of mining, by six men at 7s. 6d. in the pound. Taken by Saml. Matthews, Henry Stephens, J. Henwood, Wm. Tyack, A Truscott, John Jewell.[16]

The account for this work was settled in January 1850:

Devonshire Great Consolidated Copper Mines.
Pay November 1849 – Paid January 5th 1850.
Samuel Matthews and partners for raising copper ores in August and September:
34 tons (21 cwts.) 11 cwts., at £5 5s 6d per ton = £182 2s 2d at 7s 6d.

= £68 5s 6d.

Deduct.			
108 lbs. candles, at 8d. per lb.	£3	12s	0d
195 lbs. powder.	£6	10s	0d
Safety fuse.	£1	9s	0d

Hilts.	£0	1s	9d
Cans.	£0	2s	6d
Locks.	£0	1s	6d
Smith cost.	£3	19s	6d
Drawing.	£3	0s	11d
Dressing cost.	£6	10s	8d
Use of grinder.	£0	8s	10d
Sampling & weighing.	£0	17s	6d
August cash on account.	£18	9s	0d
September ditto.	£18	9s	0d
	£63	18s	2d
Leaves.	£4	7s	7d[17]

Whilst the men only received £4 7s 7d they had already received "cash on account". With this taken into account the average annual income would be £41 5s 7d. This figure is considerably higher than the annual income of a tutworker; this is due to the greater skill required by tributer and also the greater financial risk they were taking.

The traditional demarcation between tutworkers and tributers appears to have been blurred at Devon Great Consols from an early stage. The component regarding raising ore built into the tutwork bargain renders it akin to tribute. Likewise the "stent" built into tribute bargains renders it akin to tutwork.

By 1860 there had been a distinct move away from extracting the lode on tribute to extracting it on tutwork, miners being paid by the running fathom.[18] D. B. Barton notes that the move from tribute to tutwork was a common feature of larger nineteenth century copper mines. He argues that tutwork was particularly well suited to "the working of the big copper lodes rather than to those of lead or tin, which were bunchy and required a tributer's hand and eye if they were not on occasion to be lost".[19] By the time of the Kinnaird Commission reported in 1864 the whole of the mine, with the exception of Wheal Maria, was being worked by tutworkers.[20]

At the time of the "five week month strike" pitches were being let for two months, rather than monthly as in the 1840s.[21] In addition to the skilled tributers and tutworkers there were unskilled or semi skilled underground workers; the trammers, fillers, landers and general labourers.[22]

Surface workers

On surface the largest class of worker were the ore dressers. During the 1840s ore dressing at the mine was fairly basic. Crop ore would have been broken down by hand and then run through a crusher to reduce it to a marketable size. Hand dressing was carried out by men and older girls and was paid for on a piece work system.

The women employed on the dressing floors were known as "bal maidens". The majority of bal maidens would have fallen into the fifteen to nineteen age bracket.[23]

The increasing importance of halvans in the late 1840s saw an increase in the handpicking of to remove "mundic, spar and other minerals".[24] This was the work of the youngest children on the mine, some of whom were as young as eight years old. As they grew older the boys would become miners whilst the girls would become cobbers.[25] Unlike other people employed on the mine who were working on variations of the piece system the pickers were paid by time, a system potentially open to misinterpretation or abuse.

In addition to the ore dressers there were a number of ancillary surface workers including smiths, carpenters and sawyers, enginemen, masons and labourers.[26] These men would be paid in accordance "the customary wages" of their trade.[27]

Employment of Children

Legally very young children could be employed at surface on mines and indeed this appears to have been the case at Devon Great Consols. For example the 1861 census contains details of the Langdon family of Devon Great Consols. The Family comprised Joseph Langdon, his wife Elizabeth and their six children including a baby aged ten months. The two oldest children: John aged nine and his sister Julia aged eleven are recorded as being employed as ore dressers, presumably, given their address, at Devon Great Consols.[28]

The Mines Act 1842 prohibited children under the age of ten from working under ground. The provisions of the act appear to have been more or less observed on the mine. In his evidence to the Kinnaird Commission in 1863 Thomas Morris noted that the age at which boys went underground varied:

> Very much depends on their size, because some boys are so big at 10 as others would be at 12; we do not let very little boys go underground.[29]

The 1861 census gives details of the Bickle family of "Devon Consols". The two eldest sons of the family: Henry aged eleven, and his younger brother aged nine, are both recorded as copper miners.[30]

Welfare

The tribute and tutworker accounts quoted by Murchison in 1850 contain an item for "Doctor and sick". The Company operated a scheme whereby one shilling and sixpence was deducted from each man's monthly wage and paid into a sick fund for the miners and their families. If a miner or his family was ill for longer than a week a weekly allowance of four shillings would be paid out by the fund.[31]

As the mine grew and developed the scheme became more sophisticated. By 1864 workers payments into the "doctor and sick club" were proportionate to their wages. The scheme as operated by Devon Great Consols appears to be one of the better schemes of the period. For example the scheme made provision for the medical "attendance" of the miners' families which was unusual for the time. Also significant for the time was the fact that members of the club were able to choose their own medical attendant themselves.[32]

If a person was injured on the mine (as opposed to being taken ill) they were paid one pound a month for as long as it was required. Should the person be rendered unfit for manual labour it was the company's policy to find them appropriate employment "on the establishment". In the case of a fatal accident occurred the deceased dependents were assisted by the company and it was the rule that in such cases that each man who was employed on the mine should make a contribution.[33] Funeral expenses were met by the Company.[34]

Whilst not a legal requirement until the introduction of the Metalliferous Mines Regulation Act 1872, drys were provided at Devon Great Consols by the early 1860s. In 1864 the Wheal Emma dry was considered to be the best on the mine. Each dry had the facility for men to dry their damp work clothes on steam pipes or flues, likewise each man had a locker and access to hot water.[35]

Throughout the mine places were provided in which "men, girls and boys may take their meals in comfort". Hot water for making tea and coffee was provided as were ovens in which food could be warmed. On the main dressing floors at

Wheal Anna Maria a large, heated room was provided where girls and boys, suitably segregated, could take their meals in relative comfort.[36]

As early as 1850 the Company decided to take upon itself some of the responsibility for the education of miners children. At the 1850 A.G.M the resolution was taken to place £100 at the disposal of Thomas Morris, the Resident Director to promote the education of miners children, the funds being distributed to schools in the vicinity,[37] a contribution made annually. By May 1863 a School, attended by nearly 100 children, had been established on the mine, the school mistress being paid for by the company, the pupils paying a nominal fee.[38,39]

The company took great pride in the moral character of its workers. If contemporary reports are to be believed Devon Great Consols was a shining example of Victorian piety:

A visitor to the mine will be most forcibly and agreeably struck with the good order and decent conduct everywhere apparent. If all accounts be true, the "bal girls" of many a mine in Cornwall might well take a lesson from the females employed at Devon Great Consols. Among hundreds of men, young women, boys and girls the writer, in the course of three days never saw an improper action, nor heard an improper word. No smoking is allowed. The girls at the dressing floor are very fond of singing psalms and hymns whilst at work, and their vocal efforts really have a very pleasing effect. Perhaps the general tendency to good which prevails can hardly be more strongly indicated by the fact that upwards of seven hundred copies of the *British Workman* and *Band of Hope* are sold on the mine monthly.[40]

Workers and Agents housing

The mining boom of the 1840s-1850s saw a huge influx of population into the Tamar Valley. Taking Tavistock as an example the population increased from 6,272 in 1841 to 8,147 in 1851, a rise of 30% occasioning a serious housing crisis.[41] The Bedford Estate recognised the problem and began building a limited number of "model" workers cottages. By 1850 the Estate had erected 86 model cottages in Tavistock and six at Gulworthy.[42]

> As early as 1851 the Company Directors were expressing concern for the housing conditions of their workforce, the matter being discussed at the 1851 A.G.M. The late rainy season has again drawn the attention of the Directors, to the inadequate lodging accommodation in the neighbourhood

> of the mines....... It is true, the Noble Duke, the Lessor of the property, has erected upwards of sixty cottages, models in their way, which, although greatly relieving the overcrowded houses in Tavistock, do not remedy the evil complained of, being situated five miles at the least, from the mines. It is unreasonable to expect, nay, almost cruel to exact "a fair day's work for a fair day's wage," from the miner who has to travel, through pouring rain, five, and in some instances, eight miles, to his work, and who after toiling in the bowels of the earth for eight long hours, has to return to his comfortless abode through the same long distance, drenched to the skin with rain. If this be the hard lot of the miner, how much more irksome and injurious must it be for the numerous women and children employed at the mines.[43]

> In spite of their heartfelt concerns about their employees' welfare the company felt that they were unable to remedy the situation; the tenure by which the Company holds the sett scarcely warrants an unconditional outlay for cottages, on sites conveniently contiguous to the mines.[44]

The Directors felt that this was the responsibility of the Duke and they hoped that he would do something about the situation. At the same meeting the decision was taken to provide a residence for Thomas Morris, the Resident Director.[45]

On 6th May 1851 W. A. Thomas wrote to Christopher Haedy drawing his attention to the housing situation.[46] At the local level the Estate recognised the need for miners housing. Writing on 24th June 1851 John Benson recommended that any housing for the mine should in fact be built at Mill Hill rather than on the mine itself. Benson felt that there was room for sixty to seventy cottages at Mill Hill and a further twenty at Ottery. His reasoning being that Mill Hill was sufficiently near the mines for the workforce without being so far from the town to deprive the inhabitants the use of the schools, the market or the places of worship, and where persons would be able to find work in either the country, the slate quarries or the town in case of the mines failing.[47]

By 17th June 1851 Theophilus Jones, the Estates architect surveyor, had prepared three estimates for "Miners cottages in the neighbourhood of Devon Great Consols" The cost ranging from £39 15s to £67 10s depending on details.[48] However delays in reaching a decision meant that the Estate's cottage building programme was shelved during 1851. Eventually the Estate gave permission to build "a few cottages as a token gesture".[49]

The Estate started work on "a few cottages" at Wheal Maria in 1852, twenty cottages having been completed by 1853.[50] Given the size of the workforce on the mine this was indeed a "token gesture", having said that it was twenty more workers cottages than the Company had constructed.

The construction of the Wheal Maria cottages did not even scratch the surface of the housing issue. The problem was considered at the 1854 A.G.M: High copper prices meant that miners could pick and choose where they worked. The Company felt that the lack of housing near the mine meant that many skilled miners had gone elsewhere; compelling them to employ less skilled workers. This had a direct impact on the mining operation the Company being forced to cease "less profitable and merely experimental operations, and devote the entire disposable force to the development of the more essential parts of the mines".[51] In an attempt to ameliorate the problem the Directors resolved to approach the Duke with a view to providing more housing. By early June 1854 W. A. Thomas had presented a paper to Haedy which argued that Company had lost an estimated £40-£50,000 as a consequence of the housing shortage. Thomas' paper argued that housing was the Estate's responsibility given that they were the landowners and that they were profiting from the dues paid by the mine. In correspondence with the Duke on June 26th 1854 Haedy wrote:

> I told him (W. A. Thomas) if the company had really lost so much through the want of cottages on or near the mines as £40 or £50,000, it was a matter of surprise that they had not saved a very large proportion of that money by laying out a moderate portion of it in building cottages. He urged the shortness of their lease, & I answered that it mattered little how short or how long their lease was when the expenditure of £4 or £6,000 would bring back so large a return as the sum he mentioned, the saving of a loss being equal to a gain. They would like to build the cottages themselves then make that a reason for having an extended lease of the mines. I do not think it advisable to agree to that. I think whatever cottages are built should be built by your grace and I think it will be best to adhere to the determination you came to of having the proposed cottages at Mill Hill built as soon as Mr Jones has time to do so, which would be expected would be next winter.[52]

Unfortunately this approach fell on deaf ears, no more Estate cottages being erected on the mine. Whilst the Company felt that they could not be reasonably expected to provide adequate housing for its workforce they had no such qualms in providing accommodation for agents of the Company.

Most impressive of the Company's houses was Abbotsfield. Abbotsfield, on the outskirts of Tavistock, was built for Thomas Morris, the resident director. Construction was sanctioned by the 1851 Annual General Meeting and by the time of the 1854 A.G.M Morris was happily ensconced.[53,54]

The various Mine Captains were provided with housing on or near the mine; the "Agents House" at Anna Maria, for example, being completed by May 1849.[55] Eventually a self contained community developed on the mine mainly clustered around Wheal Anna Maria and Wheal Josiah. As with any community came the usual round of births, deaths and marriages many of which were recorded in the pages of the *Tavistock Gazette*, the following being typical examples:

> Births – Clemo – January 1st, at Devon Great Consols, the wife of Mr. F. Clemo, of a daughter.[56]
>
> Births - Rodda, June 24th, at Devon Great Consols Mine, the wife of Captain Rodda of a Son – Stillborn.[57]
>
> Deaths – Clemo. – July 1st, at Wheal Maria Cottages, Susanna, wife of Mr. Francis Clemo, aged 45 years.[58]

One of the more intriguing episodes in the life of the Devon Great Consols community was recorded in the *Tavistock Gazette*:

> Notice – I, the undersigned, John Mitchell, of the Devon Great Consolidated Mines, Tavistock, Engineer, and Foreman at the Devon Great Consolidated Mines Iron Works, having found it expedient and requisite to purchase for myself all necessaries and goods for my wife and family, and establishment, and not to suffer Mrs. Mary Mitchell, my wife, to purchase goods or to contract on my behalf.
>
> I do hereby give notice that Mrs. Mary Mitchell, my wife, is no longer authorised by me to make purchases, or contract on my behalf, and that I will not be responsible for the performance of any engagements she may enter into.
>
> Dated the 11th day of September, one thousand eight hundred and sixty. John Mitchell Witness – Edward Chilcott, Solicitor, Tavistock.[59]

One can only speculate what the unfortunate Mrs Mitchell had done to incur the wrath of her husband!

The recreational diversions of the community which grew up on the mine were typical of their age:

> Those who reside on the mine are in some measure a community to themselves, and find great solace for their isolation in music. Vocal and instrumental concerts of a very pleasant character are sustained among the families of the principal agents; and a brass band, known as the Devon Great Consols Brass Band, meets regularly for practice under the direction of Captain Cock. Sacred music is also very much practiced in connection with the choirs of places of worship in the vicinity.[60]

> The occasion is rare when so much musical talent can be found in so small a compass as here exists. Vocal or instrumental, combined or separate, there is abundant material for either. The chief regret is that more frequent opportunities are not offered to the public for appreciating the really good music, which can be produced by the agents of this mine, and their friends. We venture to say that had the concert which was given at Wheal Josiah, on Wednesday evening, taken place in the Town Hall, a very large audience would have been the result..... The musical portions were most effectively rendered. Few who were present, will easily forget the chorus "See the conquering hero comes," for it was as perfect as human voices could make it. One meed of praise we must not omit. The difficult and often unappreciated post of conductor, was filled by Miss Mitchell, and it is but an act of simple justice to that young lady, to say that she manifested all the qualities necessary to make a true musician.[61]

Appendix 1 references

1. *Mining Journal* 26 October 1850
2. 6th Annual report 1850
3. *Mining Journal* 7 July 1860
4. *Tavistock Gazette* 27 June 1879
5. *Mining Journal* 14 June 1879
6. *Tavistock Gazette* 30 May 1879
7. *Mining Journal* 26 October 1850
8. *Mining Journal* 7 July 1860
9. *Mining Journal* 27 May 1871

10. *Mining Journal* 26 October 1850
11. *Mining Journal* 26 October 1850
12. *Mining Journal* 26 October 1850
13. *Mining Journal* 26 October 1850
14. *Mining Journal* 26 October 1850
15. *Mining Journal* 26 October 1850
16. *Mining Journal* 26 October 1850
17. *Mining Journal* 26 October 1850
18. *Mining Journal* 28 July 1860
19. Barton D. B., 1968, *Essays in Cornish Mining History*, Vol. 1, Bradford Barton Ltd.
20. Kinnaird, 1864, *Report of the Commissioners appointed to inquire into the condition of all mines in Great Britain*, H.M.S.O.
21. *Mining Journal* 27 April 1878
22. *Mining Journal* 26 October 1850
23. Mayers, L., 2004, *Balmaidens*, Hypathia Trust.
24. *Mining Journal* 26 October 1850
25. *Tavistock Gazette* 2-23 December 1864
26. *Mining Journal* 26 October 1850
27. *Tavistock Gazette* 2-23 December 1864
28. Census 1861
29. Kinnaird Report, 1864, *op. cit.*
30. Census, 1861
31. *Mining Journal* 28 October 1850
32. *Tavistock Gazette* 2-23 December 1864
33. *Tavistock Gazette* 2-23 December 1864
34. Wm. Woolcock MSS
35. *Mining Journal* 2-23 December 1864
36. *Mining Journal* 2-23 December 1864
37. 6th Annual Report, 1850
38. *Mining Journal* 16th May 1863
39. *Tavistock Gazette* 2-23 December 1864
40. *Tavistock Gazette* 2-23 December 1864
41. Brayshaw M, 1982, The Duke of Bedford's Model Cottages in Tavistock 1840 – 1870, Transactions of the Devonshire Association Vol. 114
42. Brayshaw M., 1982, *op. cit.*
43. 7th Annual Report 1851
44. 7th Annual Report 1851
45. 7th Annual Report 1851
46. Devon Record Office document L1258 E11 – 130

47. Devon Record Office document L1258 E11 – 130
48. Devon Record Office document L1258 E11 – 130
49. Brayshaw M., 1982, *op. cit.*
50. Brayshaw M., 1982, *op. cit.*
51. 10th Annual Report, 1854
52. Devon Record Office document L1258 E11 – 130
53. 7th Annual Report 1851
54. 10th Annual Report, 1854
55. 5th Annual Report, 1849
56. *Tavistock Gazette* 7 January 1870
57. *Tavistock Gazette* 17 May 1872
58. *Tavistock Gazette* 5 July 1872
59. *Tavistock Gazette* 14 September 1860
60. *Tavistock Gazette* 2-23 December 1864
61. *Tavistock Gazette* 20 November 1874

Appendix 2

Geology and lodes

"The largest sulphide lode in the west of England" – H. G. Dines.

Country Rock

The majority of the lodes at Devon Great Consols are found in the killas; described by J. A. Phillips, writing in the 1880s, as "mottled killas"[1] and by J. H. Collins as "lousy killas".[2] Killas is a fairly imprecise, generic, term which in the context of Devon Great Consols refers to a metamorphosed Devonian slate known as the Kate Brook Slate. Whilst the killas is a product of regional metamorphism during the Variscan Orogeny, the mottling, which Phillips attributed to "minute and imperfect crystals of andalusite",[3] appears to be a product of later contact metamorphism. The contact metamorphism is related to the emplacement of the granite around 290 million years ago. The granite, which comprises the Frementor section of the mine, is part of the Gunnislake granite mass which itself is part of the larger, multiphase, granite batholith which dominates the geology of a large part of south west England.

The deep drilling program undertaken in 1979 by Cominco demonstrated that the granite underlies the killas at Devon Great Consols (see chapter 11).

Mineralisation

At its simplest economic mineralisation at Devon Great Consols consists of a shallow tin (cassiterite) zone underlain by copper (chalcopyrite and chalcocite) and arsenic (arsenopyrite) probably becoming barren at depth.

The Main Lode, for example, is characterised by a notable gossan or "iron hat"; gossan being the weathered and decomposed upper section of a lode. Typically

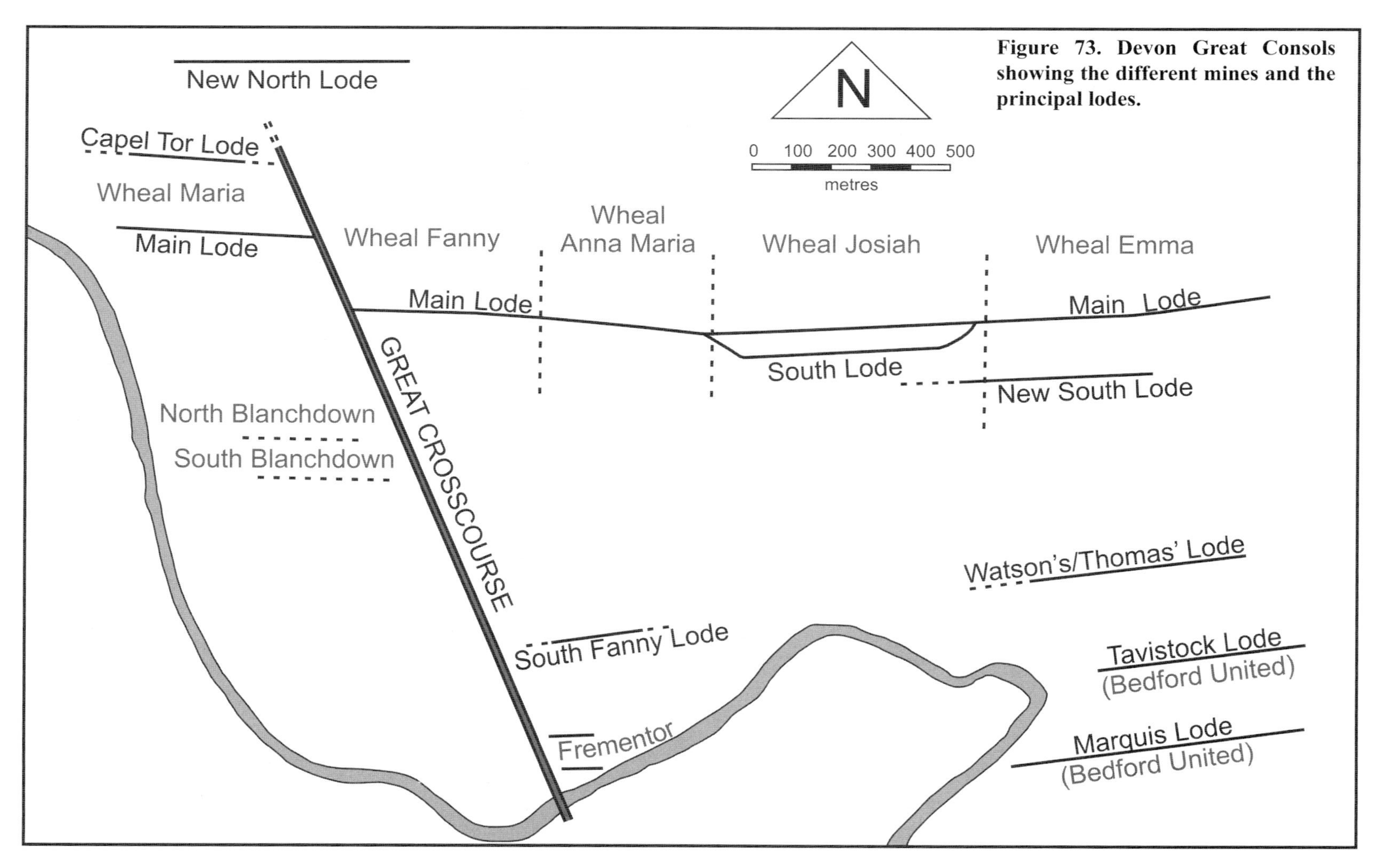

Figure 73. Devon Great Consols showing the different mines and the principal lodes.

sulphides such as chalcopyrite are leached out of this part of the lode to be re-deposited at greater depths, creating a "zone of secondary enrichment"; the mineral chalcocite is notable in this section. Cassiterite, in contrast to the sulphide ores, is not water soluble and consequently remains present in the upper, weathered section of the lode creating the shallow tin zone which attracted the early tinners.

Below the shallow tin zone lies the copper and arsenic on which the nineteenth century wealth of the mine was based. Typical of Tamar Valley lodes the Devon Great Consols lodes exhibit a horizontal zonation or banding. In crude terms arsenopyrite occurs on the outside of the lode, chalcopyrite on the inside. This was extremely fortuitous: during the 1840s-1860s the miners were able to extract the copper ore leaving the arsenic *in situ.* When arsenic production began in earnest "all" the miners needed to do was extract the arsenopyrite from the developed copper stopes without having to resort to additional development work; the infrastructure such as levels, shafts and plant already being in place.

That said, lode structure is much more complex than the simplistic model outlined above: mineralisation at Devon Great Consols appears to be the result of a multiple phase process of mineral emplacement giving rise to horizontal zoning. With regard to the Tamar Valley ore field as a whole, a four phase model of emplacement has been suggested. Whilst details differ from lode to lode Beer and Scrivener outline the process as follows:

> The earliest phase caused intensive tourmalinisation of the slates around the main fractures and the introduction of disseminated fine cassiterite. This was followed by a fracture infilling of quartz and a little chlorite with cassiterite, wolframite, some pyrite and abundant arsenopyrite. After reopening the fractures were again filled by a third phase, depositing chlorite with some quartz, pyrite, chalcopyrite, further arsenopyrite and a little galena, sphalerite and siderite. The final episode was the infilling of all fractures and cavities by fluorite and siderite with a little pyrite and chalcopyrite.[4]

The "classic" vertical zonation model of West Country mineralisation suggests that tin should occur below the copper as was amply demonstrated in the Camborne-Redruth district. The belief the tin would be discovered at depth drove exploration at Devon Great Consols during the 1870s and 1880s as demonstrated by the sinking of both Richards' and Railway Shafts. Whilst neither shaft penetrated the granite they did demonstrate the absence of tin at depth in the killas. It will be recalled that the cores recovered from Cominco's D3 borehole, which did penetrate the granite, contained no traces of tin. Whilst the evidence of one borehole will not give a definitive answer, on present evidence it is arguable that tin does not occur at depth at Devon Great Consols. It is also interesting to note that the D3 cores from the granite showed traces of "low temperature minerals" such as lead and zinc.

Lodes

A considerable number of lodes received attention during the history of the mine. Most important was the Main Lode, followed by South Lode and New South Lode. Thomas'-Watson's Lode proved important particularly in the latter days of the nineteenth century. Capel Tor Lode, New North Lode and the Frementor Lodes are also worthy of note having received a reasonable amount of attention. In addition a number of minor lodes were worked by tinners or cut during nineteenth century exploratory work.

The lode geology is complicated by a number of crosscourses or faults which run through the sett striking roughly north-south being part of the Tamar fault zone. There are at least ten crosscourses at Devon Great Consols, chief of which is the Great Crosscourse.[5]

The Devon Great Consols lodes tend to strike roughly east-west; J. A. Phillips comments that they "have a direction varying from 12 degrees to 20 degrees S. of E. and N. of W. They underlie S., and is on an average about two feet in a fathom".[6]

With regard to the composition of the Devon Great Consols lodes Phillips notes:

> The lodes are, for the most part, composed of quartz associated with iron pyrites, with sometimes a little tin ore; while the carbonate of iron is usually present, not infrequently forming the material cementing together the brecciated portions of the veins. One of the lodes in these mines sometimes assumes the appearance of having been filled with fragments of crushed killas, united by crystalline iron pyrites. When a lode of this description has been opened up by levels or otherwise, the pyrites becomes heavily oxidised, the veinstone is quickly disintegrated, and very heavy timbering is required to keep the ground open.[7]

Main Lode: Main Lode has been described as "the largest sulphide lode in the west of England".[8] The Main Lode may well have been worked as early as the sixteenth century, albeit for tin (see chapter 1).

With regard to the Main Lode, J. H. Collins observed that:

> Before the great importance of these lodes was known cindery masses of gozzan had been noticed in the woods and plantations at several points by miners for many years, but it was only in the year 1844 that permission was obtained to work on any considerable scale. ..... Occasionally this gozzan was found to extend to great depths, but in general at depths of from 20 to 40 fathoms it passed into compact masses of copper and iron pyrites with much quartz and fluor spar, and on the walls thick bodies of arsenical pyrites...... The lode was often 20, 30, and even 40 feet wide, and copper ore pitches yielding 10 to 15 tons to the cubic fathom, and others of arsenical pyrites yielding from 15 to 30 tons to the fathom.[9]

After the 1844 discovery Main Lode was quickly opened up eastwards. The miners soon encountered "The Great Crosscourse", which became the boundary between Wheal Maria and Wheal Fanny. The Great Crosscourse heaved the lode approximately 80 fathoms to the right.[10] The crosscourse is chiefly composed of fluccan with occasional gossan at surface. Phillips observed that "spots of lead ore have also occasionally been found in the fluccan".[11] By the time the mine "closed" thirteen hundredweight of lead had been sold realising £4.[12]

As Main Lode was explored it soon became clear that the further east the lode was explored the deeper it went. Thus at Wheal Maria payable copper was cut at seventeen and a half fathoms whilst at Wheal Emma ore ground was not encountered until the 47 fathom level. Likewise at Wheal Maria very little if any good ore was raised below the 60 fathom level whilst at Wheal Emma the lode was productive to at least the 190.

As noted Richards' Shaft at Wheal Josiah was sunk to the 300 fathom level in an attempt to prove the tin ground which was postulated to lie below the copper. Unfortunately the lode at depth proved poor being composed of "hard capel, quartz and a little mundic (see chapter 7).[13] The deep drilling of 1979 cut the main lode at depth, proving that it continued into the granite where the lode comprised of quartz with traces copper, lead and zinc sulphides.[14]

South Lode: South Lode, discovered in 1848,[15] branches off from the Main Lode in Wheal Anna Maria at an angle of twenty degrees and rejoins it after three hundred fathoms in Wheal Josiah. The lode runs roughly parallel to Main Lode separated by a maximum of forty seven fathoms. Collins notes that South Lode was "extremely rich" over its whole length. The richest parts of South Lode lay between the 40 and 115 fathom levels, although the lode was explored to the 144.[16,17]

The junction between Main and South Lodes was particularly rich:

> The great course of ore formed by the junction of these two principal lodes was composed largely of rich copper pyrites in a brecciated and cavernous vein - stone composed mainly of quartz, fluor spar, carbonate of iron, and iron pyrites, capels of mispickel containing traces of cassiterite bordered by and, often enough, chalcopyrite sprinkled through it – "dredge ore" as it was called – sometimes 40 feet in width, and rich enough in copper to pay for treatment. This enormous ore body continued to yield largely until the year 1866.[18]

New South Lode: Discovered in 1864 New South Lode was mainly worked at Wheal Emma although it does extend into the eastern end of Wheal Josiah. It runs parallel to the Main Lode at a distance of about 100 fathoms. Three main shafts were sunk on the lode: Counthouse Shaft at Wheal Josiah, New Shaft and Railway Shaft at Wheal Emma.

Attempts, including the sinking of Blackwell's Shaft and the driving of Jeffery's crosscut were made to prove the western end of the lode during the 1870s (See chapter 7). These attempts were abortive, the lode containing little of interest or value west of Counthouse Shaft.

New South Lode was explored down to the 260 via Railway Shaft in the hope of discovering tin. Unfortunately the lode proved unproductive at depth and by the summer of 1889 all the workings below the 190 on New South Lode were allowed to flood (see chapter 8).

Thomas'/Watson's Lode: The lode worked at Thomas'/Watson's lies about a third of a mile south of the Main Lode and 280 yards to the north of North or Tavistock Lode of Bedford United mine.

It is possible that this lode had been worked for tin at shallow depths and may equate to the Hanging Cleave and Hanging Cleave Beam tinworks mentioned in the 1778 list of the late Mr Warne's tinworks.[19]

During the nineteenth century phase of working the most productive ground lay between the 136 and the 172 fathom levels in the vicinity of Watson's Engine Shaft.[20] The lode was explored to the 172 fathom level with limited stoping below that level.[21]

Capel Tor Lode: Capel Tor Lode lies to the north of the Main lode at Wheal Maria. The lode was extensively tried in the late 1880s and early 1890s (see chapters 8 and 9), although the lode was explored to a depth of 112 fathoms it never met expectations.

New North Lode: To the north of Capel Tor Lode in the Wheal Maria section of the mine lies New North Lode. This lode was tried in the 1870s with little result in spite of having been explored down to a 54 fathom level.

Frementor Lodes: The Frementor lodes lie in or in close proximity the granite. Unlike the other Devon Great Consols lodes these lodes mainly contained cassiterite and wolfram.

Dines describes Frementor thus:

> The small part of the Gunnislake granite mass that extends across the Tamar opposite Clitters Wood contains bunches or strings of cassiterite – wolfram ore, 2 to 6 in. wide in greisenized country. These unite locally in depth to form a lode of up to 6 ft. wide.....[22]

In terms of economic geology there are two lodes of significance, the **Main Lode** which lies wholly in the granite and **North Lode** which lies on the immediate north or killas side of the granite/killas contact.

Whilst Frementor has received attention since the latter part of the sixteenth century (see Chapter 1) little if any work was carried out between 1844 and the beginning of the twentieth century, presumably due to the complex nature of the ore. The majority of work was carried out by the Bedford Estate during 1917-1918 and 1924-1930 (see Chapter 11).

Minor Lodes

Wheal Maria North Lode was cut by a crosscut on 28 fathom level.[23] During the year ending May 1856 a further crosscut on the 60 had been driven north from Gard's Shaft, cutting North Lode at 17 fathoms. North Lode proved to comprise hard capel quartz. As far as is known no other work was carried out on North Lode.

Woolridge's Lode lies to the south of Main Lode and was explored in both Wheal Maria and Wheal Fanny, the lode being heaved by the great crosscourse. Unlike the majority of Devon Great Consols lodes Woolridge's appears to dip to the north. In Wheal Maria Woolridge's was explored on the 20, 28 and 40 fathom levels. The 7th annual report of 1851 notes that the lode had been cut on the 28 and 40 by crosscuts driven south from Main Lode. At Wheal Fanny the working plan shows that Woolridge's had been explored on the 45 fathom level by Doidge's cross cut and on the 55 by Pryor's crosscut.[24] Dines wrongly identifies this lode as "Worbridge's.[25]

Anna Maria Middle Lode: This lies to the south of Main Lode, and was cut by Jeffery's cross cut driven from the 80 fathom level.

Frank's Lode: Lies between Main and South Lode, worked during 1863.[26]

No.2 South Lode: Lies between inclined Shaft on the Main Lode and New South Lode.

Blanchdown Wood Lodes: At least two lodes appear to have received attention in Blanchdown Wood These lodes occur on the high ground to the south of the nineteenth century workings on Main Lode Given that neither structure has been developed to any depth both lodes may be considered primarily tin lodes. It is possible that the references to "the late Mr Warne's tinworks" in North and South Blanchdown Wood outlined in "Mrs Cake's list" of September 1778 might refer to workings on these lodes.

Placer tin deposits

In addition to what might be termed "lode works", tin was also recovered from alluvial deposits. The stream, known as the Cat, flowing south from Scrub Tor along the western boundary of the nineteenth century sett has already been mentioned in this context (see Chapter 1). At least two further valleys cross the sett; the Hele Valley runs east-west, parallel and south of the Main Lode separating

Grenoven and Blanchdown woods whilst the Rubbytown Valley (also known as Rubbytown Bottom) runs north-south roughly paralleling the boundary between Wheal Josiah and Wheal Emma. Whilst nineteenth century works appear to have obliterated earlier archaeology it seems highly probable that alluvial tin deposits were worked in both valleys.

Appendix 2 references

1. Phillips J. A., 1884, *A Treatise on Ore Deposits*, Macmillan & Co.
2. Beer K. E and Scrivener R. *in* Durrance and Laming, 1982, *The Geology of Devon*, University of Exeter Press.
3. Phillips J. A., 1884, *op. cit.*
4. Phillips J. A,, 1884, *op. cit.*
5. Phillips J. A., 1884, *op. cit.*
6. Dines H. G, 1956, *The Metalliferous Mining Region of South West England*, H.M.S.O.
7. Collins J. H., 1912, *Observations on the West of England Mining Region*, (reprint 1988), Cornish Mining Classics, Truro.
8. Phillips J. A., 1884, *A Treatise on Ore Deposits*, Macmillan & Co.
9. Phillips J. A, 1884, *op. cit.*
10. Bedford Estate Mining Reports 1899 – 1906
11. *Mining Journal* 12 January 1878
12. British Geological Survey, Online borehole record viewer
13. 4th Annual report, 1848
14. Collins J. H., 1912, *op. cit.*
15. Dines H. G, 1956, *op. cit.*
16. Collins J. H., 1912, *op. cit.*
17. Devon Record Office document LM1258: particulars of the late Mr Warne's Tinworks (a fourth part) in the parish of Tavistock (now in possession of Mrs Elizabeth Cake), September 1778
18. Dines H. G, 1956, *op. cit.*
19. Devon Record Office document T158m/E44 a-b, 1901 Mining Report
20. Dines H. G, 1956, *op. cit.*
21. 9th Annual Report, 1853
22. Devon Great Consols working plan, author's collection
23. Dines H. G, 1956, *op. cit.*
24. Kinnaird Report, 1864, *op. cit.*

Appendix 3

Production figures 1845-1904

Unless otherwise stated these figures are taken from Burt R. *et al*, 1984, *Devon & Somerset Mines*, University of Exeter.

Copper

Date	Ore (tons)	Metal (tons)	Value (£)
1845*	11,288	1,484.337	1,000,971.75
1846*	15,684	1,578.358	95,469.5
1847*	14,175	1,474.565	95,873
1848	16,374	1,684.1	93,418.3
1849	15,777	1,508.6	96,934.7
1850	17,089	1,632.3	109,989.6
1851	18,921	1,708.6	111,082
1852	20,802	1,659.1	134,224.1
1853	24,120	1,583.5	147,281.3
1854	23,174	1,486.1	143,224.2
1855	23,467	1,385.6	131,294.1
1856	29,425	1,687.1	143,045
1857	25,726	1,472.6	139,770.4
1858	23,102	1,352.1	116,772.1
1859	22,832	1,243.9	114,033.7
1860	21,920	1,224.5	109,326.8
1861	20,801	1,178.5	103,072.5
1862	24,615	1,448.1	116,942.2
1863	26,756	1,672.6	128,799.1
1864	25,956	1,613.3	136,189.8
1865	25,259	1,593	126,932.3

1866	22,671	1,453.5	98,144.7
1867	20,067	1,333.3	94,510.8
1868	20,955	1,391.5	92,590.3
1869	16,379	1,010.4	62,014.9
1870	16,332	884.1	49,729.7
1871	17,413	868.8	50,116.6
1872	15,315	717.1	52,942.2
1873	8,716	428.5	24,615.3
1874	5,676	364.3	22,920.6
1875	7,085	416.5	29,668
1876	9,420	521.7	33,626.1
1877	11,383.6	613.4	34,250.4
1878	9,394.8	425	18,307.2
1879	10,261	518.8	22,598.7
1880	10,116.2	512.1	24,929.9
1881	10,922.7	442.1	201,13.5
1882	11,970.8	598.5	41,897
1883	11,127.2	459	20,815
1884	10,520		16,827
1885	9,778		11,988
1886	6,077		6,498
1887	4,209		6,538
1888	4,784		8,501
1889	637		958
1890	4,368		4,310
1891	3,142		3,960
1892	2,572		3,475
1893	2,312		3,716
1894	2,211		2,907
1895	2,008		3,807
1896	1,530		2,971
1897	1,417		2,773
1898	1,084	71	2,188
1899	837		2,224
1900	1,015		2,737
1901	842		1,948
1902	268		370

*Figures for 1845 to 1847 derived from ticketings, information kindly supplied by A. Neill.

Arsenic

Date	Refined arsenic (tons)	Value
1868	473.5	2,802.6
1869	1,372.6	7,726.7
1870	2,237.1	12,557.9
1871	2,220.2	1,1863
1872	2,222.3	11,400.9
1873	1,878.6	12,100.5
1874	1,842	12,762
1875	1,212	10,503.1
1876	1,521.6	14,705.4
1877	2,327.9	19,029.5
1878	2,481.8	16,966
1879	3,252.5	22,848.3
1880	3,148.8	26,283.5
1881	2,851.5	23,325
1882	2,760.1	22,080
1883	2,974.2	23,793
1884	3,198.3	25,586
1885	3,333.1	23,588
1886	2,428.2	18,342
1887	2,082	16,659
1888	2,100	19,275
1889	2,150	21,137
1890	2,615	28,523
1891	2,241	24,406
1892	2,174	20,567
1893	3,966	41,372
1894	2,212	24,608
1895	1,673	20,268
1896	1,521	20,453
1897	1,731	33,835
1898	1,723	24,240
1899	1,309	21,669
1900	1,408	25,504
1901	1,153	15,000.7
1902	879	10,513
1903	212	2,403
1904	149	1,406

Appendix 4

The mine today

The whole of the former Devon Great Consols sett in is private ownership. Whilst there are public access agreements in place it should be borne in mind that there is no public access to large parts of the site. Much of the site is a working woodland and potentially dangerous.

Neither the author, the publishers of this book nor landowners accept any responsibility or liability for any accidents however caused.

For those who do want to explore the mine there are a number of permissive paths both on the sett and utilising the course of the railway. These are part of the "Tamar Trails" initiative of the Tamar Valley AONB. Details of trails may be found on their website.

Morwellham Quay operates as a visitor attraction for which an admission fee is charged.

Whilst small underground sections of the mine are accessible to experienced mine explorers the decision has been taken not to include details in the gazetteer. Would be mine explorers are strongly recommended to contact the Plymouth Caving Group who have been exploring the mine since the late 1960s.

The Industrial Archaeology of Devon Great Consols

For such a large site Devon Great Consols is, at least at first glance, somewhat disappointing. The reclamation of the site in the first years of the 20th century means that there are few standing structures with the exception of housing which is still in occupation. The Wheal Anna Maria floors and associated dumps have been particularly hard hit, the destruction of the nineteenth century floors compounded by later working, particularly during the 1970s. In a similar vein the comparatively recent landscaping of Richards' Shaft is a significant loss.

That said, on closer inspection there is still much to be seen. For example the site is rich in "early" mining remains. Much also remains from the nineteenth century including large portions of the water power and associated flat rod systems, the trackbed of the railway, most of the shafts and remains of wheel pits and engine houses. The jewel in the crown, however, is the 20th century archaeology, the 1920s arsenic works, which has survived largely intact, is one of the best examples in the country. That the 1970s ball mill has survived, given the voracity of the scrap metal trade, is nothing short of miraculous.

Since the mid 2000s the mine has received the dubious attentions of the "Tamar Valley Mining Heritage Project" (the timber decking at the 1920s arsenic works being particularly objectionable). Under their auspices miles of fencing have appeared on the mine significantly altering the character of the site. Many standing structures have been "conserved" and consolidated. Whilst this will at least preserve them for future generations the manner in which some of the work has been carried out is questionable.

Selected Gazetteer

Inclusion of a site in this gazetteer does not imply that there is any public access to that particular site. **In particular please respect residents' privacy.**

It should be noted that as a former mine parts of the site are exceedingly dangerous. Putting a foot wrong, particularly in the vicinity of the Fremen-tor gunnises, could have fatal consequences.

The following gazetteer is not intended to be comprehensive but should give the intending explorer a good introduction to the mine.

"Early" mining

Grid Reference	Type	Notes
SX 414 743 to SX 415 742	Stream work	Cat stream running north – south along western boundary of sett.
SX 4170 7409	Open work	Capel Tor Lode
SX 4165 7345	Open work	North Blanchdown Wood Lode .
SX 4180 7317	Open work	South Blanchdown Wood Lode . Lode back works to both the east and west
SX 4229 7264	Open work	Frementor, odd alignment
SX 7234 7251	Open work	Frementor North Lode. Reworked during WWI. On north side of granite killas contact
SX 4238 7248	Open work	Frementor "Main Lode". Extensively worked during WWI and late 1920s. **A slip here could be fatal.**
SX 4320 7338	Open work	Wheal Josiah, possibly on New South Lode

Shafts

Name	Grid reference	Lode	Comments
Western	SX 4180 7436	New North Lode	Shaft not positively identified. However there is a granite marker at this point.
Eastern	SX 418 743	New North Lode	Extensive dump. Exact location of shaft not located.
Western (formerly Boundary Shaft)	SX 4152 7409	Capel Tor	Marked with granite post
Eastern	SX 4160 7409	Capel Tor	Large dump
Gard's	SX 4177 7391	Main Lode (Wheal Maria)	Marked with Granite post. Also well preserved capstan round.
Morris'	SX 4190 7390	Main Lode (Wheal Maria)	Large crater
Western Engine	SX 4212 7370	Main Lode (Wheal Fanny)	Marked with granite post
Eastern Engine	SX 4224 7370	Main Lode (Wheal Fanny)	Marked with granite post. Adjacent capstan excavated 2011.
"1920s"	SX 4227 7371	Main Lode (Wheal Fanny)	Vestiges of head frame
Ventilating	SX 4235 7371	Main Lode (Wheal Fanny)	Marked with granite post
Engine Shaft	SX 4267 7365	Main Lode (Wheal Anna Maria)	Marked with granite post
Field Shaft	SX 4283 7363	Main Lode (Wheal Anna Maria)	Surface depression
South Lode (?)	In vicinity of SX 429 736	South Lode (?) (Wheal Josiah)	Large surface depression and (small) dump.

Richards'	SX 430 736	Main Lode (Wheal Josiah)	The site has been obliterated, capped and landscaped
Hitchins'	SX 4315 7365	Main Lode (Wheal Josiah)	Man engine used in this shaft.
"New" Agnes	SX 4351 7358	Main Lode (Wheal Josiah)	Bob pit on west side
"Old" Agnes	SX 4350 7363	Main Lode (Wheal Josiah)	Fenced depression
Inclined	SX 4386 7370	Main Lode (Wheal Emma)	Marked with granite post. Man engine used in this shaft.
Thomas'	SX 440 737	Main Lode (Wheal Emma)	Rough depression
Blackwell's	SX 4233 7343	New South Lode	Prominent dump
Counthouse	SX 4341 7346	New South Lode	Marked with granite post.
Railway	SX 4367 7355	New South Lode	Difficult to locate with exact certainty
New	SX 4379 7355	New South Lode	Marked with Granite post.
South Fanny Engine (a.k.a. South Lode Shaft)	SX 4255 7287	South Fanny Lode (a.k.a. South Lode)	Open to blockage. Fine flat rod cutting.
Bawden's	SX 4372 7303	Watson's	Marked with granite post
Watson's Engine	SX 4391 7304	Watson's	Extensive dumps.
"Skip"	In vicinity of SX 424 724	Frementor	Within open gunnis.
Plunger	SX 4329 7311		Surface depression

Ore Dressing

Site	Grid reference	Remains
Wheal Maria floor	SX 416 738	Reasonably unmolested
Wheal Fanny floor	SX 421 736	Fragments of cobbing floor and small ponds.
Anna Maria "old" floor	SX 426 735	Little survives. Fragments of cobbing floors
Anna Maria "new"	SX 425 734	Incoherent
New Dressing Plant (1920s)	SX 4262 7330	Obliterated by 1970s reworking
Wheal Josiah	SX 431 735	Reasonably unmolested
Wheal Maria precipitation works	SX 41517375	Traces of launders
Anna Maria precipitation works	SX 426 731	Obliterated by 1970s working
Blanchdown precipitation works	SX 434 731	Ochreous
South Fanny precipitation works	SX 429 728	Remains of collapsed launders
1970s mill	SX 4268 7318	Ball mill survives intact

Arsenic works

Site	Grid reference	Remains
1860s works	SX 424 731	Works levelled on closure. Although some structures survive above ground level.
1920s works	SX 426 731	Survives in good condition
1920s waterfall chamber	SX 4259 7347	Survives in fair condition
1920s stack	SX 4259 7356	Survives to full height.

Water Wheels

Wheel pits which do survive tend to survive as earthwork features, having been robbed of their stone.

Location	Grid reference	Size	Notes	Remains
Wheal Maria	SX 4160 7394	50′ x 4′. Hoisting and pumping	1848	Pit identifiable, robbed of stone
Wheal Fanny	SX 4208 7370	35′ x 4′. Crusher (?)	1846, 1850 also driving stamps	Traces of masonry
Wheal Fanny	SX 4215 7372	32′ x 4′. Hoisting (?)	1848	Pit identifiable, robbed of stone
Blanchdown	SX 434 730	40′ x 12′. Pumping	1849 by Nicholls, Williams. Fed by "Great leat". Pumping Richards' Shaft & Hitchins' Shaft	Pit identifiable, robbed of stone. 1849 date stone *in situ*.
Blanchdown	SX 434 730	40′ x 12′. Pumping	1850 by Nicholls, Williams. Fed by "Great leat". Pumping Anna Maria Engine Shaft & Field Shaft	Pit identifiable, robbed of stone
Wheal Thomas / Blanchdown	SX 4359 7501	36′ hoisting and pumping	1850	
Great Leat	SX 4330 7288	30′ x 16′ Pumping	1852. "Plunger Wheel", pumping from river to Wh. Josiah	Pile of demolition rubble (possibly of later date).
Wheal Josiah	SX 4343 7362	30′ x 4′ hauling	1855, hauling from Agnes Shaft	Pile of demolition rubble.
Great Leat	SX 43327288	32′ x 10′	By 1860 Pumping Agnes Shaft	Pit identifiable, robbed of stone
Foundry	SX 4163 738	35′ x 4′	By 1860. Presumably driving blower for cupola.	Pit survives in fair condition.

Steam Engines

Location	Grid reference	Size/type	Notes	Remains
Wheal Maria, Morris' Shaft	SX 4189 7389	40″ pumping, by Perran Foundry.	1845. Moved to Wheal Anna Maria in 1864	Granite cylinder bed survives, not in situ (SX 4187 7389) . Vestiges of boiler house and stack.
Anna Maria, New halvans floor	SX 4260 7331	30″ Rotary by Nicholls, Matthews	1855	Fly wheel loading survives incorporated into 1920s arsenic works. Includes adjacent remains of roller crusher.
Wheal Josiah, Hitchins' Shaft	SX 4309 7363	16″/30″ Sims compound. By Perran Foundry	1845 pumping, 1849 converted to whim, 1865 driving man engine.	House levelled but site identifiable
Wheal Emma, Inclined Shaft.	SX 4390 7370	22″ Whim	1851, 1865/1866 driving man engine on Inclined Shaft (?)	Flywheel /drum loadings in fine condition. Boiler house identifiable
New Shaft	SX 4373 7353	Whim	Post 1864 and by 1867	House ruinous, but identifiable, flywheel drum loadings
Morwellham incline.	SX 4427 7068	22″. Largely constructed in DGC foundry	1858	Archaeologically excavated

Bibliography

Archival material.

Archival sources are referenced in the text.

Extensive use was made of the Bedford Estate papers (DRO ref: L1258/) held in the Devon Record Office; these include leases, agents correspondence and mine reports. In 2011 a mass of Bedford Estate mine reports resurfaced in the hands of a Bristol antiquarian book dealer. These have since been acquired by the Devon Record Office (DRO ref: 8187) and are set to revolutionise our understanding of mining in West Devon. The Devon record office also holds copies of mine plans deposited as part of the abandoned mines record (AMR). The Devon Great Consols workbooks dating from 1922-1925 are also held here. (The workbooks were the basis of much of P. H. G. Richardson's work on the mine).

Limited use has been made of the Cornwall Record Office by the current author mainly due to its incredibly slow and frustrating procedures. Of the material consulted the Toll papers (CRO AD159) are of particular interest.

Reference has also been made to material held in the Public Record Office/ National Archive.

William Woolcock's notebook (Woolcock MSS) is part of the archive at Morwellham Quay. This has been transcribed and it is hoped to make the text more widely available.

Books, Journals etc.

Andre, G. G., 1878, *A Descriptive Treatise on Mining Machinery, Tools and other Appliances used in Mining*, Vol. 2, E. & F. N. Spon.

Barclay, C. F., 2004, *Mines of the Tamar & Tavy* (ed. Stewart R. J.), Tamar Mining Press.

Barton, D. B., 1964, *A Historical Survey of the Mines and Mineral Railways of East Cornwall and West Devon*, D. Bradford Barton Ltd.

Barton, D. B., 1966, *The Cornish Beam Engine*, D. Bradford Barton Ltd.

Barton D. B., 1966, *The Redruth and Chacewater Railway 1824 – 1915*. 2nd ed., Bradford Barton

Barton D. B., 1968, *Essays in Cornish Mining History*, Vol. 1, Bradford Barton Ltd.

Barton, D. B., 1970, *Essays in Cornish Mining History*, Vol. 2, D. Bradford Barton Ltd.

Bawden, M., 1914, Mines and mining in the Tavistock district, *Transactions of the Devonshire Association*, Vol. XLVI. pp. 256 - 264.

Beament, R., 2008, Devon Great Consols in the 1970s, Unpublished manuscript.

Booker, F., 1967, *Industrial Archaeology of the Tamar Valley*, David & Charles.

Brayshay, M., 1982, The Duke of Bedford's model cottages in Tavistock 1840-1870, *Transactions of the Devonshire Association*, Vol. 114. pp. 115-131.

Brooke, J., New Light on Devon Great Consols, *Devon & Cornwall Notes & Queries*, Vol. XXX, 1965-7 Part VII, p.183.

Brooke, J., 1977, Who discovered Devon Great Consols? *Northern Cavern & Mine Research Society Memoirs*, Vol. 5., pp 21-22

Brooke, J., 1982, The last years of Devon Great Consols, *Journal of the Trevithick Society*, No. 9, pp. 69-72.

Browne, W., 1848. *The Cornish Engine Reporter* No.12, January 1848. St Austell.

Browning, P. A, (n.d.), The exploration of Devon Great Consols by Plymouth Caving Group, Unpublished manuscript.

Buck, C., 2002, *Devon Great Consols, archaeological assessment*, Cornwall Archaeological Unit.

Burt, R. (ed.), 1969, *Cornish Mining, Essays on the Organisation of Cornish Mines and the Cornish Mining Economy*, David & Charles.

Burt R., 1977, *John Taylor, mining entrepreneur and engineer 1779-1863*. Moorland Publishing.

Burt, R. *et al*, 1984, *Devon & Somerset Mines*, University of Exeter.

Burt, R., 1988, Arsenic - Its significance for the survival of South Western metal mining in the late nineteenth and early twentieth centuries, *Journal of the Trevithick Society*, No. 15, pp. 5 – 26.

Chowen, G, 1863, *Some Accounts of the rise and progress of mining in Devonshire: from the time of the Phoenicians to the present,* Thomas S. Chave.

Collins, J. H., 1912, *Observations on the West of England Mining Region*, (reprint 1988), Cornish Mining Classics, Truro.

Cook, D. & Kirk, D., 2009, *Turner in the Tamar Valley*, Tamar Valley AONB.

Cornish Mining Development Association, Annual reports 1974-1978.

Darlington, J., 1878, *On the Dressing of Ores* (reprint 2002), extracted from Ure's dictionary of arts, manufactures and mines, Dragonwheel Books.

Darlington, J. & Phillips, J. A. 1857, *Records of Mining and Metallurgy*, E. & F. N. Spon.

Davies, E. H, 1902, *Machinery for Metalliferous Mines*, Crosby Lockwood.

Devonshire Great Consolidated Copper Mining Company, Annual reports 1846-1856.

Dewey, H., 1920, *Arsenic and Antimony Ores*, Special Reports on the Mineral Resources of Great Britain, Vol. XV, H.M.S.O.

Dewey, H., 1923, *Copper Ores of Cornwall and Devon*, Special Reports on the Mineral Resources of Great Britain, Vol. XXVIII, H.M.S.O.

Dewey, H, Dines, H. G. *et al*, 1923, *Tungsten and Manganese Ores*, Special Reports on the Mineral Resources of Great Britain, Vol. I, H.M.S.O

Dickinson, M, 1985, The Duke's men, the Wheal Maria people and others, the story of the Devon Great Consols "ghost railways", *Tamar Journal*, Vol. 7, pp. 31-40

Dines, H. G, 1956, *The Metalliferous Mining Region of South West England*, H.M.S.O.

Dixon, D., 1995, The Richards family of Tavistock and Mary Tavy, *Journal of the Trevithick Society*, No. 22, pp. 67-76.

Dixon, T. & Pye, A. R., 1989, The arsenic works at Devon Great Consols Mine, *Devon Archaeological Society Proceedings*, No. 47.

Durrance, *et al.*, 1982, *The Geology of Devon*, University of Exeter Press.

Goodridge, J. C., 1964, Devon Great Consols, a study of Victorian mining enterprise, *Transactions of the Devonshire Association*, Vol. XCVI. pp. 228-268.

Greeves, T. A. P., 1981, *The Devon tin industry 1450-1750*, Doctoral thesis, University of Exeter.

Harvey, C. & Press, J., 1990, The city and mining enterprise: The making of the Morris family fortune, *Journal of the William Morris Society*, 91, pp. 3-14.

Hateley R., 1977. *Industrial locomotives of South West England.* Industrial Railway Society.

Jenkin, A. K. H,. 1974, *Mines of Devon*, Vol. 1 The Southern Area, David &

Charles.

Jenkin, A. K. H., 2005, *Mines of Devon*, Landmark.

The Jurist Vol. XVI, Part II, 1853.

Kinnaird, 1864, *Report of the Commissioners appointed to inquire into the condition of all mines in Great Britain*, H.M.S.O.

The Mining Journal (References in text).

Mayers, L., 2004, *Balmaidens*, Hypathia Trust.

Mayers, L., 2011, *The Tamar Bal Maidens*, Blaize Bailey Books.

Messenger M., 2001, *Caradon and Looe – The canal, railways and mines.* Twelveheads Press.

Miners' Association of Cornwall and Devonshire, Annual reports. (References in text).

Murchison, J. H., 1858, *Review of the Progress of British mining.....*1857, Murchison.

Phillips, J. A., 1884, *A Treatise on Ore Deposits*, Macmillan & Co.

Pharmaceutical Journal and *Transactions of the Royal Pharmaceutical Society of Great Britain*,1852-1853 Vol. 12, page 308

Richardson, P. H. G., 1979-1980, Mining at Devon Great Consols after 1903, *Tamar Journal* 1979-1980, pp. 33 – 42.

Richardson, P. H. G., 1995, *Mines of Dartmoor & the Tamar Valley after 1913*, (2nd edition) Devon Books.

The Royal Cornwall Gazette. (References in text)

Shambrook, R., 1982, The Devonshire Great Consolidated Copper Mining Company, *Journal of the Trevithick Society*, No. 9, pp. 62-68.

Spargo T, 1864, *Statistics and observations on the mines of Cornwall and Devon,* Vincent & Skeen London.

Stewart, R. J. (ed.), 2003, *Devon Great Consols*, a collection of contemporary articles & reports, Tamar Mining Press.

Stewart, R. J., 2005, *Ore Dressing at Devon Great Consols*, Tamar Mining Press.

The Tavistock Gazette. (References in text).

Taylor J., 1829, *Records of Mining,* Part 1. John Murray London.

Thomas, C., 1867, *Mining Fields of the West*, (1967 Reprint), D. Bradford Barton Ltd.

Thomas, H., 1896, *Cornish Mining Interviews*, Camborne Printing and Stationary Co.

Toll, R. W. 1938, The arsenic industry in the Tavistock district of Devon, *Sands Clays & Minerals*, April 1938, pp 224-227.

Toll, R. W., 1947, Copper recovery from low-grade waters, *Mining Magazine*, August 1947, pp. 83-84.

Toll, R. W., 1953, Arsenic in West Devon & East Cornwall, *Mining Magazine*, August 1953, pp. 83-88.

Waterhouse, R. E, (in prep.), *The Tavistock Canal.*

Watson, J. Y., 1843, *Compendium of British Mining*, London.

The West Briton. (References in text).

Williams, R. 2011, The Tolgarrick tin streams of Cornish Tin and Engineering Ltd., *Journal of the Trevithick Society*, No. 38, pp. 40 – 74.

Index